Germline Stem Cells

T0076013

Edited by

Steven X. Hou
National Cancer Institute
Frederick, MD

Shree Ram Singh
National Cancer Institute
Frederick, MD

☀ Humana Press

Editors
Steven X. Hou
National Cancer Institute
Frederick, MD
shou@mail.ncifcrf.gov

Shree Ram Singh
National Cancer Institute
Frederick, MD
singhshr@mail.ncifcrf.gov

Series Editor
John M. Walker, Professor Emeritus
School of Life Sciences
University of Hertfordshire Hatfield
Hertfordshire AL10 9AB, UK.

ISBN 978-1-60327-213-1   e-ISBN 978-1-60327-214-8
DOI: 1.1007/978-1-60327-214-8

Library of Congress Control Number: 2007938052

Cover illustration: Figure 1A from Chapter 7, "Identification and Characterization of Spermatogonial Subtypes and Their Expansion in Whole Mounts and Tissue Sections from Primate Testes" by Jens Ehmcke and Stefan Schlatt.
A conventional hematoxylin staining of adult rhesus monkey testicular tissue showing a central seminiferous tubule at stage II of the seminiferous epithelial cycle.

Printed on acid-free paper

9 8 7 6 5 4 3 2 1

springer.com

Preface

Stem cells are responsible for maintaining tissue homeostasis. They possess unlimited self-renewal potency. Through asymmetric cell division, a stem cell in adult tissues can produce an offspring that will maintain the stem cell populations and a daughter cell that will differentiate into various short-lived cell types to replace damaged or dying cells. Tumors may originate from a few transformed cancer stem cells. Stem cells have immense potential for therapeutic use in regenerative medicine and for developing anticancer therapies that specifically eliminate cancer stem cells. Germline stem cells are the source of human and animal reproduction. The knowledge gained from studying germline stem cells may find immediate application in preserving endangered wildlife, managing commercial livestock, overcoming fertility problems in humans, and treating testicular and ovary tumors.

To make use of this potential, we have to learn how to isolate, characterize, and maintain germline stem cells. To isolate, characterize, and maintain stem cells, we must first understand the molecular parameters that define a germline stem cell and the mechanisms that regulate stem cell behavior. The protocols in *Germline Stem Cells* are intended to present selected genetic, molecular, and cellular techniques used in germline stem cell research. The book is divided into two parts. Part I covers germline stem cell identification and regulation in model organisms. Part II covers current techniques used in in vitro culture and applications of germline stem cells. Each chapter begins with a brief overview of the topic, list of necessary equipment and reagents, step-by-step laboratory protocols, and tips on troubleshooting and avoiding known pitfalls. We hope that *Germline Stem Cell* protocols provides basic techniques to cell and molecular biologists, tissue engineers, clinicians, geneticists, and students involved in various aspects of germline stem cell research and application.

We would like to thank Prof. John M. Walker and the staff at Humana Press for their invitation, editorial guidance, and assistance throughout preparation of this book for publication. We also would like to express our sincere appreciation and gratitude to the contributors for sharing their precious laboratory expertise with the germline stem cell research community.

Steven X. Hou
Shree Ram Singh

Contents

Contributors

Maria E. Ayala, PhD
Unidad de Investigacion en Biologia de la Reproduccion, FES Zaragoza,
UNAM, Mexico

Antonin Bukovsky, MD, PhD, DSc
Laboratory of Development, Differentiation and Cancer, Department
of Obstetrics and Gynecology, The University of Tennessee Graduate School
of Medicine, Knoxville, TN

Karen M. Chapman, BS
Howard Hughes Medical Institute, University of Texas, Southwestern Medical
Center, Dallas, TX

Hélio Chiarini-Garcia, PhD
Laboratory of Structural Biology and Reproduction, Department
of Morphology—ICB Federal University of Minas Gerais, Belo Horizonte
MG Brazil

Pleas Copas, MD
Department of Obstetrics and Gynecology, The University of Tennessee
Graduate School of Medicine, Knoxville, TN

Sarah L. Crittenden, PhD
Howard Hughes Medical Institute and Department of Biochemistry,
University of Wisconsin–Madison, Madison, WI

Christina Tenenhaus Dann, PhD
University of Texas Southwestern Medical Center, Department
of Pharmacology and Cecil H. and Ida Green Center for Reproductive
Biology Sciences, Dallas, TX

David A. Dansereau, PhD
Department of Biology, McGill University, Montreal, Quebec, Canada

Ina Dobrinski, M.V.Sc., Ph.D., Dip. ACT
Center for Animal Transgenesis and Germ Cell Research, School of Veterinary
Medicine, University of Pennsylvania, Philadelphia, PA

Roberto Dominguez, PhD
Unidad de Investigacion en Biologia de la Reproduccion, FES Zaragoza,
UNAM, Mexico

August Dorn, PhD
Institute of Zoology, Johannes Gutenberg University, Mainz, Germany

David C. Dorn, MD, PhD
Laboratory of Developmental Hematopoiesis, Cell Biology Program, Memorial
Sloan-Kettering Cancer Center, New York, NY

Jens Ehmcke, PhD
University of Pittsburgh School of Medicine, Center for Research
in Reproductive Physiology, Department of Cell Biology and Physiology,
Pittsburgh, PA

Ilaria Falciatori, PhD
Howard Hughes Medical Institute, University of Texas Southwestern Medical
Center, Department of Pharmacology, Dallas, TX

David L. Garbers, PhD
Howard Hughes Medical Institute, University of Texas Southwestern Medical
Center, Department of Pharmacology and Cecil H. and Ida Green Center
for Reproductive Biology Sciences, Dallas, TX

Satish K. Gupta, PhD
Gamete Antigen Laboratory, National Institute of Immunology, New Delhi, India

F. Kent Hamra, PhD
University of Texas Southwestern Medical Center, Department of Pharmacology
and Cecil H. and Ida Green Center for Reproductive Biology Sciences, Dallas, TX

Jason D. Heaney, PhD
Case Western Reserve University, Department of Genetics, Cleveland, OH

Steven X. Hou, PhD
Mouse Cancer Genetics Program, National Institutes of Health, National Cancer
Institute at Frederick, Frederick, MD

Tiina Immonen, PhD
Institute of Biomedicine, Biochemistry and Developmental Biology,
University of Helsinki, Finland

Judith Kimble, PhD
Howard Hughes Medical Institute and Department of Biochemistry,
University of Wisconsin–Madison, Madison, WI

Paul Lasko, PhD
Department of Biology, McGill University, Montreal, Quebec, Canada

Kate Lillard-Wetherell, PhD
Howard Hughes Medical Institute, University of Texas Southwestern Medical
Center, Department of Pharmacology, TX

Wei Liu, PhD
Mouse Cancer Genetics Program, National Institutes of Health, National Cancer
Institute at Frederick, Frederick, MD

Derek J. McLean, PhD
Department of Animal Sciences, Center for Reproductive Biology, Washington
State University, Pullman, WA

Marvin L. Meistrich, PhD
Department of Experimental Radiation Oncology, The University of Texas,
M.D. Anderson Cancer Center, Houston, TX

Durga Prasad Mishra, PhD
Department of Pharmacology, School of Medicine, University of California,
Irvine, Irvine, CA

Joseph H. Nadeau, PhD
Case Western Reserve University, Department of Genetics, Cleveland, OH

Rahul Rathi, BVSc, MSc, PhD
Center for Animal Transgenesis and Germ Cell Research School of Veterinary
Medicine, University of Pennsylvania, Philadelphia, PA

Hannu Sariola, MD
Institute of Biomedicine, Biochemistry and Developmental Biology,
University of Helsinki, Finland

Paolo Sassone-Corsi, PhD
Department of Pharmacology, School of Medicine, University of California,
Irvine, Irvine, CA

Stefan Schlatt, PhD
University of Pittsburgh School of Medicine, Center for Research
in Reproductive Physiology, Department of Cell Biology and Physiology,
Pittsburgh, PA

Shree Ram Singh, PhD
Mouse Cancer Genetics Program, National Institutes of Health,
National Cancer Institute at Frederick, Frederick, MD

Marta Svetlikova, BSc
Laboratory of Development, Differentiation and Cancer, Department
of Obstetrics and Gynecology, The University of Tennessee Graduate
School of Medicine, Knoxville, TN

Nirmala B. Upadhyaya, MD
Department of Obstetrics and Gynecology, The University of Tennessee
Graduate School of Medicine, Knoxville, TN

Stuart E. Van Meter, MD
Department of Pathology, The University of Tennessee
Graduate School of Medicine, Knoxville, TN

Irma Virant-Klun, PhD
Department of Obstetrics and Gynecology, University Medical Centre
Ljubljana, Ljubljana, Slovenia

Zhuoru Wu, PhD
University of Texas Southwestern Medical Center, Department
of Pharmacology, Dallas, TX

Part I
Identification and Regulation of Germline Stem Cells in Model Organisms

Chapter 1
The Development of Germline Stem Cells in *Drosophila*

David A. Dansereau and Paul Lasko

Contents

Summary Germline stem cells (GSCs) in *Drosophila* are a valuable model to explore of how adult stem cells are regulated in vivo. Genetic dissection of this system has shown that stem cell fate is determined and maintained by the stem cell's somatic microenvironment or niche. In *Drosophila* gonads, the stem cell niche—the cap cell cluster in females and the hub in males—acts as a signaling center to recruit GSCs from among a small population of undifferentiated primordial germ cells (PGCs). Short-range signals from the niche specify and regulate stem cell fate by maintaining the undifferentiated state of the PGCs next to the niche. Germline cells that do not receive the niche signals because of their location assume the default fate and differentiate. Once GSCs are specified, adherens junctions maintain close association between the stem cells and their niche and help to orient stem cell division so that one daughter is displaced from the niche and differentiates. In females, stem cell fate depends on bone morphogenetic protein (BMP) signals from the cap cells; in males, hub cells express the cytokine-like ligand Unpaired, which activates the Janus kinase-signal transducers and activators of transcription (Jak-Stat) pathway in stem cells. Although the signaling pathways operating between the niche and stem cells are different, there are common general features in both males and females, including the arrangement of cell types, many of the genes used, and the logic of the system that maintains stem cell fate.

Keywords *Drosophila*; germline stem cell; GSC; PGC; primordial germ cell; stem cell fate; stem cell niche.

From: *Methods in Molecular Biology, Vol. 450, Germline Stem Cells*
Edited by S. X. Hou and S. R. Singh © Humana Press, Totowa, NJ

1.1   Introduction

Stem cells are defined by their capacity to self-renew and to generate daughters that differentiate into one or more terminal cell types. Proper regulation of this property is critical for animal development, growth control, and reproduction. Understanding stem cell function is potentially very important for future developments in gene therapies and regenerative medicine. Research in *Drosophila* on germline stem cells (GSCs) has been instrumental in defining the important function of the stem cell's somatic microenvironment or niche in the control of its division and self-renewal *(1,2)*. The *Drosophila* germline is an excellent model of stem cell biology because the system is genetically tractable, the sterility and rudimentary gonads resulting from GSC loss are easily recognized, and the stem cells can be easily identified based on molecular markers and position in the gonad. Germline morphology and development, from embryogenesis to the differentiation of gametes in adult flies, has been well characterized and offers a solid basis for the study of GSC fate determination, maintenance, and differentiation.

Drosophila GSCs are derived from embryonic pole cells, the first cell type defined in the embryo. They migrate from the posterior to meet the somatic gonadal precursors (SGPs) and form the embryonic gonad, a simple structure made up of about ten primordial germ cells (PGCs) intermingled with and surrounded by mesodermal cells. PGCs divide and remain in an undifferentiated state until the stem cell niche develops in the anterior of the gonad. The niche acts as a signaling source that maintains the undifferentiated state of PGCs near the niche, thereby maintaining their capacity to populate the niche as GSCs. During the final phase of niche development, adherens junctions develop between the niche and the GSCs, thereby orienting stem cell division and holding the stem cells close to the maintenance signals of the niche.

1.2   Primordial Germ Cell Formation

Germline stem cells are derived from a population of PGCs, a distinctive lineage of cells set aside from the soma early in development. Germ cells are distinguished from somatic cells by the expression of germline-specific molecular markers, the most widely used of which is the DEAD-box RNA helicase Vasa *(3)*. Although the mechanisms used to define the germline vary, the determinant role of localized germplasm has been described in many animal species, including *Drosophila*, *Caenorhabditis elegans*, *Xenopus*, and zebrafish *(4–6)*. In mammals, PGC fate is specified in a subset of the epiblast through a fundamentally different mechanism, involving an inductive bone morphogenetic protein (BMP) signal from the extraembryonic ectoderm *(7,8)*. Epigenetic modes of germ cell specification may be the more widely distributed in evolution *(6)*.

PGCs in *Drosophila* begin their lives as pole cells, products of the first cellularization event in the syncytial embryo. Pole cells are committed to the germ cell fate at the time of their formation through the cytoplasmic inheritance of maternally

deposited pole plasm, or germplasm, which is sufficient for germline determination
(4,9). The pole plasm contains polar granules, electron-dense fibrous aggregates of
RNA, and protein, which are closely associated with dividing pole bud nuclei and
are specifically inherited by the PGCs on cellularization (*see* **Fig. 1.1**).

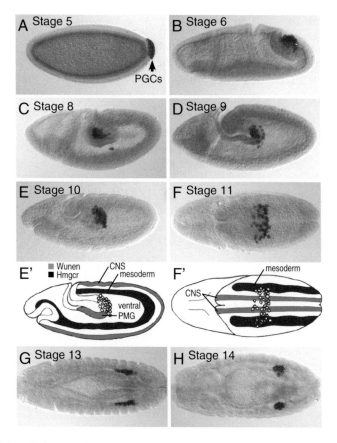

Fig. 1.1 Primordial germ cell migration. Lateral view, dorsal up, anterior left (**A–E**). The germline is the first distinct cell lineage established in the *Drosophila* embryo when primordial germ
cells (PGCs), or pole cells, cellularize at the posterior pole of the embryo (**A**). PGC fate is determined by their inheritance of polar granules, which are located in the posterior cytoplasm of the
embryo. A polar granule component, Vasa, is used here to mark the germline (**A–H**). PGCs are
carried dorsally and anteriorly as the germ band extends, and during gastrulation, the posterior
midgut (PMG) invaginates and carries the PGCs into the PMG lumen (**B** and **C**). PGCs then
migrate through the epithelium of the PMG into the interior of the embryo (**D** and **E**). PGCs
migrate away from Wunen expression in the ventral PMG (grey in **E'**) and toward the 3-hydroxy-
3-methylglutaryl coenzyme A (HMG-CoA) reductase (Hmgcr)-expressing mesoderm (black in
E') and thus move dorsally. Dorsal view, anterior left (**F–H**). PGCs migrate toward the Hmgcr-
expressing mesoderm (black in **F'**) and split into two groups as they are repelled from the midline
by Wunen expression in the central nervous system (CNS; gray in **F'**). Within the mesoderm,
PGCs associate with the somatic gonadal precursors (SGPs) in parasegments 10–12 (**G**), before
migrating anteriorly and coalescing to form embryonic gonads in parasegment 10 (**H**)

Genetic screens have revealed many important constituents of the pole plasm. Loss of some polar granule components can cause maternal lethality, because the pole plasm is essential for stable localization of the essential posterior morphogen Nanos. Other mutations affecting pole plasm components cause a grandchildless phenotype, in which viable embryos are produced that lack germ cells. A critical component for polar granule assembly in *Drosophila* is Oskar (Osk), with accumulation in the posterior oocyte cytoplasm that is necessary and sufficient for the recruitment of Vasa and Tudor, thus initiating the polar granule assembly pathway *(4,9)*. A second class of pole plasm components does not affect somatic development and is involved only in germ cell development. This class includes the RNAs *gcl* and *pgc*, which are discussed in **Subheading 1.2.1**, and Piwi, which has been implicated in gene silencing by RNA interference *(10,11)*. Reducing the maternal contribution of Piwi, or the RNA-induced silencing complex (RISC) components Dicer1 or Fragile X mental retardation protein (FMRP), drastically reduces the number of pole cells without affecting posterior somatic patterning *(11)*. In addition to their roles in maintaining GSC fate and controlling GSC division in the adult *(12–14)*, Piwi and other components of the microRNA machinery may regulate germline fate determination through microRNA-mediated translational control in the pole plasm.

Polar granules are thought to regulate the translation of maternal messenger RNAs (mRNAs) that are required for germ cell determination and early germ cell behavior. The pole plasm is assembled and stabilized at the posterior pole of the developing oocyte before fertilization *(4,9)*. Its constituents are produced by nurse cells during oogenesis and pass into the oocyte via large cellular interconnections called ring canals. Polar granules share components (including the translational regulators Vasa and Aubergine) with the perinuclear nuage, electron-dense ribonucleoprotein bodies near the germline cell nuclear pores. Regardless of how they are specified, a common characteristic of germ cells is the presence of these extranuclear organelles (e.g., nuage, P-granules, or chromatoid bodies), which are possibly sites of messenger ribonucleoprotein (mRNP) biosynthesis and germline-specific translational control.

1.2.1   *Transcriptional Quiescence of PGCs*

The PGCs are transcriptionally quiescent until the onset of gastrulation, when they initiate zygotic transcription of germline-specific genes such as Vasa *(15)*. PGC transcriptional quiescence correlates with reduced levels of active RNA polymerase II (RNAP II) that is phosphorylated on its C-terminal domain (CTD), suggesting that transcriptional repression might be established through regulating RNA Pol II function *(16,17)*. The maternally supplied mRNAs *gcl*, *nos*, and *pgc* contribute to the transcriptional quiescence of PGCs, apparently through independent mechanisms that repress the transcription of different sets of genes *(18)*.

nanos mRNA is maternally loaded into the pole plasm and is translated after fertilization. It forms a posterior-to-anterior protein gradient and directs posterior somatic patterning by shaping the anterior Hunchback morphogen gradient. Nos protein is inherited by the PGCs, in which it maintains transcriptional quiescence until they begin to migrate through the midgut epithelium *(15,16,19)*. In embryos produced by *nos* mutant females, PGCs abnormally transcribe a subset of somatic genes, such as *Sex-lethal (Sxl)*, and the segmentation genes *fushi tarazu* and *even skipped (19–21)*. They do not complete their migration into the embryonic gonad, fail to repress mitosis, lose germ cell fate, and are eliminated by apoptosis *(20,22– 24)*. The cell cycle effects caused by loss of Nos can be explained by a direct role for Nos and the RNA binding protein Pumilio (Pum) in repressing the translation of *CyclinB* mRNA *(25,26)*. *CyclinB* mRNA is another pole plasm component, and its translation is normally repressed in migrating PGCs until they reach the gonad. CyclinB normally promotes the division of PGCs, and its premature translation in *nos* or *pum* mutants allows precocious mitosis in migrating PGCs *(25,27)*. Because Nos is a translational repressor, it presumably acts indirectly to repress transcription in wild-type PGCs by regulating the translation of another, unidentified factor.

Another gene that affects transcriptional quiescence of PGCs is *gcl*. Maternally supplied *gcl* mRNA is translated at the posterior pole of the embryo and is specifically inherited by the PGCs *(28)*. The level of RNAP II CTD phosphorylation is increased in pole buds (precursors to PGCs) of embryos derived from *gcl* mutant mothers, bringing it to a level similar to that of the surrounding somatic nuclei. This correlates with the misexpression of somatic genes, such as *sisterless-a* and *scute*, in the pole buds of *gcl* mutant embryos *(17)*. Conversely, ectopic *gcl* expression at the anterior of an otherwise wild-type embryo is sufficient to downregulate somatic expression of *sisterless-a* and *tailless* in anterior cells *(17)*. Thus, *gcl* is implicated in establishing transcriptional quiescence in newly formed pole buds. Although its mechanism of action remains to be determined, *gcl* protein associates with the nucleoplasmic surface of pole cell nuclear envelopes, where it might affect chromatin compaction *(29)*. Even though most *gcl* mutant embryos do not form PGCs, *gcl* is not absolutely required for this process; some embryos form a few PGCs and develop into fertile adults, suggesting redundancy in the mechanisms that silence transcription in the germline.

Similar to *gcl*, PGCs in embryos that lack maternally contributed *pgc* have elevated levels of RNAP II CTD phosphorylation and express somatic genes such as *zerknüllt*, *tailless*, and *slam (18,30,31)*. Loss of *pgc* leads to increased levels of methylated histone H3, which is associated with increased transcriptional activity *(18)*. *pgc* mRNA is localized to the germplasm and is inherited by PGCs, and its disappearance correlates with the onset of zygotic transcription in the germline *(31)*. A small 71-amino acid protein that is transiently expressed in the PGCs of stage 4–6 embryos is encoded by *pgc* (A. Nakamura, personal communication). Although the functions of *gcl* and *pgc* in transcriptional repression have not been defined, these results suggest that the undifferentiated state of germline precursors is maintained during embryogenesis by preventing transcriptional responses in PGCs to somatic patterning signals.

1.3   The Embryonic Gonad

1.3.1   Primordial Germ Cell Migration

In *Drosophila* as well as many vertebrate systems, PGCs and the somatic cells of the gonad are specified at different locations, so the germ cells must migrate to reach the somatic part of the gonad (*see* **Fig. 1.1**). PGC migration is a process involving several discrete steps, controlled by genetically separable mechanisms *(32–34)*. During gastrulation, the germ band extends dorsally, and the PGCs are carried into the lumen of the invaginating posterior midgut (PMG). Once inside the PMG lumen, the PGCs lose close association with one another, become amoeboid, and migrate through the PMG epithelium. Migrating PGCs use filopodia to contact each other and their substrate; in the absence of Jak-Stat (Janus kinase-signal transducer and activator of transcription) signaling, the filopodia are reduced, and PGC are frequently mislocalized *(35)*. This is the first active step in PGC migration, and it involves the G protein-coupled receptor (GPCR) encoded by *trapped in endoderm 1* (tre1; *36*). *tre-1* mRNA is maternally encoded and acts in the germline, but the ligand for this receptor has yet to be identified. Once the PGCs traverse the PMG epithelium, they move dorsally along the midgut and split into two groups that move laterally, away from the midline. PGCs migrate toward the mesoderm to contact three bilateral clusters of SGPs in parasegments 11 and 12 *(37)*. During germ band retraction, the resulting groups of somatic and germline cells migrate anteriorly until they coalesce into a tight ball to form the embryonic gonad (*see* **Subheading 1.3.2**).

Migration of PGCs away from the ventral side of the PMG, and later away from the dorsal midline, is driven by repulsive cues provided by two phospholipid phosphatases, Wunen and Wunen2 (*see* **Fig. 1.1** and **refs. 38** and *39*). The two Wunens, which are believed to have similar functions, are expressed together in tissues flanking the migration route, where they appear to provide a repulsive migration cue to PGCs. Wunens are expressed ventrally in the PMG, guiding PGC migration dorsally (*see* **Fig. 1.1E** and **refs. 38** and *39*). Expression of Wunens in the central nervous system (CNS) then drives PGC migration laterally, away from the midline (*see* **Fig. 1.1F** and **ref. 40**). Although Wunen expression in the mesoderm is sufficient to override the normal attractive cues of the SGPs, in the absence of both Wunens, germ cells migrate randomly through the embryo. The lipid phosphate hydrolysis activity of the two Wunens is localized to the extracellular environment. Although Wunen substrates in the embryo are unknown, their predicted enzymatic activity has led to the suggestion that Wunens act as an attractant sink instead of a repellant source *(41,42)*. Wunens could locally hydrolyze and thus destroy a phospholipid attractant, and the PGCs would migrate toward higher levels of that attractant.

In addition to the Wunen cues, PGCs are guided toward the SGPs by attractive signals, including Hedgehog (Hh) and one or more products of a lipid metabolism pathway involving 3-hydroxy-3-methylglutaryl coenzyme A (HMG-CoA) reductase

(Hmgcr). Localized overexpression of either Hh or Hmgcr is sufficient to alter the migration path of PGCs, and both are expressed at high levels in SGPs as well as in other tissues, where they play vital roles in development *(43–46)*. Although the Hmgcr pathway attractant itself has yet to be identified, geranyl-geranyl diphosphate, an isoprenoid product of the pathway that is attached to a variety of proteins, is crucial, implicating a protein modified by this lipid as the PGC guidance cue *(45)*. The Hh signaling pathway is potentiated by Hmgcr, suggesting that the Hh signal emanating from the SGPs may be modified by Hmgcr, providing a mechanism for the migrating germ cells to distinguish Hh produced by the SGPs *(46,47)*. Hmgcr and geranyl-geranyl protein modifications have also been implicated in zebrafish germ cell migration, suggesting evolutionary conservation of signals that guide migrating germ cells in spite of wide divergence in anatomical structure *(48,49)*.

1.3.2 Gonad Coalescence

Gonad development involves the coordination of precursors from two distinct lineages: the SGPs of mesodermal origin and the PGCs that completed migration from the embryonic posterior pole. The SGPs are specified in bilateral clusters from part of the dorsolateral mesoderm of parasegments 10–12 *(50)*. Differential expression of *abdominal A (abdA)* and *Abdominal B (AbdB)*, homeotic genes that specify regional identity along the anterior–posterior axis of the embryonic abdomen, applies anterior, posterior, and male-specific SGP (msSGP) identities among the adjacent parasegmental clusters *(37)*. SGPs can be identified by their expression of the 68–77 LacZ line or the nuclear proteins Eyes Absent (Eya) or Zfh-1 *(51–53)*. On adherence of migratory PGCs and the gonadal precursors, they migrate anteriorly to form bilateral groups of cells in ps10, each including on average 12 PGCs and 30 mesodermal cells that coalesce into a round and compact gonad *(37,54,55)*. After coalescence, SGPs intermingle with and individually ensheath PGCs in a process that requires the cell adhesion molecule E-Cadherin (E-Cad) and a novel transmembrane protein Fear-of-Intimacy *(56,57)*. During coalescence, the PGCs are round and lack cellular extensions that would suggest an active role in the process; in contrast, the SGPs send cytoplasmic extensions that contact germ cells and other SGPs.

The Maf family transcription factor encoded by *traffic jam (tj)* is a key regulator of the morphogenetic movements that occur after gonad coalescence *(58)*. Tj protein is expressed in the SGPs of embryonic gonads and in the somatic cells in contact with the germline at later stages of development. In *tj* mutants, SGPs are specified normally and ensheath PGCs but are restricted to the outside of the germline cluster instead of intermingling with the PGCs *(58)*. This defect leads to the formation of small, disorganized gonads and sterility *(58,59)*. Tj regulates the expression of the cell adhesion molecules E-Cad, Fasciclin 3 (Fas3), and Neurotactin, and the two-layer gonad formed in *tj* mutants is consistent with altered adhesive properties between the soma and germline causing a cell-sorting phenotype.

1.4   Female GSC Niche Development

The germarium houses three types of stem cells: GSCs, escort stem cells (ESCs), and somatic stem cells (SSCs) that produce follicle cells (*see* **Table 1.1**). The core of the female GSC niche is a group of five to seven nondividing somatic cap cells that physically anchor two or three GSCs to the anterior of each germarium (*60–62*). Anterior to each cap cell cluster, a single-file array of about eight terminal filament (TF) cells make up the TF that connects the germarium with the sheath cells that surround the ovariole. Both the cap cells and TF, which can be visualized with a *hedgehog-lacZ* enhancer trap (*60, see* **Table 1.2** for a list of markers), are the sources of important stem cell maintenance signals such as Hh and Dpp, and the survival of GSCs is dependent on close association with these cells (*see* **Subheading 1.6.4** and **ref.** *63*). To the posterior of the cap cells and surrounding the GSCs and newly formed cystocytes are the inner germarial sheath (IGS) cells, which can be identified by *ptc-lacZ* expression (*64*). IGS cells line the surface of the anterior half of the germarium and are thought to act as a niche for the SSCs (*65*). Cystoblast differentiation is supported by a subpopulation of IGS cells called *escort cells*, derived from ESCs that send thin cytoplasmic processes to insulate cystoblasts and cystocyte clusters prior to follicle formation, when they die and are replaced by follicle cells (*66*).

During larval stages, the female gonad can be divided into three regions: the anterior somatic cells, where the TF and cap cell precursors reside; the medial region, housing the PGCs and somatic interstitial cells; and the posterior region, housing the precursors of the basal stalk cells (*see* **Fig. 1.2D**). During the third larval instar, cells are added to each TF gradually and in a progressive manner from medial to lateral across the ovary until early pupal development (*67,68*). Cap cell differentiation occurs in early pupae as TF formation is completed. The niche develops by cell rearrangement and cell recruitment, so that an individual niche is polyclonal in origin, technically limiting the ability to perform genetic loss-of-function

Table 1.1   Cell types in the male and female germline stem cell (GSC) niches

Cell type	Testis	Ovary
Primary GSC niche	Hub cells	Cap cells (CpC)
Primary niche–GSC pathway	Jak-Stat	Dpp-Bam
Differentiating GSC daughter	Gonialblast (GB)	Cystoblast (CB)
Somatic cells supporting differentiation of cysts	Somatic cyst cells (SCCs)	Inner germarial sheath (IGS) cells/escort cells (ECs)
Progenitor of somatic cyst cells	Somatic cyst progenitor stem (SCP) cells	Escort stem cells (ESCs)
Follicle cell progenitors (female specific)		Somatic stem cells (SSCs) or follicle stem cells (FSCs)

Table 1.2 Useful niche and germline stem cell (GSC) markers

LacZ P-element insertion lines	
842-lacZ	Hub cells, SSCs and cyst cells *(87)*
1444-lacZ	IGS cells *(129)*
3914-lacZ	Cap cells (weak: TF, IGS) and hub cells (weak: somatic cyst progenitors) *(21)*
68-77-lacZ	Gonadal mesoderm *(37)*
bab-lacZ	Somatic cells of the pupal niche *(68)*
bluetail-lacZ	Male-specific SGPs *(86)*
esg-lacZ	Male anterior SGPs and hub *(88)*
hh-lacZ	Hub cells and cap cells/TF *(60)*
M5-4-lacZ	Hub cells, GSCs and gonialblasts *(130)*
ptc-lacZ	Cap cells (weak: TF, IGS) *(60)*
Stat-lacZ	Cap and IGS cells *(131)*
wingless-lacZ	Cyst progenitor cells and cyst cells (17-en-40) *(64)*
Antibodies	
α-Spectrin	Spectrosome and fusome *(70)*
Bag of Marbles	Dividing cysts *(106,119)*
E-Cadherin	Cap cells *(77)*, hub cells *(83)*
Eyes absent	Male-specific somatic gonadal precursors (msSGPs) *(86)*
Engrailed	Cap cells, terminal filament *(64)*
Nanos	Highest in GSCs and cystoblasts (female) *(24)*
pMad (phosphorylated Mothers against *Dpp*)	GSCs (weak in dividing cysts) *(112,118)*
Vasa	Germline *(132)*

studies in the niche *(68,69)*. Advances in available genetic tools, such as the recent development of a method to target clone formation to the somatic gonadal cells *(69)*, hold promise to expand the genetic analysis of niche morphogenesis and GSC maintenance.

GSCs can be defined as the only mitotically active germline cells that remain in the gonad and that continue to divide long after they are genetically marked. They are most commonly identified as the germline cells in contact with the cap cells (female) or hub cells (male) of the stem cell niche and that contain a particular cytoplasmic organelle known as a *spectrosome*, which in the female is invariably attached to the site of contact with the niche (*see* **Fig. 1.2E** and **refs. 70** and **71**). Viewed with antibodies recognizing an adducin-like protein (MAb1B1) or α-Spectrin, the spectrosome is an apical sphere that elongates and reaches into the daughter cell during mitosis, before it is partitioned unequally on cytokinesis, with the stem cell retaining a greater portion *(72)*. Dividing GSCs are easily observed because of a prolonged G1 phase of mitosis, during which the mother and daughter cells are connected by an elongated spectrosome through a temporary cytoplasmic bridge. After division, the spectrosome retracts to the anterior of the GSC and regains its spherical shape. The spectrosome is the progenitor of the fusome, a meshwork of membranes and membrane skeletal proteins representing a germ cell-specific

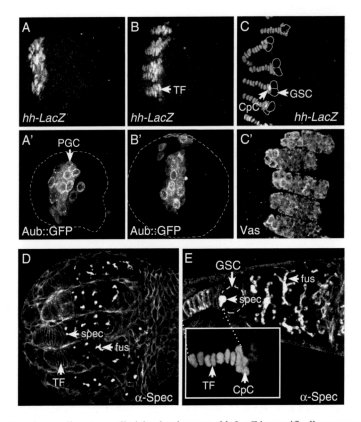

Fig. 1.2 Female germline stem cell niche development. *hh-LacZ* is specifically expressed in the developing GSC niche in the anterior of the developing ovary (**A–C**). Niche development commences in early third larval instar with the identity of somatic terminal filament (TF) cells in the anterior of the gonad. Terminal filament cells intercalate and separate into clusters (**A**), then extend by cell shape changes and by cell recruitment during late third larval instar (**B**). The niche develops anterior to the primordial germ cell (PGC) population, which is located in the medial region of the ovary, and contacts a subset of PGCs, marked here with Aub::GFP (**A', B'**). Cap cells (CpC) are formed in pupal ovaries after TF formation is complete (**C**). GSC fate is restricted to anterior PGCs during the early pupal stage based on their juxtaposition to cap cells (**D**, early pupae; **E**, late pupae). Antibodies to α-Spectrin mark the cell membranes of TF cells, the spectrosomes of PGCs and GSCs (spec), and the fusomes of differentiating germline cells (fus). Anterior PGCs next to the developing niche maintain a single spectrosome that becomes attached to the CpC–GSC contact site. The differentiation of germline cells further away from the niche is marked by the presence of branched fusomes (**D, E**)

modification of the endoplasmic reticulum *(73,74)*. During cyst divisions, the fusome grows and fuses to become an elongated, branched structure that extends throughout the cyst, where it coordinates mitosis and microtubule polarity.

In the female, PGCs proliferate during larval stages to produce a population of over 100 undifferentiated germline cells *(75)*. PGCs proliferate with random division orientation, and clonal analysis indicated dispersion of daughter cells, so that there is little detectable relationship between clonal history and position *(76)*. Lineage analysis of single, marked PGCs did not indicate early restriction of GSC fate among PGCs, although there is a tendency for PGCs in the anterior of the embryonic gonad to populate the adult niche *(76)*. GSCs are anchored to the cap cells by adherens junctions, and the division plane of the GSC division is such that one daughter cell loses close contact with the cap cells and differentiates *(77)*. Adherens junctions between presumptive GSCs and newly formed cap cells appear during early pupal stages, at the same time that differentiation is first apparent in PGCs away from the niche (using Bam expression or branched fusomes as markers) *(77,78)*. The spectrosomes of the PGCs in closest proximity to the cap cells become anchored to the site of the newly forming adherens junction, thereby adopting the architecture that we use to recognize adult GSCs *(78)*. Thus, after the completion of the stem cell niches, the nearby anterior PGCs become GSCs, and the rest of the PGCs directly enter the cyst differentiation pathway *(77–79)*.

1.5 Male GSC Niche Development

In the testis, there are two types of stem cells that maintain spermatogenesis: the GSCs and the cyst progenitor cells, which produce somatic cyst cells (SCCs) to encapsulate differentiating germ cells (*see* **Table 1.1**). Approximately 12 nondividing somatic hub cells at the apical tip of the testis, commonly marked with Fas3 antibodies *(80)*, maintain five to nine spectrosome-containing GSCs in a characteristic rosette pattern (*see* **Fig. 1.3** and **refs.** *81* and *82*). The hub cells make up the GSC niche, the cellular microenvironment that regulates, induces, and supports the development of GSCs. Each GSC is surrounded by two somatic cyst progenitor cells, which are also attached directly to the hub. During male GSC division, one spindle pole associates with the GSC–niche interface, and the GSC divides radially with respect to the hub. This orientation ensures that one of the two progeny cells remains in the niche as a GSC, and the other, the gonialblast, is displaced from the niche and differentiates *(83)*. Each gonialblast is surrounded by two SCCs that support its differentiation. The behavior of the fusome and ring canals during cyst divisions in testes is highly similar to that in ovaries, except that the male GSC spectrosomes are not specifically localized to the GSC–niche interface as in the female (*see* **Fig. 1.3** and **ref.** *83*).

Although striking similarities exist between the GSC niches of *Drosophila* males and females *(66,84)*, sexual dimorphism in gonad morphogenesis is present soon after gonad coalescence. Markers such as Eyes absent (Eya) and the *Drosophila* homolog of Sox9 (Sox100B) are expressed at highest levels in the msSGPs at the posterior of the gonad *(85,86)*. The msSGPs are initially specified in ps13 of both males and females but are subsequently lost in the female by apoptosis *(85)*. At the anterior of the gonad, *escargot (esg)-LacZ* marks a different subset of SGPs in which the precursors of the male stem cell niche are located *(87,88)*. The expression of *esg-LacZ* becomes further restricted to a subset of

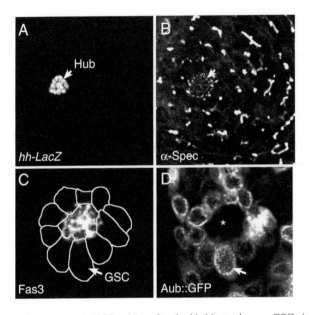

Fig. 1.3 The germline stem cell (GSC) niche of male third instar larvae. GSCs in the testis are maintained by a cluster of somatic hub cells that specifically express an *hh-lacZ* enhancer trap (**A**) and Fas3 protein (**C**). The position of the hub from **A** is noted in **B** (arrow). Antibodies to α-Spectrin mark GSC spectrosomes next to the hub and the fusomes of differentiating gonial-blasts and dividing cysts (**B**). GSCs are arranged around the hub in a characteristic rosette pattern (**C**). The positions of GSCs from **D** are noted around the Fas3-expressing hub in **C**. Germline cells arranged around the hub (*) express an Aub::GFP (green fluorescent protein) fusion protein (**D**)

anterior SGPs when development of the male stem cell niche commences at the end of embryogenesis *(88)*. It is not known how *esg-LacZ* expression and hub cell fate are restricted to a subset of anterior SGPs.

The embryonic male stem cell niche closely resembles that of the adult male: a tight cluster of SGPs that express *esg-LacZ,* Fas3, and the Jak-Stat ligand Unpaired, which is associated with a group of anterior PGCs in the same rosette pattern as that observed in the adult *(81,88)*. The restriction of hub cell fate and the rosette pattern of PGCs around the hub suggest that a functional GSC niche forms in male embryos, much earlier than niche development in the female. The presence of sper-matogonia in newly hatched larvae and cysts of spermatocytes before the second larval instar indicate that spermatogenesis has commenced before hatching *(89)* and suggest that the PGCs organized around the embryonic hub are already specified as GSCs. Although it is not clear how the newly formed hub recruits GSCs from the PGC population, the major GSC maintenance signals in the male, Unpaired and Glass-bottom boat (Gbb; *see* **Subheadings 1.6.3** and **1.6.4**), are both short-range signals, suggesting that anterior PGCs first need to be in close contact with the hub to receive these signals.

Part of the mechanism used to capture GSCs involves the localization of high levels of E-Cad, a large homophyllic transmembrane adhesion molecule, to the hub. E-Cad expression is initially increased in all SGPs at the time of gonad coalescence, but high-level E-Cad expression is refined to a pattern resembling that of *esg-lacZ*: first to the anterior portion of the gonad *(57)* and then to the hub cells *(90,91)*. Although not yet demonstrated in the niche, *esg* positively regulates E-Cad expression during tracheal development *(92)*. E-Cad is an integral component of the adherens junctions that maintain GSC fate in adults by ensuring physical contact between GSCs and the niche and that polarize GSC divisions perpendicular to the hub *(77,83,84)*.

Drosophila gef26 encodes a guanine nucleotide exchange factor for the Rap guanine triphosphatase (GTPase) pathway that localizes to the hub–GSC interface *(93,94)*. In *gef26* mutants, the hub is specified normally, but GSC fate is not maintained because of the absence of adherens junctions between hub cells and GSCs *(93)*. Experiments in mammalian cell culture and *Drosophila* epithelia indicated a direct role for the Rap pathway in localizing E-Cad to newly formed intercellular contact sites, facilitating the development of mature adherens junctions *(95,96)*. The interaction between the Rap pathway and E-Cad suggests that a Rap-mediated signal from the hub guides the formation of adherens junctions specifically at the interface of hub cells and nearby PGCs, effectively capturing GSCs from among the population of PGCs near the hub *(93)*.

1.6   Niche–PGC Interactions

1.6.1   Egfr Signaling

After the gonad forms, regulatory interactions between the PGCs and gonadal mesoderm ensure that a proper ratio of germline to soma is maintained through larval stages. An example illustrating the importance of this mechanism is the phenotype of *germcell-less (gcl)* mutants. Although most embryos produced by *gcl* mutant females do not form PGCs, some do form a few PGCs and develop into fertile adults *(97)*. Examination of *gcl* mutant larval gonads, which in L1 have an average of one to three PGCs, revealed an increase in PGC division rate through larval development until their numbers reach near wild type *(75)*. PGC number is likely to be regulated by the BMP family member Dpp because overexpressing Dpp in the somatic cells of larval gonads increases PGC number by promoting proliferation *(78)*. PGC number is also regulated by the somatic activity of the Epidermal growth factor receptor (Egfr), which is probably activated by the Egf ligand Spitz from PGCs *(75)*. Decreasing Egfr pathway activity in the soma, or decreasing Spitz expression in the PGCs, yields an expanded number of PGCs and a concomitant decrease in the number of somatic cells intermingled with the PGCs *(58,75)*. Conversely, constitutively high Egfr activity in the soma reduces the number of PGCs through an undefined inhibitory mechanism. Thus, a dividing population of PGCs appears to positively regulate the number of somatic support cells, which in turn keep PGC numbers in check during larval development by inhibiting further PGC division.

In addition to regulating mitosis, germline–soma communication affects the encapsulation of germline cells by somatic support cells. Stem cell tumor (Stet), a rhomboid-like transmembrane protease, is required in male and female germline cells to promote their enclosure by somatic cells *(98)*. Rhomboid cleaves and activates the membrane-tethered Egfr ligand Spitz in the signaling cell, implicating Stet in Egfr pathway regulation. Without germline Stet, somatic support cells do not surround germ cells properly, leading to the accumulation of cysts at early stages of differentiation in adult gonads *(98)*. The *stet* phenotype is consistent with disrupted signaling between the germline and somatic support cells because of the absence of normal intercellular contacts between the two cell types.

1.6.2   Notch Signaling

Notch signaling is involved in a feedback regulatory mechanism between GSCs and niche cells in the female. Notch and its ligands Delta and Serrate are large transmembrane proteins with extracellular EGF-like repeats *(99)*. Notch signaling operates by a direct mechanism by which, on activation, the intracellular domain of Notch is released from the membrane and accesses the nucleus. In the nucleus, it interacts with the DNA-binding protein Suppressor of Hairless [Su(H)] to activate expression of target genes, such as genes of the *Enhancer of split Complex* [*E(spl)-C*]. GSCs mutant for Delta or Serrate are not maintained in the niche, and Notch and Su(H) are dispensable in GSCs, indicating that Notch ligands signal from the GSCs to another cell type to maintain GSC fate *(100)*. One target of the germline Delta signal are the cap cells because germline overexpression of Delta induces up to a tenfold increase in the number of cap cells at the anterior of the germarium *(100)*. This increase in cap cell number increases the capacity of the stem cell niche because more cells with GSC markers, such as increased pMAD levels (*see* **Subheading 1.6.4**), are present after germline Delta overexpression. These results illustrate that a germline-to-niche signal affects the normal functioning of the niche, and that the flow of information between niche and stem cells is not unidirectional. The ESCs are also possible targets for the germline Delta signal, and further study is likely to result in a more elaborate model of signaling cross talk among the cells of the niche and the GSCs.

1.6.3   Jak-Stat Signaling

The Jak-Stat pathway is the major signaling pathway required for GSC maintenance in the male. Upd, a short-range signal expressed in the hub cells, maintains

GSC fate by activating the Jak-Stat signaling pathway in GSCs *(101,102)*. In *Drosophila*, Jak-Stat signaling initiates when Upd binds to the Domeless receptor and activates Hopscotch (the Janus kinase). Hopscotch activity results in phosphorylation of the Stat92E transcription factor, which translocates into the nucleus and activates target gene transcription *(103)*. GSCs mutant for *hop* or *Stat92E* cannot transduce the Upd signal and lose their ability to self-renew, whereas Upd overexpression is sufficient to induce an excess of GSC fate *(101,102)*. These results indicate that Upd is both necessary and sufficient for GSC fate in the male. Understanding the mechanism by which GSC fate is controlled by the Jak-Stat pathway awaits the identification of relevant target genes.

In the female, the Jak-Stat pathway plays a less-important role in GSC maintenance. Upd overexpression in the cap cells and escort cells causes an increase in GSC proliferation without a striking effect on GSC number *(66)*. Reducing *Stat92E* function specifically in the ESCs and escort cells affects the normal organization of the anterior germarium, leading to a reduction in stem cell number *(66)*. These findings illustrate the complexity of the GSC–niche signaling network because the niche signals affect the stem cells as well as their support cells, and the support cells affect stem cell proliferation and differentiation.

1.6.4   Dpp Signaling and Bag of Marbles

Dpp signaling is the major GSC-maintaining pathway in females, fulfilling a function analogous to the Jak-Stat pathway in the male. Dpp, the *Drosophila* homolog of human BMP2, acts nonautonomously over the short range from the stem cell niche to the GSCs to repress the transcription of *bag of marbles (bam) (104,105)*. *Bam* is the cystoblast determinant that is both necessary and sufficient for GSC differentiation in the female *(106,107)*. As Bam contains no well-characterized functional motifs, the biochemical mechanism by which it regulates cyst formation remains unclear. Bam works together with Benign Gonial Cell Neoplasm (Bgcn), an unusual RNA helicase-like protein, suggesting that a Bam:Bgcn complex might regulate the translation of cystoblast promoting mRNAs *(108)*.

Loss of *dpp* signaling in GSCs allows *bam* expression and cyst differentiation, and conversely, constitutive *dpp* signaling prevents *bam* transcription and blocks the differentiation of GSC daughter cells *(104,109)*. It appears that *bam* is the principal Dpp target because Bam expression is sufficient to drive differentiation of GSCs or cystoblasts even in the presence of high levels of Dpp *(109)*. The repression of *bam* by Dpp is direct as the Dpp pathway transcriptional corepressors Mothers against Dpp (Mad) and Medea (Med) bind to transcriptional silencer elements in the *bam* promoter *(104,105,109)*.

In PGCs, *bam* is not expressed until early pupal stages, when the PGCs not in contact with the cap cells begin to differentiate. Forced expression of *bam* in PGCs during larval stages is sufficient to drive the differentiation of all PGCs, effectively eliminating PGC fate *(110)*. Similarly, preventing Dpp signal transduction in PGCs by removing the Dpp receptor Tkv induces differentiation of PGCs, beginning as early as late embryogenesis *(110)*. These data strongly support a model in which Dpp is continuously required throughout embryonic and larval development to repress Bam expression and PGC differentiation. In fact, active BMP signaling, measured with an antibody specific to the active phosphorylated form of the transcription factor MAD (pMAD), can be observed in embryonic pole cells and in all PGCs throughout larval stages *(110–113)*. Once niche formation is completed in early pupae, a Dpp signal from the cap cells prevents anterior PGCs from differentiating, thereby maintaining their potential to become GSCs *(78)*.

bam mutant ovaries fill with undifferentiated germline cells that resemble GSCs, confirming the critical role for *bam* in the differentiation of the stem cell's descendants *(106,107)*. *bam* mutant germline cells can be maintained for long periods in the presence of Dpp and cultured somatic cells from the germarium *(114)*. Under these conditions, they maintain their spectrosomes and even develop adherens junctions with somatic cells *(114)*. Undifferentiated cells isolated from *bam* mutant ovaries, when injected into the dorsal mesoderm of embryos, are able to populate the embryonic gonad and establish a dividing population of *bam* mutant stem cells in the adult *(115)*. These cells retain the capacity to differentiate on transgenic rescue of *bam* expression. These results show that the *bam* mutant GSC-like cells are not committed to a post-stem-cell fate because they are able to revert to PGC fate and establish a stem cell population in the adult. They also imply a close relationship between the undifferentiated state of GSCs and PGCs.

As in the female, the BMP ligands Dpp and Gbb play a role in male GSC maintenance *(116–118)*. Loss of BMP pathway components leads to Bam expression in GSCs and loss of GSC fate. As in the female, overexpressing Bam is sufficient to drive GSC differentiation. However, unlike results with Upd in the male or Dpp in the female, overexpression of Dpp or Gbb is not sufficient to block the differentiation of gonialblasts in the male—Bam is not required for gonialblast differentiation but instead restricts the number of cyst divisions to four *(116–119)*. These results indicate that Bam repression in both male and female GSCs is required to maintain GSC fate, but that Bam differentially affects the differentiation of cystoblasts and gonialblasts. The BMP and Jak-Stat signaling pathways control GSC self-renewal in both males and females, but the relative importance of the two pathways is reversed, perhaps reflecting differences in gonadal development.

1.6.5 *Nanos*

Nanos (Nos) is an evolutionarily conserved translational repressor that maintains the fate of germline progenitors in *C. elegans*, zebrafish, and mouse embryos *(120–122)*. In *Drosophila*, Nos functions in a complex with Pumilio (Pum), the founding member of a widely conserved class of RNA-binding proteins (Puf domain proteins). By preventing their differentiation, *nos* promotes PGC fate during larval stages and GSC fate in adults *(24,123,124)*. If *nos* or *pum* function is removed during larval stages, a large proportion of PGCs, even those in contact with the stem cell niche, prematurely differentiate into cysts *(124)*. Differentiation of *nos* mutant PGCs begins as early as the first larval instar, so that by the third instar, 16-cell cysts are already present. Similarly, loss of *nos* or *pum* from adult GSCs allows them to differentiate *(110)*. Analogous to its role as a translational repressor of *hunchback* mRNA during embryogenesis *(123)*, a Nos:Pum heterodimer likely prevents the activation of a PGC–GSC differentiation pathway by repressing the translation of factors important for germ cell differentiation.

Remarkably, loss of *pum* is sufficient to allow the differentiation of *bam* mutant germline cells, suggesting that Bam represses the Nos:Pum complex to allow the translation of differentiation-promoting mRNAs *(125,126)*. In this model, Bam prevents the function of the Nos:Pum complex in cystoblasts and dividing cysts, allowing cells to differentiate if low Dpp signaling permits Bam expression. In GSCs, it is proposed that Dpp signaling prevents Bam expression, thus activating the Nos:Pum complex and maintaining an undifferentiated state *(125,126)*.

If applied to PGCs, however, there are several observations that are not easily explicable within the framework of this model. First, in *nos* mutant larval gonads, differentiating cysts show active *bam* transcription *(110)*. Bam is not normally expressed before pupation, and this is an unexpected result if Nos acts downstream of Dpp and Bam. Second, loss of *nos* or *pum* does not allow the differentiation of PGCs that are constitutively active for Dpp signaling *(110)*. These results indicate that the Nos:Pum complex does not prevent PGC differentiation in a simple linear pathway downstream of Dpp signaling, underscoring the complexity of interactions between extrinsic somatic signals and factors intrinsic to PGCs that prevent differentiation.

1.7 Conclusion

Once the stem cell niche is established, it functions as a signaling center to recruit GSCs. A remarkable example illustrates this point. In the female germline, loss of Dpp signaling allows the GSCs to begin the cystoblast differentiation program. However, if Dpp signaling is reactivated in the cystoblast daughters, their limited differentiation can be reversed, and they can repopulate the niche as stem cells *(127)*. Similarly, early stages of gonialblast differentiation can be reversed in the male on the return of Jak-Stat activity to the differentiating male germline cysts *(128)*. Direct physical interaction between GCSs and their niche is essential for recruiting GSCs and for stem cell maintenance. GSCs are held in the niche by adherens junctions, and GSCs mutant for the cell adhesion molecules E-Cad or Armadillo (β-Catenin) lose their adherens junctions and differentiate *(65,83)*. E-Cad accumulates between cap cells and putative GSCs from the beginning of niche establishment, and this is important in recruiting the GSCs to the niche and maintaining their proximity to stem cell maintenance factors. Localizing high levels of E-Cad expression to the niche might play a central role in recruiting nearby PGCs into the niche. Once recruited, the undifferentiated state of the PGCs would then be maintained by niche signals into adulthood as GSCs. The remainder of the PGCs, and the daughters of the newly formed GSC, do not receive the niche signals, assume the default fate, and differentiate.

Acknowledgments We are grateful to Akira Nakamura for sharing unpublished results. We thank T. Xie for providing *hh-lacZ* flies; P. Macdonald for flies expressing Aub::GFP; and the Developmental Studies Hybridoma Bank, University of Iowa, for α-Spectrin and Fas3 antibodies. P.L. acknowledges financial support from the National Institute of Child Health and Human Development, the Canadian Institutes of Health Research, the Natural Sciences and Engeneering Research Council of Canada, and the National Cancer Institute of Canada.

References

1. Fuchs, E., Tumbar, T., and Guasch, G. (2004) Socializing with the neighbors: stem cells and their niche. *Cell.* **116,** 769–778.
2. Wong, M. D., Jin, Z., and Xie, T. (2005) Molecular mechanisms of germline stem cell regulation. *Annu. Rev. Genet.* **39,** 173–195.
3. Saffman, E. E., and Lasko, P. (1999) Germline development in vertebrates and invertebrates. *Cell Mol. Life Sci.* **55,** 1141–1163.
4. Mahowald, A. P. (2001) Assembly of the *Drosophila* germ plasm. *Int. Rev. Cytol.* **203,** 187–213.
5. Seydoux, G., and Schedl, T. (2001) The germline in *C. elegans*: origins, proliferation, and silencing. *Int. Rev. Cytol.* **203,** 139–185.
6. Extavour, C. G., and Akam, M. (2003) Mechanisms of germ cell specification across the metazoans: epigenesis and preformation. *Development.* **130,** 5869–5884.
7. Tsang, T. E., Khoo, P. L., Jamieson, R. V., et al. (2001) The allocation and differentiation of mouse primordial germ cells. *Int. J. Dev. Biol.* **45,** 549–555.
8. Saitou, M., Barton, S. C., and Surani, M. A. (2002) A molecular programme for the specification of germ cell fate in mice. *Nature.* **418,** 293–300.
9. Houston, D. W., and King, M. L. (2000) Germ plasm and molecular determinants of germ cell fate. *Curr. Top. Dev. Biol.* **50,** 155–181.
10. Pal-Bhadra, M., Bhadra, U., and Birchler, J. A. (2002) RNAi related mechanisms affect both transcriptional and posttranscriptional transgene silencing in *Drosophila. Mol. Cell.* **9,** 315–327.
11. Megosh, H. B., Cox, D. N., Campbell, C., and Lin, H. (2006) The role of PIWI and the miRNA machinery in *Drosophila* germline determination. *Curr. Biol.* **16,** 1884–1894.
12. Cox, D. N., Chao, A., Baker, J., Chang, L., Qiao, D., and Lin, H. (1998) A novel class of evolutionarily conserved genes defined by *piwi* are essential for stem cell self-renewal. *Genes Dev.* **12,** 3715–3727.
13. Hatfield, S. D., Shcherbata, H. R., Fischer, K. A., Nakahara, K., Carthew, R. W., and Ruohola-Baker, H. (2005) Stem cell division is regulated by the microRNA pathway. *Nature.* **435,** 974–978.
14. Forstemann, K., Tomari, Y., Du, T., et al. (2005) Normal microRNA maturation and germ-line stem cell maintenance requires Loquacious, a double-stranded RNA-binding domain protein. *PLoS Biol.* **3,** e236.
15. Van Doren, M., Williamson, A. L., and Lehmann, R. (1998) Regulation of zygotic gene expression in Drosophila primordial germ cells. *Curr. Biol.* **8,** 243–246.
16. Seydoux, G., and Dunn, M. A. (1997) Transcriptionally repressed germ cells lack a subpopulation of phosphorylated RNA polymerase II in early embryos of *Caenorhabditis elegans* and *Drosophila melanogaster. Development.* **124,** 2191–2201.

17. Leatherman, J. L., Levin, L., Boero, J., and Jongens, T. A. (2002) *germ cell-less* acts to repress transcription during the establishment of the *Drosophila* germ cell lineage. *Curr. Biol.* **12**, 1681–1685.

18. Martinho, R. G., Kunwar, P. S., Casanova, J., and Lehmann, R. (2004) A noncoding RNA is required for the repression of RNApolII-dependent transcription in primordial germ cells. *Curr. Biol.* **14**, 159–165.

19. Deshpande, G., Calhoun, G., Yanowitz, J. L., and Schedl, P. D. (1999) Novel functions of Nanos in downregulating mitosis and transcription during the development of the *Drosophila* germline. *Cell.* **99**, 271–281.

20. Kobayashi, S., Yamada, M., Asaoka, M., and Kitamura, T. (1996) Essential role of the posterior morphogen Nanos for germline development in *Drosophila*. *Nature.* **380**, 708–711.

21. Asaoka, M., Sano, H., Obara, Y., and Kobayashi, S. (1998) Maternal Nanos regulates zygotic gene expression in germline progenitors of *Drosophila melanogaster. Mech. Dev.* **78**, 153–158.

22. Hayashi, Y., Hayashi, M., and Kobayashi, S. (2004) Nanos suppresses somatic cell fate in *Drosophila* germ line. *Proc. Natl. Acad. Sci. U. S. A.* **101**, 10338–10342.

23. Coffman, C. R. (2003) Cell migration and programmed cell death of *Drosophila* germ cells. *Ann. N. Y. Acad. Sci.* **995**, 117–126.

24. Forbes, A., and Lehmann, R. (1998) Nanos and Pumilio have critical roles in the development and function of *Drosophila* germline stem cells. *Development.* **125**, 679–690.

25. Asaoka-Taguchi, M., Yamada, M., Nakamura, A., Hanyu, K., and Kobayashi, S. (1999) Maternal Pumilio acts together with Nanos in germline development in *Drosophila* embryos. *Nat. Cell. Biol.* **1**, 431–437.

26. Dalby, B., and Glover, D. M. (1992) 3' non-translated sequences in *Drosophila cyclin B* transcripts direct posterior pole accumulation late in oogenesis and peri-nuclear association in syncytial embryos. *Development.* **115**, 989–997.

27. Wang, Z., and Lin, H. (2005) The division of *Drosophila* germline stem cells and their precursors requires a specific cyclin. *Curr. Biol.* **15**, 328–333.

28. Jongens, T. A., Hay, B., Jan, L. Y., and Jan, Y. N. (1992) The *germ cell-less* gene product: a posteriorly localized component necessary for germ cell development in *Drosophila. Cell.* **70**, 569–584.

29. Jongens, T. A., Ackerman, L. D., Swedlow, J. R., Jan, L. Y., and Jan, Y. N. (1994) *germ cell-less* encodes a cell type-specific nuclear pore-associated protein and functions early in the germ-cell specification pathway of *Drosophila. Genes Dev.* **8**, 2123–2136.

30. Deshpande, G., Calhoun, G., and Schedl, P. (2004) Overlapping mechanisms function to establish transcriptional quiescence in the embryonic *Drosophila* germline. *Development.* **131**, 1247–1257.

31. Nakamura, A., Amikura, R., Mukai, M., Kobayashi, S., and Lasko, P. F. (1996) Requirement for a noncoding RNA in *Drosophila* polar granules for germ cell establishment. *Science.* **274**, 2075–2079.

32. Muller, H. A. (2002) Germ cell migration: as slow as molasses. *Curr. Biol.* **12**, R612–R614.

33. Santos, A. C., and Lehmann, R. (2004) Germ cell specification and migration in *Drosophila* and beyond. *Curr. Biol.* **14**, R578–R589.

34. Molyneaux, K., and Wylie, C. (2004) Primordial germ cell migration. *Int. J. Dev. Biol.* **48**, 537–544.

35. Li, J., Xia, F., and Li, W. X. (2003) Coactivation of STAT and Ras is required for germ cell proliferation and invasive migration in *Drosophila*. *Dev. Cell.* **5**, 787–798.

36. Kunwar, P. S., Starz-Gaiano, M., Bainton, R. J., Heberlein, U., and Lehmann, R. (2003) Tre1, a G protein-coupled receptor, directs transepithelial migration of *Drosophila* germ cells. *PLoS Biol.* **1**, E80.

37. Boyle, M., and DiNardo, S. (1995) Specification, migration and assembly of the somatic cells of the *Drosophila* gonad. *Development.* **121**, 1815–1825.

38. Starz-Gaiano, M., Cho, N. K., Forbes, A., and Lehmann, R. (2001) Spatially restricted activity of a *Drosophila* lipid phosphatase guides migrating germ cells. *Development.* **128**, 983–991.

39. Zhang, N., Zhang, J., Purcell, K. J., Cheng, Y., and Howard, K. (1997) The *Drosophila* protein Wunen repels migrating germ cells. *Nature*. **385**, 64–67.
40. Sano, H., Renault, A. D., and Lehmann, R. (2005) Control of lateral migration and germ cell elimination by the *Drosophila melanogaster* lipid phosphate phosphatases Wunen and Wunen 2. *J. Cell Biol*. **171**, 675–683.
41. Renault, A. D., Sigal, Y. J., Morris, A. J., and Lehmann, R. (2004) Soma-germ line competition for lipid phosphate uptake regulates germ cell migration and survival. *Science*. **305**, 1963–1966.
42. Hanyu-Nakamura, K., Kobayashi, S., and Nakamura, A. (2004) Germ cell-autonomous Wunen2 is required for germline development in *Drosophila* embryos. *Development*. **131**, 4545–4553.
43. Deshpande, G., Swanhart, L., Chiang, P., and Schedl, P. (2001) Hedgehog signaling in germ cell migration. *Cell*. **106**, 759–769.
44. Van Doren, M., Broihier, H. T., Moore, L. A., and Lehmann, R. (1998) HMG-CoA reductase guides migrating primordial germ cells. *Nature*. **396**, 466–469.
45. Santos, A. C., and Lehmann, R. (2004) Isoprenoids control germ cell migration downstream of HMGCoA reductase. *Dev. Cell*. **6**, 283–293.
46. Deshpande, G., and Schedl, P. (2005) HMGCoA reductase potentiates hedgehog signaling in *Drosophila melanogaster*. *Dev. Cell*. **9**, 629–638.
47. Besse, F., Busson, D., and Pret, A. M. (2005) Hedgehog signaling controls Soma-Germen interactions during *Drosophila* ovarian morphogenesis. *Dev. Dyn*. **234**, 422–431.
48. Thorpe, J. L., Doitsidou, M., Ho, S. Y., Raz, E., and Farber, S. A. (2004) Germ cell migration in zebrafish is dependent on HMGCoA reductase activity and prenylation. *Dev. Cell*. **6**, 295–302.
49. Kunwar, P. S., Siekhaus, D. E., and Lehmann, R. (2006) In vivo migration: a germ cell perspective. *Annu. Rev. Cell Dev. Biol*. **22**, 237–265.
50. Warrior, R. (1994) Primordial germ cell migration and the assembly of the *Drosophila* embryonic gonad. *Dev. Biol*. **166**, 180–194.
51. Boyle, M., Bonini, N., and DiNardo, S. (1997) Expression and function of *clift* in the development of somatic gonadal precursors within the *Drosophila* mesoderm. *Development*. **124**, 971–982.
52. Broihier, H. T., Moore, L. A., Van Doren, M., Newman, S., and Lehmann, R. (1998) *zfh-1* is required for germ cell migration and gonadal mesoderm development in *Drosophila*. *Development*. **125**, 655–666.
53. Simon, J., Peifer, M., Bender, W., and O'Connor, M. (1990) Regulatory elements of the *bithorax* complex that control expression along the anterior-posterior axis. *EMBO J*. **9**, 3945–3956.
54. Poirie, M., Niederer, E., and Steinmann-Zwicky, M. (1995) A sex-specific number of germ cells in embryonic gonads of *Drosophila*. *Development*. **121**, 1867–1873.
55. Sonnenblick, B. P. (1941) Germ cell movements and sex differentiation of the gonads in the *Drosophila* embryo. *Proc. Natl. Acad. Sci. U. S. A*. **27**, 484–489.
56. Van Doren, M., Mathews, W. R., Samuels, M., Moore, L. A., Broihier, H. T., and Lehmann, R. (2003) *fear of intimacy* encodes a novel transmembrane protein required for gonad morphogenesis in *Drosophila*. *Development*. **130**, 2355–2364.
57. Jenkins, A. B., McCaffery, J. M., and Van Doren, M. (2003) *Drosophila* E-Cadherin is essential for proper germ cell-soma interaction during gonad morphogenesis. *Development*. **130**, 4417–4426.
58. Li, M. A., Alls, J. D., Avancini, R. M., Koo, K., and Godt, D. (2003) The large Maf factor Traffic Jam controls gonad morphogenesis in *Drosophila*. *Nat. Cell Biol*. **5**, 994–1000.
59. Schupbach, T., and Wieschaus, E. (1991) Female sterile mutations on the second chromosome of *Drosophila melanogaster*. II. Mutations blocking oogenesis or altering egg morphology. *Genetics*. **129**, 1119–1136.
60. Forbes, A. J., Lin, H., Ingham, P. W., and Spradling, A. C. (1996) Hedgehog is required for the proliferation and specification of ovarian somatic cells prior to egg chamber formation in *Drosophila*. *Development*. **122**, 1125–1135.

61. Spradling, A. C., de Cuevas, M., Drummond-Barbosa, D., et al. (1997) The *Drosophila* germarium: stem cells, germ line cysts, and oocytes. *Cold Spring Harb. Symp. Quant. Biol.* **62,** 25–34.

62. Xie, T., and Spradling, A. C. (2000) A niche maintaining germ line stem cells in the *Drosophila* ovary. *Science.* **290,** 328–330.

63. Spradling, A., Drummond-Barbosa, D., and Kai, T. (2001) Stem cells find their niche. *Nature.* **414,** 98–104.

64. Forbes, A. J., Spradling, A. C., Ingham, P. W., and Lin, H. (1996) The role of segment polarity genes during early oogenesis in *Drosophila. Development.* **122,** 3283–3294.

65. Song, X., and Xie, T. (2002) DE-cadherin-mediated cell adhesion is essential for maintaining somatic stem cells in the *Drosophila* ovary. *Proc. Natl. Acad. Sci. U. S. A.* **99,** 14813–14818.

66. Decotto, E., and Spradling, A. C. (2005) The *Drosophila* ovarian and testis stem cell niches: similar somatic stem cells and signals. *Dev. Cell.* **9,** 501–510.

67. Sahut-Barnola, I., Godt, D., Laski, F. A., and Couderc, J. L. (1995) *Drosophila* ovary morphogenesis: analysis of terminal filament formation and identification of a gene required for this process. *Dev. Biol.* **170,** 127–135.

68. Godt, D., and Laski, F. A. (1995) Mechanisms of cell rearrangement and cell recruitment in *Drosophila* ovary morphogenesis and the requirement of *bric a brac. Development.* **121,** 173–187.

69. Bolivar, J., Pearson, J., Lopez-Onieva, L., and Gonzalez-Reyes, A. (2006) Genetic dissection of a stem cell niche: the case of the *Drosophila* ovary. *Dev. Dyn.* **235,** 2969–2979.

70. de Cuevas, M., and Spradling, A. C. (1998) Morphogenesis of the *Drosophila* fusome and its implications for oocyte specification. *Development.* **125,** 2781–2789.

71. Lin, H., Yue, L., and Spradling, A. C. (1994) The *Drosophila* fusome, a germline-specific organelle, contains membrane skeletal proteins and functions in cyst formation. *Development.* **120,** 947–956.

72. Deng, W., and Lin, H. (1997) Spectrosomes and fusomes anchor mitotic spindles during asymmetric germ cell divisions and facilitate the formation of a polarized microtubule array for oocyte specification in *Drosophila. Dev. Biol.* **189,** 79–94.

73. Büning, J. (1994) *The insect ovary: ultrastructure, previtellogenic growth, and evolution,* Chapman and Hall, New York.

74. Pepling, M. E., and Spradling, A. C. (1998) Female mouse germ cells form synchronously dividing cysts. *Development.* **125,** 3323–3328.

75. Gilboa, L., and Lehmann, R. (2006) Soma-germline interactions coordinate homeostasis and growth in the *Drosophila* gonad. *Nature.* **443,** 97–100.

76. Asaoka, M., and Lin, H. (2004) Germline stem cells in the *Drosophila* ovary descend from pole cells in the anterior region of the embryonic gonad. *Development.* **131,** 5079–5089.

77. Song, X., Zhu, C. H., Doan, C., and Xie, T. (2002) Germline stem cells anchored by adherens junctions in the *Drosophila* ovary niches. *Science.* **296,** 1855–1857.

78. Zhu, C. H., and Xie, T. (2003) Clonal expansion of ovarian germline stem cells during niche formation in *Drosophila. Development.* **130,** 2579–2588.

79. Bhat, K. M., and Schedl, P. (1997) Establishment of stem cell identity in the *Drosophila* germline. *Dev Dyn.* **210,** 371–382.

80. Brower, D. L., Smith, R. J., and Wilcox, M. (1981) Differentiation within the gonads of *Drosophila* revealed by immunofluorescence. *J. Embryol. Exp. Morphol.* **63,** 233–242.

81. Hardy, R. W., Tokuyasu, K. T., Lindsley, D. L., and Garavito, M. (1979) The germinal proliferation center in the testis of *Drosophila melanogaster. J. Ultrastruct. Res.* **69,** 180–190.

82. Gonczy, P., and DiNardo, S. (1996) The germ line regulates somatic cyst cell proliferation and fate during *Drosophila* spermatogenesis. *Development.* **122,** 2437–2447.

83. Yamashita, Y. M., Jones, D. L., and Fuller, M. T. (2003) Orientation of asymmetric stem cell division by the APC tumor suppressor and centrosome. *Science.* **301,** 1547–1550.

84. Gilboa, L., and Lehmann, R. (2004) How different is Venus from Mars? The genetics of germ-line stem cells in *Drosophila* females and males. *Development.* **131,** 4895–4905.

85. DeFalco, T. J., Verney, G., Jenkins, A. B., McCaffery, J. M., Russell, S., and Van Doren, M. (2003) Sex-specific apoptosis regulates sexual dimorphism in the Drosophila embryonic gonad. *Dev. Cell.* **5,** 205–216.

86. DeFalco, T., Le Bras, S., and Van Doren, M. (2004) Abdominal-B is essential for proper sexually dimorphic development of the *Drosophila* gonad. *Mech. Dev.* **121,** 1323–1333.

87. Gonczy, P., Viswanathan, S., and DiNardo, S. (1992) Probing spermatogenesis in *Drosophila* with P-element enhancer detectors. *Development.* **114,** 89–98.

88. Le Bras, S., and Van Doren, M. (2006) Development of the male germline stem cell niche in *Drosophila. Dev. Biol.* **294,** 92–103.

89. Fuller, M. T. (1993) *in The development of* Drosophila melanogaster (Bate, M., and Martinez Arias, A., Eds.), Cold Spring Harbor Laboratory Press, Cold Spring Harbor, NY, Vol. 1, pp. 71–147.

90. Kiger, A. A., White-Cooper, H., and Fuller, M. T. (2000) Somatic support cells restrict germline stem cell self-renewal and promote differentiation. *Nature.* **407,** 750–754.

91. Tazuke, S. I., Schulz, C., Gilboa, L., et al. (2002) A germline-specific gap junction protein required for survival of differentiating early germ cells. *Development.* **129,** 2529–2539.

92. Tanaka-Matakatsu, M., Uemura, T., Oda, H., Takeichi, M., and Hayashi, S. (1996) Cadherin-mediated cell adhesion and cell motility in *Drosophila* trachea regulated by the transcription factor Escargot. *Development.* **122,** 3697–3705.

93. Wang, H., Singh, S. R., Zheng, Z., et al. (2006) Rap-GEF signaling controls stem cell anchoring to their niche through regulating DE-Cadherin-mediated cell adhesion in the *Drosophila* testis. *Dev. Cell.* **10,** 117–126.

94. Lee, J. H., Cho, K. S., Lee, J., et al. (2002) *Drosophila* PDZ-GEF, a guanine nucleotide exchange factor for Rap1 GTPase, reveals a novel upstream regulatory mechanism in the mitogen-activated protein kinase signaling pathway. *Mol. Cell Biol.* **22,** 7658–7666.

95. Price, L. S., Hajdo-Milasinovic, A., Zhao, J., Zwartkruis, F. J., Collard, J. G., and Bos, J. L. (2004) Rap1 regulates E-Cadherin-mediated cell-cell adhesion. *J. Biol. Chem.* **279,** 35127–35132.

96. Knox, A. L., and Brown, N. H. (2002) Rap1 GTPase regulation of adherens junction positioning and cell adhesion. *Science.* **295,** 1285–1288.

97. Robertson, S. E., Dockendorff, T. C., Leatherman, J. L., Faulkner, D. L., and Jongens, T. A. (1999) *germ cell-less* is required only during the establishment of the germ cell lineage of *Drosophila* and has activities which are dependent and independent of its localization to the nuclear envelope. *Dev. Biol.* **215,** 288–297.

98. Schulz, C., Wood, C. G., Jones, D. L., Tazuke, S. I., and Fuller, M. T. (2002) Signaling from germ cells mediated by the *rhomboid* homolog *stet* organizes encapsulation by somatic support cells. *Development.* **129,** 4523–4534.

99. Artavanis-Tsakonas, S., Matsuno, K., and Fortini, M. E. (1995) Notch signaling. *Science.* **268,** 225–232.

100. Ward, E. J., Shcherbata, H. R., Reynolds, S. H., Fischer, K. A., Hatfield, S. D., and Ruohola-Baker, H. (2006) Stem cells signal to the niche through the Notch pathway in the *Drosophila* ovary. *Curr. Biol.* 16, 2352–2358.

101. Kiger, A. A., Jones, D. L., Schulz, C., Rogers, M. B., and Fuller, M. T. (2001) Stem cell self-renewal specified by JAK-STAT activation in response to a support cell cue. *Science.* **294,** 2542–2545.

102. Tulina, N., and Matunis, E. (2001) Control of stem cell self-renewal in *Drosophila* spermatogenesis by JAK-STAT signaling. *Science.* **294,** 2546–2549.

103. Rawlings, J. S., Rosler, K. M., and Harrison, D. A. (2004) The JAK/STAT signaling pathway. *J. Cell Sci.* **117,** 1281–1283.

104. Song, X., Wong, M. D., Kawase, E., et al. (2004) Bmp signals from niche cells directly repress transcription of a differentiation-promoting gene, *bag of marbles,* in germline stem cells in the *Drosophila* ovary. *Development.* **131,** 1353–1364.

105. Chen, D., and McKearin, D. M. (2003) A discrete transcriptional silencer in the *bam* gene determines asymmetric division of the *Drosophila* germline stem cell. *Development.* **130,** 1159–1170.

106. McKearin, D., and Ohlstein, B. (1995) A role for the *Drosophila* Bag-of-Marbles protein in the differentiation of cystoblasts from germline stem cells. *Development.* **121,** 2937–2947.
107. Ohlstein, B., and McKearin, D. (1997) Ectopic expression of the *Drosophila* Bam protein eliminates oogenic germline stem cells. *Development.* **124,** 3651–3662.
108. Ohlstein, B., Lavoie, C. A., Vef, O., Gateff, E., and McKearin, D. M. (2000) The *Drosophila* cystoblast differentiation factor, *benign gonial cell neoplasm*, is related to DExH-box proteins and interacts genetically with *bag of marbles*. *Genetics.* **155,** 1809–1819.
109. Chen, D., and McKearin, D. (2003) Dpp signaling silences *bam* transcription directly to establish asymmetric divisions of germline stem cells. *Curr. Biol.* **13,** 1786–1791.
110. Gilboa, L., and Lehmann, R. (2004) Repression of primordial germ cell differentiation parallels germ line stem cell maintenance. *Curr. Biol.* **14,** 981–986.
111. Kai, T., and Spradling, A. (2003) An empty *Drosophila* stem cell niche reactivates the proliferation of ectopic cells. *Proc. Natl. Acad. Sci. U. S. A.* **100,** 4633–4638.
112. Gilboa, L., Forbes, A., Tazuke, S. I., Fuller, M. T., and Lehmann, R. (2003) Germ line stem cell differentiation in *Drosophila* requires gap junctions and proceeds via an intermediate state. *Development.* **130,** 6625–6634.
113. Dorfman, R., and Shilo, B. Z. (2001) Biphasic activation of the BMP pathway patterns the *Drosophila* embryonic dorsal region. *Development.* **128,** 965–972.
114. Niki, Y., Yamaguchi, T., and Mahowald, A. P. (2006) Establishment of stable cell lines of *Drosophila* germ-line stem cells. *Proc. Natl. Acad. Sci. U. S. A.* **103,** 16325–16330.
115. Niki, Y., and Mahowald, A. P. (2003) Ovarian cystocytes can repopulate the embryonic germ line and produce functional gametes. *Proc. Natl. Acad. Sci. U. S. A.* **100,** 14042–14045.
116. Shivdasani, A. A., and Ingham, P. W. (2003) Regulation of stem cell maintenance and transit amplifying cell proliferation by TGF-beta signaling in *Drosophila* spermatogenesis. *Curr. Biol.* **13,** 2065–2072.
117. Schulz, C., Kiger, A. A., Tazuke, S. I., et al. (2004) A misexpression screen reveals effects of Bag of Marbles and TGF-beta class signaling on the *Drosophila* male germ-line stem cell lineage. *Genetics.* **167,** 707–723.
118. Kawase, E., Wong, M. D., Ding, B. C., and Xie, T. (2004) Gbb/Bmp signaling is essential for maintaining germline stem cells and for repressing bam transcription in the *Drosophila* testis. *Development.* **131,** 1365–1375.
119. Gonczy, P., Matunis, E., and DiNardo, S. (1997) *bag of marbles* and *benign gonial cell neoplasm* act in the germline to restrict proliferation during *Drosophila* spermatogenesis. *Development.* **124,** 4361–4371.
120. Subramaniam, K., and Seydoux, G. (1999) *nos-1* and *nos-2*, two genes related to *Drosophila nanos*, regulate primordial germ cell development and survival in *Caenorhabditis elegans*. *Development.* **126,** 4861–4871.
121. Koprunner, M., Thisse, C., Thisse, B., and Raz, E. (2001) A zebrafish *nanos*-related gene is essential for the development of primordial germ cells. *Genes Dev.* **15,** 2877–2885.
122. Tsuda, M., Sasaoka, Y., Kiso, M., et al. (2003) Conserved role of Nanos proteins in germ cell development. *Science.* **301,** 1239–1241.
123. Parisi, M., and Lin, H. (2000) Translational repression: a duet of Nanos and Pumilio. *Curr. Biol.* **10,** R81–R83.
124. Wang, Z., and Lin, H. (2004) Nanos maintains germline stem cell self-renewal by preventing differentiation. *Science.* **303,** 2016–2019.
125. Szakmary, A., Cox, D. N., Wang, Z., and Lin, H. (2005) Regulatory relationship among *piwi*, *pumilio*, and *bag of marbles* in *Drosophila* germline stem cell self-renewal and differentiation. *Curr. Biol.* **15,** 171–178.
126. Chen, D., and McKearin, D. (2005) Gene circuitry controlling a stem cell niche. *Curr. Biol.* **15,** 179–184.
127. Kai, T., and Spradling, A. (2004) Differentiating germ cells can revert into functional stem cells in *Drosophila melanogaster* ovaries. *Nature.* **428,** 564–569.
128. Brawley, C., and Matunis, E. (2004) Regeneration of male germline stem cells by spermatogonial dedifferentiation in vivo. *Science.* **304,** 1331–1334.

129. Margolis, J., and Spradling, A. (1995) Identification and behavior of epithelial stem cells in the *Drosophila* ovary. *Development*. **121,** 3797–3807.
130. Tran, J., Brenner, T. J., and DiNardo, S. (2000) Somatic control over the germline stem cell lineage during *Drosophila* spermatogenesis. *Nature*. **407,** 754–757.
131. Silver, D. L., and Montell, D. J. (2001) Paracrine signaling through the JAK/STAT pathway activates invasive behavior of ovarian epithelial cells in *Drosophila*. *Cell*. **107,** 831–841.
132. Lasko, P. F., and Ashburner, M. (1988) The product of the *Drosophila* gene *vasa* is very similar to eukaryotic initiation factor-4A. *Nature*. **335,** 611–617.

Chapter 2
Analysis of the *C. elegans* Germline Stem Cell Region

Sarah L. Crittenden and Judith Kimble

Contents

Summary We present methods for characterizing the mitotic and early meiotic regions of the *Caenorhabditis elegans* germline. The methods include examination of germlines in living and fixed worms, cell cycle analysis, analysis of markers, and initial characterization of mutants that affect germline proliferation.

Keywords BrdU; *C. elegans*; cell cycle; germline; meiosis; mitosis; proliferation; stem cells.

2.1 Introduction

Identification of stem cells and the pathways that regulate them is important for both clinical research and more basic biomedical science. The *Caenorhabditis elegans* germline is a simple and well-studied model for understanding the genetic and molecular regulation of stem cells *(1–3)*. First, all stages of germ cell development, from stem cell to differentiated gamete, are present in the adult gonad at one time (*see* **Fig. 2.1A**). Second, establishment and maintenance of *C. elegans* germline stem cells (GSCs) is controlled by a single somatic cell, the distal tip cell (DTC) (*see* **Fig. 2.1A**), which provides the niche for the stem cells *(4)*. Finally, regulators of stem cell self-renewal have been identified and analyzed in depth (*see* **Fig. 2.1B**). Both the Notch signaling pathway *(5)* and the PUF (for Pumilio and FBF) family of RNA regulators maintain stem cells in *C. elegans* *(3,5)*. These regulators are also likely to control stem cells in other systems. For example, Notch signaling has been suggested to play roles in stem cell regulation in vertebrates (for review, *see* **ref. 6**).

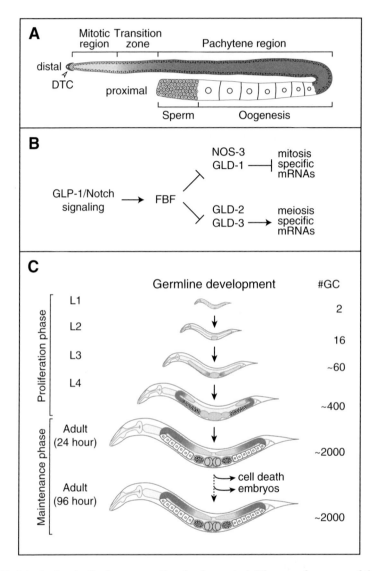

Fig. 2.1 Introduction to *C. elegans* germline development. **A** Diagram of one arm of the adult hermaphrodite germline. The somatic distal tip cell (DTC), *arrowhead*, is located at the distal end and provides a niche for proliferating germline stem cells. Germ cells differentiate as they move away from the DTC, toward the proximal end of the germline. **B** Pathway controlling the proliferation/differentiation switch in the *C. elegans* germline. **C** Diagram showing the stages of germline development from the larval proliferative phases through the adult maintenance phase. Approximate germ cell numbers for each stage are given on the *right*

PUF proteins are required for GSC self-renewal in flies (*see* **ref. 7** and references therein) and have been found in human spermatogonia and ES cells (*8*).

We describe methods for studying *C. elegans* GSCs. Criteria to identify stem cells can vary; thus, it is crucial to define what is known in the system under consideration and to define the criteria for identifying cell types. In the *C. elegans* germline, mitotic cells are at the distal end, adjacent to the DTC (*see* **Fig. 2.1A**). The mitotic cells self-renew and contribute to the more proximal meiotic population, replenishing the germline as mature gametes are lost (*see* **Fig. 2.1A**). The GSCs reside within the mitotic region of the germline; however, their position and numbers have not been unambiguously identified. A subset of mitotic germ cells in the few rows closest to the DTC expresses high levels of the mitotic activators GLP-1, FBF-1, and FBF-2 and low levels of the meiotic activator GLD-1 (*see* **Figs. 2.1B** and **2.6**) *(3,5)*. Because of their expression pattern and their position near the DTC, this subset is likely to include the GSCs. Mitotic germ cells further from the DTC begin to express markers of differentiation, such as GLD-1 and HIM-3 *(9–11)*. These germ cells may be analogous to transit-amplifying cells in other systems. Based on analyses of mitotic index, S-phase index, and molecular markers, some of these more proximal cells appear to be in premeiotic S-phase *(10,12)*. All mitotic germ cells, including stem cells, cycle at similar rates and divide with a random orientation *(12–14)*.

We present methods for identifying and characterizing undifferentiated and proliferating germ cells, including identification of the proliferative region, premeiotic S-phase region and putative stem cells. We then discuss how we characterize new mutants using procedures for wild-type germlines. There are excellent chapters about other useful techniques freely available on the WormBook web site (http://www.wormbook.org) (*see* **Subheading 2.2**).

2.2   Materials

2.2.1   *Reagents*

1. 4% agarose in distilled water (dH_2O) for microscopy of live *C. elegans*.
2. M9: 22 mM KH_2PO_4, 22 mM Na_2HPO_4, 85 mM NaCl, 1 mM $MgSO_4$.
3. M9 plus 0.25 mM levamisole.
4. Slides and coverslips.
5. Subbing solution: Bring 200 mL dH_2O to 60°C, then add 0.4 g gelatin; cool to 40°C. Add 0.04 g chrome alum and sodium azide to 1 mM. Add poly-L-lysine (Sigma, cat. no. P1524) to 1 mg/mL. Store subbing solution at 4°C. To sub slides, put subbing solution on slide for 10 min at room temperature. Wick off excess liquid. Dry in 65°C oven for approx. 30 min. Slides can be stored in the oven or at room temperature.
6. Paraformaldehyde: 16% stock (Electron Microscopy Sciences, cat. no. 15710).
7. PBSB: Phosphate-buffered saline (PBS; for 1 L: 8 g NaCl, 0.2 g KCl, 1.44 g Na_2HPO_4, 0.24 g KH_2PO_4, pH to 7.2 with NaOH) containing 0.5% bovine serum albumin (BSA).

8. DAPI (4′,6-diamidino-2-phenylindole) (Molecular Probes, Invitrogen cat. no. D1306).
9. Vectashield (Vector Labs, cat. no. H-1000).
10. Hoechst 33342 (Molecular Probes, Invitrogen cat. no. H3570).
11. SYTO-12 (Molecular Probes, Invitrogen cat. no. S7574).
12. Rabbit antiphosphohistone H3 (PH3) polyclonal antibody (Upstate Biotechnology, cat. no. 06–570).
13. Mouse anti-PH3 monoclonal antibody (Cell Signaling Technology, cat. no. 97065).
14. M9–agar plates: 1.2% agar, 0.6% agarose in M9 salts containing 0.1 mg/mL ampicillin *(15)*.
15. BrdU (bromodeoxyuridine) (BD-Pharmingen, cat. no. 550891). Store aliquots at −70°C. Once an aliquot is thawed, keep it at 4°C for no more than 1 wk.
16. Anti-BrdU monoclonal antibody (B44, Becton-Dickinson, cat. no. 347580).
17. TO-PRO-3 (Molecular Probes, Invitrogen cat. no. T3605).
18. Thymidine-deficient *Escherichia coli* MG1693 (*E. coli* stock center, http://cgsc.biology.yale.edu/top.html, CGSC#: 6411).
19. *lag-2*::GFP (green fluorescent protein) (*Caenorhabditis* Genetics Center (CGC), http://www.cbs.umn.edu/CGC/strains, strain #JK2868).
20. *lim-7*::GFP (*Caenorhabditis* Genetics Center (CGC), http://www.cbs.umn.edu/CGC/strains, strain #DG1575).

2.2.2 Web Resources

1. Reinke microarray data: http://cmgm.stanford.edu/~kimlab/germline/.
2. *in situ* RNA expression database: http://nematode.lab.nig.ac.jp/.
3. *C. elegans* site: http://elegans.swmed.edu/.
4. WormBase: http://www.wormbase.org/.
5. *C. elegans* strain collection (CGC): http://www.cbs.umn.edu/CGC/.
6. WormBook: http://www.wormbook.org/.
7. WormAtlas: http://www.wormatlas.org/index.htm.

2.3 Methods

2.3.1 Identification of Proliferating Cells in Wild-Type C. elegans Hermaphrodites

The *C. elegans* germline is composed of a U-shaped tube containing approx. 1000 germ cells in different states of differentiation. The mitotic germ cells, including GSCs, reside at one end, adjacent to the somatic DTC. The germline can be observed in living animals (*see* **Subheading 2.3.1.1** and **Fig. 2.2A**). The organization of proliferative and meiotic cells is similar in hermaphrodites and males. The *C. elegans* hermaphrodites are self-fertile XX animals that first make sperm, then oocytes, whereas males are XO animals that make only sperm.

In wild-type young adult hermaphrodites (*see* **Note 1**) the mitotic region is approx. 20 cell diameters in length and contains approx. 225–250 germ cells. The length of the mitotic region is defined as the number of cell diameters between the DTC and the transition zone (TZ) *(10,12)*. The mitotic region was initially defined by the positions of mitoses; however, now we know it is likely also to contain germ cells that have switched to early stages of meiosis (*see* **Subheading 2.3.3**, **Fig. 2.6**, and **refs.** *10* and *12*). The TZ contains nuclei in early meiotic prophase; when stained with DAPI, the DNA in these nuclei has a distinctive crescent shape *(16)* (crescents; *see* **Fig. 2.2**, **Subheadings 2.3.1.2** and **2.3.1.4**). Most nuclei between the DTC and the TZ are proliferating; however, approx. 50 germ cells in the most

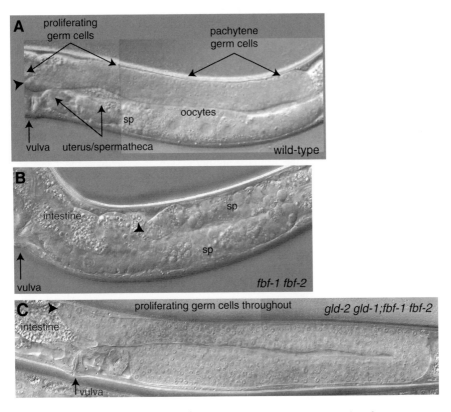

Fig. 2.2 Microscopy of wild-type and mutant germ lines. **A** Differential interference contrast (DIC) micrograph of a wild-type adult hermaphrodite. **B** DIC micrograph of an *fbf-1 fbf-2* double-mutant germline. Mature sperm are seen at the distal end. **C** DIC micrograph of a *gld-2 gld-1; fbf-1 fbf-2* mutant germline containing only mitotic germ cells

proximal rows appear to be in premeiotic S-phase *(12)*. In wild-type germlines, apoptosis occurs in some oogenic cells in the late pachytene stage of meiotic prophase (*see* **Subheading 2.3.1.3** and **ref.** *17*).

2.3.1.1 Mounting Live Animals for Light Microscopy (*see* **Notes 2 and 3**)

1. To make an agarose pad on slide: Drop melted 4% agar on glass slide, quickly put another slide on top at a 90° angle, and press to create a thin layer of agarose (about 0.4-mm thick) (*see* **ref.** *18*, p. 596). Slides with lab tape can be used as spacers for the top slide.
2. Pick animals onto agarose pad next to a drop of M9 containing 0.25 mM levamisole (*see* **Note 4**). Levamisole will prevent animals from moving.
3. Cover with coverslip, avoiding bubbles. For short-term observation, it is not necessary to seal the edges of the coverslip.
4. Observe germline using differential interference contrast (DIC) microscopy.

2.3.1.2 Immunohistochemistry of Extruded Gonads

The protocol described next works for staining with anti-GLP-1, -FBF-1, -GLD-1, -HIM-3, and other antibodies on extruded gonads (somatic tissues remain associated with the germline). Anti-FBF-2 requires a different fixation (*see* **Note 5**). Other protocols for staining germlines are available *(10)* (*see* **Note 6**). If only DAPI staining is required, skip the primary antibody step and incubate the fixed and blocked germlines with DAPI. Continue from there (*see* **Note 7**).

1. Put 7–10 μL M9 containing 0.25 mM levamisole on subbed slide.
2. Pick 10–20 animals into M9.
3. Cut the animals with a syringe needle or scalpel behind the pharynx or at the tail. The germline and intestine will extrude from the cuticle (*see* **Notes 8 and 9**).
4. Add 100 μL 1–2% paraformaldehyde and incubate for 10 min at room temperature in a humidified chamber (e.g., a box with a wet paper towel taped to the lid) (*see* **Note 5**).
5. Remove paraformaldehyde and replace with 100 μL 0.1% Triton X-100 in PBSB for 5 min at room temperature.
6. Block in 100 μL PBSB for 30 min at room temperature.
7. Incubate with primary antibody. Dilution, incubation time, and temperature should be determined for each antibody.
8. Wash three times with 100 μL PBSB for 15 min each.
9. Incubate with 100 μL secondary antibody. Dilution, incubation time, and temperature should be determined for each antibody in PBSB (*see* **Note 10**). Include 0.1 μg/mL DAPI to stain DNA (*see* **Note 11**).
10. Wash three times with 100 μL PBSB for 15 min each.

11. Add 7 μL Vectashield, place coverslip over germlines, and seal with nail polish.
12. Observe using fluorescence microscopy.

2.3.1.3 Visualization of Germline Apoptosis with SYTO-12 *(17)*

1. Incubate animals in 33 μ*M* SYTO-12 (taken up by dying cells) or 50 μg/mL Hoechst 33342 (DNA dye) in M9 with some *E. coli* in a microfuge tube for 4–5 h (*see* **Notes 12** and **13**).
2. Transfer to seeded plate for 30–60 min.
3. Mount on agarose pad for DIC microscopy.
4. Observe live animals on agarose pads (*see* **Subheading 2.3.1.1**) using fluorescence microscopy and DIC. SYTO-12 emits green fluorescence and can be observed using fluorescein isothiocyanate (FITC) filter sets. Hoechst can be observed using DAPI filter sets.

2.3.1.4 Scoring Position of Transition Zone and Counting Mitotic Cell Number

1. Using DAPI-stained germlines (*see* **Subheading 2.3.1.2**), identify the distal end by locating the DTC nucleus (oval with diffuse DAPI staining; *see* **Fig. 2.3**) (*see* **Note 14**).

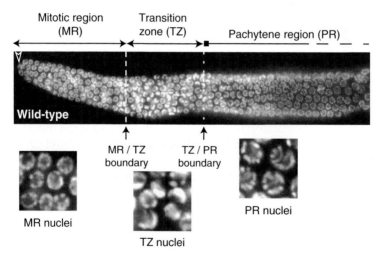

Fig. 2.3 Identification of mitotic and meiotic germ cells. DAPI (4′,6-diamidino-2-phenylindole)-stained distal arm of a wild-type adult hermaphrodite. Mitotic region (MR), transition zone (TZ), and pachytene region (PR) nuclei are enlarged below. Dotted lines indicate the MR–TZ and TZ–PR boundaries. Note the crescents of DNA in the TZ nuclei and the condensed strands of chromatin in the PR nuclei. *Arrowhead* points to DTC nucleus

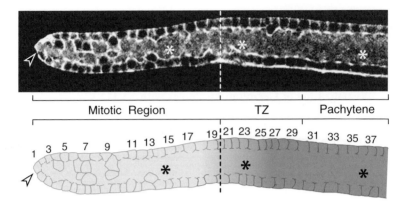

Fig. 2.4 Counting cell diameters from distal end of germline. *Top,* Actin-stained distal arm of a wild-type adult hermaphrodite. *Dotted line* indicates MR/TZ boundary. *Arrowhead* points to position of DTC. *Bottom,* diagram numbered starting with the row of germ cells immediately adjacent to the DTC

2. The start of the TZ is identified as the most distal row of cells containing more than approx. 30% crescents (*see* **Note 15**). More proximal rows should also contain more than approx. 30% crescents and will include some cells that have progressed into pachytene (*see* **Fig. 2.3**).
3. Count germ cell rows and total number of cells (*see* **Note 16**) starting with the row immediately adjacent to the DTC nucleus and going to the TZ (*see* **Fig. 2.4**).
4. Mitotic regions in young adult wild-type hermaphrodite germlines average 19–20 cell diameters (range 15–24) and contain 225–250 germ cells (range ~ 200 to ~ 300) (*see* **Note 1**).

2.3.2 Useful Markers

A number of markers are helpful for identifying mitotic region germ cells, meiotic germ cells, and the somatic gonad cells that affect proliferation (*see* **Table 2.1** and **Fig. 2.6**). Mitotic region markers include the Notch receptor GLP-1, which is abundant in membranes of mitotic germ cells *(19)*; the PUF family RNA regulators FBF-1 and FBF-2, which are abundant in the cytoplasm of mitotic germ cells *(20,21)*; the p53 homolog CEP-1, which is present in mitotic nuclei *(22)*; and the meiotic cohesin REC-8, which is enriched in the cytoplasm of mitotic region cells under certain fixation conditions *(10)*. All of these markers are also expressed either at lower levels in early meiosis or in other parts of the germline (*see* **Table 2.1** and **Fig. 2.6**). Meiotic markers include the maxi KH/STAR domain-containing RNA regulator GLD-1, which increases in the cytoplasm of more proximal mitotic germ cells and peaks in the meiotic region *(9,23)*, and the meiotic protein HIM-3, which is present in the

Table 2.1 Markers for identifying mitotic region germ cells, meiotic germ cells, and the somatic gonad cells that affect proliferation in *Caenorhabditis elegans*

Marker	Molecular identity	Pattern
GLP-1	Notch receptor	Membrane associated, strong in mitotic germ cells, lower levels in early meiotic region (*19*)
FBF-1 and FBF-2	PUF family RNA regulators	Cytoplasmic, strong in mitotic region; levels decrease as germ cells progress through meiosis (*20,21*)
REC-8	Cohesin	Nuclear in mitotic cells under specific fixation conditions (*see* **ref.** *10* and **Fig. 6**); on chromosomes in meiotic prophase (*47*)
CEP-1	*C. elegans* p53	Nuclear in mitotic region and proximal cells (*22*)
PH3	Histone H3 phosphorylated on serine 10	Chromosomes in late prophase (mitotic and meiotic) and M phase (*10,12,24*)
GLD-1	maxi-KH/STAR domain RNA-binding protein	Cytoplasmic, increasing in proximal mitotic cells, strong in meiotic germ cells (*9,23*)
HIM-3	Component of central element of synaptonemal complex	Cytoplasmic in proximal MR, on chromosomes in meiotic prophase (*11*)
PGL-1	P-granule component	Only in germ cells, not in somatic cells (*44*)
SP56	Sperm-specific protein	Spermatocytes and mature sperm (*45*)
RME-2	*C. elegans* yolk receptor	Oocyte membranes (*46*)
lag-2::GFP	*lag-2* promoter driving GFP	Somatic DTC (*12,48*)
lim-7::GFP	*lim-7* promoter driving GFP	Somatic sheath cells 1–4 (*34,49*)

cytoplasm of proximal mitotic germ cells and becomes localized to chromosomes in early meiosis (*10,11*). Hansen and colleagues (*10*) used a combination of REC-8 and HIM-3 staining to identify a region of overlap called the *meiotic entry region*. The meiotic entry region has substantial overlap with the premeiotic S-phase region identified by cell cycle analysis (*10,12*) (*see* **Fig. 2.6**). There are currently no markers that specifically label the different mitotic cell stages proposed to be present in the mitotic region such as stem cells, transit-amplifying cells, or premeiotic S-phase cells.

2.3.3   Cell Cycle Analysis

Cell cycle analysis allows estimation of cell cycle lengths, identification of quiescent or slow-cycling cells, and identification of cells entering meiotic S-phase (*see* **Fig. 2.6**). Anti-PH3 staining, which labels condensed chromosomes in late G2 and M phase nuclei (*24*), is used to determine mitotic index (% of cells in M phase) (*see* **Subheading 2.3.3.1**) (*10,12*). Alternatively, mitotic figures can be scored by DAPI staining (*13*). BrdU labeling is used to determine the S-phase labeling index (% of cells in S-phase) and to estimate total cell cycle length (*see* **Subheadings 2.3.3.2** and **2.3.3.5**). The proximal region of the mitotic region, where mitotic index is low and labeling index is the same as the rest of the mitotic region, is likely composed largely of germ cells in premeiotic S-phase (*see* **Subheading 2.3.3.6**). Some premeiotic S-phase cells are also found in the TZ.

2.3.3.1 Mitotic Index with Anti-PH3

1. Prepare animals for antibody staining (*see* **Subheading 2.3.1.3**).
2. Stain with anti-PH3.
3. Score position and number of PH3-positive nuclei in mitotic region (*see* **Fig. 2.5B** and **Note 16**).
4. Count number of nuclei/row for normalizing.
5. Calculate mitotic index by dividing number of PH3-positive cells by the total number of cells in the region of interest.

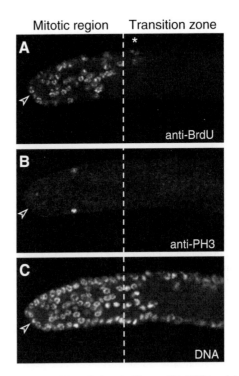

Fig. 2.5 Bromodeoxyuridine (BrdU) and phosphohistone H3 (PH3) staining. Adult hermaphrodite germline stained with anti-BrdU (**A**), anti-PH3 (**B**), and TO-PRO-3 (**C**) after a 15-min pulse with BrdU-labeled bacteria. The mitotic region (MR)/transition zone (TZ) boundary is indicated by a *dotted line*. The distal end is indicated by an *arrowhead*. BrdU-labeled nuclei within the TZ are indicated with an *asterisk*

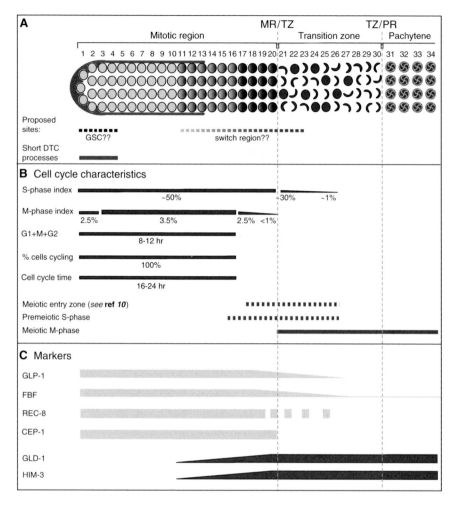

Fig. 2.6 Summary of cell cycle characteristics and markers for mitotic and meiotic regions of the adult hermaphrodite germline. **A** Diagram of the distal arm of the germline. *Dark dotted line* indicates region likely to contain germline stem cells (GSCs). *Graded dotted line* indicates region where cells are making the transition from mitosis into meiosis. **B** Summary of cell cycle characteristics *(10,12)*. **C** Summary of where markers are expressed in the distal part of the gonad. See text, **Table 2.1**, and references therein for detailed descriptions of patterns and staining methods

2.3.3.2 BrdU Labeling

Labeling on plates (modified from **ref. *15***) is the best protocol in our hands. *Caenorhabditis elegans* can also be labeled in solution (*see* **Notes 12 and 17**).

1. Grow an overnight culture of *E. coli* MG1693 in LB.
2. To 100 mL M9, add 3 mL MG1693 overnight culture, 1–2.5 mL 40% glucose (0.4%–1%, usually 1%), 100 μL 1.25 mg/mL thiamine (1.25 μg/mL), 100 μL 0.5 m*M* thymidine (0.5 μ*M*), 100 μL 1*M* MgSO$_4$ (1 m*M*), and 33 μL 30 m*M*

BrdU (Pharmingen) (10 μM). Grow approx. 48 h at 37°C. Spin to pellet *E. coli*. Resuspend in about 1 mL M9.
3. Seed M9 agar plates with 200–400 μL labeled MG1693. Try to cover most of the plate so that animals are continuously feeding on labeled bacteria. Plates can be stored at 4°C for a couple of weeks.
4. Put animals on plates for desired time period.
5. To chase, wash animals off plate with M9 containing approx. 0.1% Triton X-100 (to prevent sticking), wash once, and place animals on plate containing unlabeled bacteria.

2.3.3.3 Fixation for BrdU Staining (*12*; modified from ref. *25*)

1. Cut animals on subbed slide, add coverslip (*see* **Note 18**), freeze for approx. 10 min at −70°C or on dry ice.
2. Remove coverslip, put in −20°C methanol 1 h to days (*see* **Note 19**).
3. Remove slide from methanol; wipe off excess.
4. Block in PBSB for 15 min or more.
5. Fix with 1% paraformaldehyde for 15 min at room temperature.
6. To costain with anti-PH3, incubate with anti-PH3 overnight at 4°C at this point, then wash and postfix with 1% paraformaldehyde for 15 min at room temperature.
7. Denature DNA with 2*N* HCl for 15–30 min at room temperature (*see* **Note 20**).
8. Neutralize with 0.1*M* borate, pH 8.5, for 15 min at room temperature.
9. Block in 100 μL PBSB for 15 min or more.
10. Incubate with anti-BrdU 1:2.5 overnight at 4°C.
11. Wash three times in 100 μL PBSB for 15 min each.
12. Add secondary antibody and wash as described in **Subheading 2.3.1.2**.
13. Mount with 7 μL of Vectashield; seal with nail polish.

2.3.3.4 Calculating Labeling Index

1. Pulse animals with BrdU for a short time (e.g., 15 min).
2. Fix and stain as described in **Subheading 2.3.3**.
3. Count number of BrdU-positive germ cells and total number of germ cells in region of interest (entire MR, 1 row, or a subregion) (*see* **Fig. 2.4A,C**).
4. Divide number of BrdU-positive germ cells by total germ cell number in the region of interest to get labeling index.

2.3.3.5 Estimating Total Length of Cell Cycle

1. Label animals with BrdU (on plates) for 15 min to approx. 12 h.
2. Fix and stain some animals every 1–2 h.

3. Determine labeling index (*see* **Subheading 2.3.3.4**) for each time point.
4. The time it takes for 100% of nuclei to become BrdU$^+$ is the time of G2 + M + G1 (the time it takes for the cells in early G2 to traverse through G2, M, and G1 and finally label in S-phase; *26*).
5. Total length of cell cycle is estimated by dividing the time of G2 + M + G1 by the fraction of cells in G2 + M + G1. The fraction of cells in G2 + M + G1 is the fraction of cells not in S-phase (one-labeling index) and represents the proportion of cell cycle time spent in these phases.

2.3.3.6 Identification of Cells in Premeiotic S-Phase

1. Calculate mitotic index at each position along the distal–proximal axis (*see* **Subheading 2.3.3.1**).
2. Calculate labeling index (15-min pulse) at each position along the distal–proximal axis (*see* **Subheading 2.3.3.4**).
3. In wild type, the proximal four to six rows of the mitotic region have a lower mitotic index, but the labeling index remains constant. Because these cells do not progress into mitosis (no mitotic cells in TZ), this indicates that these cells are in premeiotic S-phase. There are S-phase nuclei in the TZ (*see* **Figs. 2.5** and **2.6**).

2.3.4 *Characterization of Germline Defects in Mutants*

The steps we take to characterize the phenotypes of new germline-defective mutants are described briefly in this section. In particular, the goal is to distinguish mutants in the stem cell regulatory pathway from those affecting more general processes such as cell cycle survival. The phenotypes studied can either be generated by RNA interference (RNAi; *see* **ref. 27**) or by chromosomal mutations. Regulatory mutants can have subtle effects, affecting only the length of the mitotic region or the number of cells within that region (e.g., *fbf-1, fbf-2, gld-2,* and *gld-3* single mutants; *21,28*). Germline defects in some regulatory genes are only seen in double-mutant combinations because of redundancy (e.g., *fbf-1* and *fbf-2; 21*). Germline phenotypes are sensitive to age of worm, temperature, and starvation. Thus, it is important to stage, grow, and score mutant animals in parallel to a wild-type control (*see* **Note 1**).

2.3.4.1 Initial Characterization of Mutants

1. Compare mutant to wild-type germlines at defined stages of development (*see* **Fig. 2.1C**). Using DIC microscopy (*see* **Subheading 2.3.3.1**), look first at adults to determine whether the germline has a normal size and pattern of cell fates. If the germline does not look normal, look progressively earlier to determine at what stage mutant differs from wild type (*see* **Note 21**).

2. Is there a defect in the proliferation/differentiation switch? If germ cells can no longer divide mitotically but can still progress through meiosis and gametogenesis, mature gametes will be present at the distal end of the germline where immature gametes are found in wild-type animals (*see* **Note 22**). This defect in germline proliferation is typical of Notch signaling and FBF loss-of-function mutants (*see* **refs. 2, 3, 20, 29** and **Fig. 2.2B**). Alternatively, if germ cells can no longer enter the meiotic cell cycle, a germline tumor can result. This is typical of *glp-1* gain-of-function mutants, which have unregulated Notch signaling *(3,30)*, and *gld-1 gld-2* loss-of-function mutants, which lack meiosis-promoting activity *(31; see* **Fig. 2.2C**).

3. Is the germline small and lacking differentiated gametes? This type of small germline can result from slowed or arrested proliferation or inappropriate cell death (*see* list of other genes in **refs. 5, 32,** and **33**).

4. If a mitotic region is present, determine its length, cell number, and mitotic index (*see* **Subheadings 2.3.1.5** and **2.3.3.1**) to identify subtle effects on proliferation.

5. Does the mutant affect germline or soma? Germline proliferation requires Notch signaling from the somatic DTC *(4)*. In addition, the distal somatic sheath cells affect germline proliferation *(34,35)*. Several approaches can be taken to determine whether the defect is germline autonomous or caused by a somatic defect.

   a. Is the messenger RNA (mRNA) present in the germline, soma, or both? First, check the in situ RNA expression database and the Reinke lab microarray data for germline-enriched transcripts (*see* **Subheading 2.2.2**). Next, perform in situ hybridizations on extruded gonads to determine whether expression is restricted to the germline or also present in the somatic cells, such as the DTC or sheath (e.g., *21,36*).

   b. Transgenic rescue is much more efficient in somatic tissues. If a mutant is easily rescued with simple arrays, it is likely that the gene functions in the soma (e.g., *see* **ref. 37**).

   c. RNAi in *rrf-1* mutant background: The *rrf-1* mutants have decreased RNAi efficiency in the soma but not in the germline *(37,38)*.

6. Where does the mutant fit in the regulatory pathway? The genetic pathway controlling the mitosis/meiosis switch has been well characterized (*see* **refs. 2, 3,** and *5* and **Fig. 2.1A**). Epistasis analysis is used to determine at what point in the pathway a gene acts *(3,10,28,39,40)*.

2.4   Notes

1. The age of the worm affects the length of the mitotic region. Most measurements have been done in young adult hermaphrodites, about 24h after L4 at 20°C. The mitotic region shortens with age *(12)*, and older animals may have lower mitotic indices (S.L.C. and J.K., unpublished observation). There are differing data about mitotic region cell numbers in older animals (compare **refs.** *12* and *34*).

2. Descriptions of worm culture and genetics are available at the WormBook Web site (http://www.wormbook.org/chapters/www_introandbasics/introandbasics.html).

3. A detailed description for using DIC microscopy to look at *C. elegans* can be found on the WormBook web site (http://www.wormbook.org/chapters/www_intromethodscellbiology/intromethodscellbiology.html).
4. Placing animals directly on the agarose pad next to the drop of M9 helps them stay in place when the coverslip is put on the slide. This is especially helpful when looking at younger animals, which tend to float in the M9 (*see* **Subheading 2.3.1.1**).
5. Paraformaldehyde fixation improves nuclear morphology. We generally use either 1% or 2%, depending on the antigen. Most antibodies work well with either concentration. If fixation with paraformaldehyde does not work, try fixing with −20°C MeOH for 10 min followed by −20°C acetone for 10 min. MeOH/acetone fixation works well for anti-FBF-2 *(21)*.
6. An alternate protocol for fixing extruded gonads used by Hansen and colleagues *(10)* calls for incubations in tubes rather than on slides. Different fixative conditions are also used. For staining germlines in younger animals (L1 to young L4), it is often easier to use a whole-mount protocol *(41,42)*. Small germlines in larvae are difficult to extrude.
7. Whole animals can be easily DAPI stained (*see* protocols in refs. *31* and *43*). DAPI staining in whole animals is useful for a general look at germline size and gamete production; however, in whole mounts it is more difficult to look at the nuclear morphology of undifferentiated germ cells or to identify germ cells in mutants with small germlines.
8. Gonad extrusion tips: More liquid on the slide helps prevent drying out. If animals are in levamisole too long, germlines will no longer extrude well, so pick fewer worms at a time if cutting takes some time. Finally, an eyelash taped to a toothpick can help position or untangle germlines.
9. There are a number of factors that can affect gonad attachment to slides. First, the slides should be freshly subbed. An increase in the amount of time the subbing solution is left on the slide can improve attachment; however, too much time can result in a thick layer, which can peel off. Second, a drop of M9 on the slide should stay rounded. If it spreads, the germlines may not stick well. Third, be sure there is little or no *E. coli* picked into the M9. Fourth, a coverslip can be laid gently on the extruded germlines to increase contact with the subbed surface. The slide and coverslip are then frozen at −70°C, or on dry ice; the coverslip is popped off, and standard fixation protocols are followed.
10. We use secondary antibodies from Jackson ImmunoResearch, preabsorbed against other species, for double and triple labeling. We typically incubate with the secondary antibody 1–2 h at room temperature at a 1:500 dilution.
11. An alternative DNA dye is TO-PRO-3, which emits a far red signal. To use TO-PRO-3, dilute approx. 1:50,000; incubate with a fixed sample for about 10 min at room temperature after the secondary antibody incubation. Then, wash as usual.
12. Add *E. coli* to live animals in solution. This can be done by collecting *E. coli* on a pick and swishing it off in the solution with the animals. The *E. coli* will ensure the animals have sufficient food and encourage them to eat the dye (*see* **Subheading 2.3.1.3**).
13. Ovulation can slow or stop in live animals kept on slides for extended periods of time. This in turn may influence the proliferation of the mitotic cells and the rate of movement of cells through the germline.
14. Sometimes, the distal end of the germline is lost or damaged during extrusion, so to get accurate counts, it is important to be sure the whole gonad is intact. To learn to identify the DTC nucleus, look at live or fixed animals carrying a *lag-2* promoter::GFP construct (*see* **Table 2.1**). The DTC has processes that embrace the germ cells in the few rows immediately adjacent to it. These processes extend approx. 12 cell diameters from the distal end of the germline in young adult hermaphrodites and lengthen with age *(12)*.
15. The start of meiotic prophase has been defined as either the position of the first crescents *(10)* or the point at which there are multiple crescents (to reduce the chance of scoring occasional anaphase chromosomes that are curved as crescent-shaped meiotic nuclei; *12,28*). These two points are usually within one to three rows of each other.
16. Germ cells do not line up in precise rows, so row counts are somewhat subjective and variable. Counting the same germline several times gives a measure of the accuracy of the counting

method. In addition, a large sample size helps increase accuracy, especially for anti-PH3 counts.

17. The *C. elegans* can be BrdU-labeled in solution by adding 5 μL 30 m*M* BrdU, worms, and some bacteria (food) to 100 μL M9. After labeling, wash with M9 containing 0.1% Triton X-100 (to prevent worms sticking to pipet), then pipet onto a seeded plate. Animals should recover on plates for about 15 min to allow better gonad extrusion. At short time points, the labeling index is similar to that for animals fed labeled bacteria; however, at later time points (e.g., >3 or 4 h), labeling is not as effective.

18. Putting a coverslip over the cut animals helps germlines make contact with the subbed slide.

19. Fresh methanol is important for good morphology. Molecular sieves can be added to methanol to keep it working well for longer periods. If germlines look shriveled, it is time to change the methanol.

20. The time for the denaturation step with HCl in the BrdU-labeling protocol can be varied to improve BrdU staining. The optimal time can be determined empirically.

21. It can be difficult to distinguish germ cells from somatic gonad cells in early development. Germ cells, but not somatic cells, will stain with anti-PGL-1 *(44)* (*see* **Table 2.1**).

22. To confirm the identity of differentiated gametes in mutant germ lines, sperm can be stained with the monoclonal antibody SP56 *(45)*, and oocytes can be stained with the polyclonal antibody anti-RME-2 *(46)* (*see* **Table 2.1**).

References

1. Kimble, J., and Crittenden, S.L (2007). Controls of germline stem cells, entry into meiosis, and the sperm/oocyte decision in *C. elegans. Annu. Rev. Cell Dev. Biol.* **23**, 405–433.

2. Crittenden, S. L., Eckmann, C. R., Wang, L., Bernstein, D. S., Wickens, M., and Kimble, J. (2003) Regulation of the mitosis/meiosis decision in the *Caenorhabditis elegans* germline. *Phil. Trans. R. Soc. Lond. B.* **358**, 1359–1362.

3. Hansen, D., and Schedl, T. (2006) The regulatory network controlling the proliferation-meiotic entry decision in the *Caenorhabditis elegans* germ line. *Curr. Top. Dev. Biol.* **76**, 185–215.

4. Kimble, J. E., and White, J. G. (1981) On the control of germ cell development in *Caenorhabditis elegans. Dev. Biol.* **81**, 208–219.

5. Kimble, J., and Crittenden, S. L. Germline proliferation and its control (August 15, 2005), in *WormBook* (The *C. elegans* Research Community, ed.), WormBook, 10.1895/wormbook.1.13.1, http://www.wormbook.org.

6. Chiba, S. (2006) Notch signaling in stem cell systems. *Stem Cells.* **24**, 2437–2447.

7. Wickens, M., Bernstein, D. S., Kimble, J., and Parker, R. (2002) A PUF family portrait: 3′ UTR regulation as a way of life. *Trends Genet.* **18**, 150–157.

8. Moore, F. L., Jaruzelska, J., Fox, M. S., et al. (2003) Human Pumilio-2 is expressed in embryonic stem cells and germ cells and interacts with DAZ (Deleted in AZoospermia) and DAZ-like proteins. *Proc. Natl. Acad. Sci. U. S. A.* **100**, 538–543.

9. Hansen, D., Wilson-Berry, L., Dang, T., and Schedl, T. (2004) Control of the proliferation vs meiotic development decision in the *C. elegans* germline through regulation of GLD-1 protein accumulation. *Development.* **131**, 93–104.

10. Hansen, D., Hubbard, E. J. A., and Schedl, T. (2004) Multi-pathway control of the proliferation vs meiotic development decision in the *Caenorhabditis elegans* germline. *Dev. Biol.* **268**, 342–357.

11. Zetka, M. C., Kawasaki, I., Strome, S., and Müller, F. (1999) Synapsis and chiasma formation in *Caenorhabditis elegans* require HIM-3, a meiotic chromosome core component that functions in chromosome segregation. *Genes Dev.* **13**, 2258–2270.

12. Crittenden, S. L., Leonhard, K. A., Byrd, D. T., and Kimble, J. (2006) Cellular analyses of the mitotic region in the *Caenorhabditis elegans* adult germ line. *Mol. Biol. Cell.* **17**, 3051–3061.
13. Maciejowski, J., Ugel, N., Mishra, B., Isopi, M., and Hubbard, E. J. A. (2006) Quantitative analysis of germline mitosis in adult *C. elegans*. *Dev. Biol.* **292**, 142–151.
14. Morrison, S. J., and Kimble, J. (2006) Asymmetric and symmetric stem-cell divisions in development and cancer. *Nature.* **441**, 1068–1074.
15. Ito, K., and McGhee, J. D. (1987) Parental DNA strands segregate randomly during embryonic development of *Caenorhabditis elegans*. *Cell.* **49**, 329–336.
16. Dernburg, A. F., McDonald, K., Moulder, G., Barstead, R., Dresser, M., and Villeneuve, A. M. (1998) Meiotic recombination in *C. elegans* initiates by a conserved mechanism and is dispensable for homologous chromosome synapsis. *Cell.* **94**, 387–398.
17. Gumienny, T. L., Lambie, E., Hartwieg, E., Horvitz, H. R., and Hengartner, M. O. (1999) Genetic control of programmed cell death in the *Caenorhabditis elegans* hermaphrodite germline. *Development.* **126**, 1011–1022.
18. Sulston, J., and Hodgkin, J. (1988) Methods, in *The nematode Caenorhabditis elegans* (W.B. Wood, ed.), Cold Spring Harbor Laboratory Press, Cold Spring Harbor, NY, Vol. 17, pp. 587–606.
19. Crittenden, S. L., Troemel, E. R., Evans, T. C., and Kimble, J. (1994) GLP-1 is localized to the mitotic region of the *C. elegans* germ line. *Development.* **120**, 2901–2911.
20. Crittenden, S. L., Bernstein, D. S., Bachorik, J. L., et al. (2002) A conserved RNA-binding protein controls germline stem cells in *Caenorhabditis elegans*. *Nature.* **417**, 660–663.
21. Lamont, L. B., Crittenden, S. L., Bernstein, D., Wickens, M., and Kimble, J. (2004) FBF-1 and FBF-2 regulate the size of the mitotic region in the *C. elegans* germline. *Dev. Cell.* **7**, 697–707.
22. Schumacher, B., Hanazawa, M., Lee, M. H., et al. (2005) Translational repression of *C. elegans* p53 by GLD-1 regulates DNA damage-induced apoptosis. *Cell.* **120**, 357–368.
23. Jones, A. R., Francis, R., and Schedl, T. (1996) GLD-1, a cytoplasmic protein essential for oocyte differentiation, shows stage-and sex-specific expression during *Caenorhabditis elegans* germline development. *Dev. Biol.* **180**, 165–183.
24. Hendzel, M. J., Wei, Y., Mancini, M. A., et al. (1997) Mitosis-specific phosphorylation of histone H3 initiates primarily within pericentromeric heterochromatin during G2 and spreads in an ordered fashion coincident with mitotic chromosome condensation. *Chromosoma.* **106**, 348–360.
25. Newmark, P. A., and Sanchez Alvarado, A. (2000) Bromodeoxyuridine specifically labels the regenerative stem cells of planarians. *Dev. Biol.* **220**, 142–153.
26. Aherne, W. A., Camplejohn, R. S., and Wright, N. A. (1977) *An introduction to cell population kinetics*, Arnold, London.
27. Ahringer, J. (ed.). Reverse genetics (April 6, 2006), in *WormBook* (The *C. elegans* Research Community, ed.), WormBook, doi/10.1895/wormbook.1.47.1, http://www.wormbook.org.
28. Eckmann, C. R., Crittenden, S. L., Suh, N., and Kimble, J. (2004) GLD-3 and control of the mitosis/meiosis decision in the germline of *Caenorhabditis elegans*. *Genetics.* **168**, 147–160.
29. Austin, J., and Kimble, J. (1987) *glp-1* is required in the germ line for regulation of the decision between mitosis and meiosis in *C. elegans*. *Cell.* **51**, 589–599.
30. Berry, L. W., Westlund, B., and Schedl, T. (1997) Germ-line tumor formation caused by activation of *glp-1*, a *Caenorhabditis elegans* member of the *Notch* family of receptors. *Development.* **124**, 925–936.
31. Kadyk, L. C., and Kimble, J. (1998) Genetic regulation of entry into meiosis in *Caenorhabditis elegans*. *Development.* **125**, 1803–1813.
32. Kraemer, B., Crittenden, S., Gallegos, M., et al. (1999) NANOS-3 and FBF proteins physically interact to control the sperm-oocyte switch in *Caenorhabditis elegans*. *Curr. Biol.* **9**, 1009–1018.

33. Subramaniam, K., and Seydoux, G. (1999) *nos-1* and *nos-2*, two genes related to *Drosophila nanos*, regulate primordial germ cell development and survival in *Caenorhabditis elegans*. *Development*. **126**, 4861–4871.
34. Killian, D. J., and Hubbard, E. J. A. (2005) *Caenorhabditis elegans* germline patterning requires coordinated development of the somatic gonadal sheath and the germ line. *Dev. Biol.* **279**, 322–335.
35. McCarter, J., Bartlett, B., Dang, T., and Schedl, T. (1997) Soma-germ cell interactions in *Caenorhabditis elegans*: multiple events of hermaphrodite germline development require the somatic sheath and spermathecal lineages. *Dev. Biol.* **181**, 121–143.
36. Lee, M.-H., and Schedl, T. RNA in situ hybridization of dissected gonads (June 14, 2006), in *WormBook* (The *C. elegans* Research Community, ed.), WormBook, 10.1895/wormbook.1.107.1, http://www.wormbook.org.
37. Killian, D. J., and Hubbard, E. J. A. (2004) *C. elegans pro-1* activity is required for soma/germline interactions that influence proliferation and differentiation in the germ line. *Development*. **131**, 1267–1278.
38. Sijen, T., Fleenor, J., Simmer, F., et al. (2001) On the role of RNA amplification in dsRNA-triggered gene silencing. *Cell*. **107**, 465–476.
39. Maine, E. M., Hansen, D., Springer, D., and Vought, V. E. (2004) *Caenorhabditis elegans atx-2* promotes germline proliferation and the oocyte fate. *Genetics*. **168**, 817–830.
40. Huang, L., and Sternberg, P. W. Genetic dissection of developmental pathways (June 14, 2006), in *WormBook* (The *C. elegans* Research Community, ed.), WormBook, 10.1895/wormbook.1.88.2, http://www.wormbook.org.
41. Epstein, H. F., and Shakes, D. C. (eds.). (1995) *Caenorhabditis elegans: modern biological analysis of an organism*, Academic Press, New York.
42. Duerr, J. S. Immunohistochemistry (June 19, 2006), in *WormBook* (The *C. elegans* Research Community, ed.), WormBook, 10.1895/wormbook.1.105.1, http://www.wormbook.org.
43. Shaham, S. WormBook: methods in cell biology (January 2, 2006), in *WormBook* (The *C. elegans* Research Community, ed.), WormBook, doi/10.1895/wormbook.1.41.1, http://www.wormbook.org.
44. Kawasaki, I., Shim, Y.-H., Kirchner, J., Kaminker, J., Wood, W. B., and Strome, S. (1998) PGL-1, a predicted RNA-binding component of germ granules, is essential for fertility in *C. elegans*. *Cell*. **94**, 635–645.
45. Ward, S., Roberts, T. M., Strome, S., Pavalko, F. M., and Hogan, E. (1986) Monoclonal antibodies that recognize a polypeptide antigenic determinant shared by multiple *Caenorhabditis elegans* sperm-specific proteins. *J. Cell Biol.* **102**, 1778–1786.
46. Grant, B., and Hirsh, D. (1999) Receptor-mediated endocytosis in the *Caenorhabditis elegans* oocyte. *Mol. Biol. Cell*. **10**, 4311–4326.
47. Pasierbek, P., Jantsch, M., Melcher, M., Schleiffer, A., Schweizer, D., and Loidl, J. (2001) A *Caenorhabditis elegans* cohesion protein with functions in meiotic chromosome pairing and disjunction. *Genes Dev.* **15**, 1349–1360.
48. Blelloch, R., Santa Anna-Arriola, S., Gao, D., Li, Y., Hodgkin, J., and Kimble, J. (1999) The *gon-1* gene is required for gonadal morphogenesis in *Caenorhabditis elegans*. *Dev. Biol.* **216**, 382–393.
49. Hall, D. H., Winfrey, V. P., Blaeuer, G., et al. (1999) Ultrastructural features of the adult hermaphrodite gonad of *Caenorhabditis elegans*: relations between the germ line and soma. *Dev. Biol.* **212**, 101–123.

Chapter 3
Immunohistological Techniques for Studying the *Drosophila* Male Germline Stem Cell

Shree Ram Singh and Steven X. Hou

Contents

Summary Stem cells are undifferentiated cells that have a remarkable ability to self-renew and produce differentiated cells that support normal development and tissue homeostasis. This unique capacity makes stem cells a powerful tool for future regenerative medicine and gene therapy. Accumulative evidence suggests that stem cell self-renewal or differentiation is controlled by both intrinsic and extrinsic factors, and that deregulation of stem cell behavior results in cancer formation, tissue degeneration, and premature aging. The *Drosophila* testis provides an excellent in vivo model for studying and understanding the fundamental cellular and molecular mechanisms controlling stem cell behavior and the relationship between niches and stem cells. At the tip of the *Drosophila* testes, germline stem cells (GSCs) and somatic stem cells (SSCs) contact each other and share common niches (known as a hub) to maintain spermatogenesis. Signaling pathways, such as the Janus kinase (JAK)/signal transducer and activator of transcription (STAT), bone morphogenetic protein (BMP), ras-associated protein–guanine nucleotide exchange factor for small GTPase (Rap-GEF), and epidermal growth factor receptor (EGFR)/mitogen-activated protein kinase (MAPK), are known to regulate self-renewal or differentiation of *Drosophila* male germline stem cells. We describe the detailed in vivo immunohistological protocols that mark GSCs, SSCs, and their progeny in *Drosophila* testes.

Keywords Antibodies; *Drosophila* testes; germline stem cell; immunostaining; niche; signaling pathways; somatic stem cell; spermatogenesis.

From: *Methods in Molecular Biology, Vol. 450, Germline Stem Cells*
Edited by S. X. Hou and S. R. Singh © Humana Press, Totowa, NJ

45

3.1   Introduction

Stem cells are unspecialized cells that have remarkable ability to self-renew, and through asymmetric cell division, they generate functional differentiated cells. These differentiated cells play a crucial role in both the maintenance of local homeostasis, by replacing dead or damaged cells, and the process of tissue remodeling. Tumors may originate from a small subset of cancer stem cells, which constitute a reservoir of self-sustaining cells with the exclusive ability to self-renew and maintain the tumor *(1)*. Mounting evidence suggests that stem cell self-renewal or differentiation is controlled by both intrinsic and extrinsic (niche) factors *(2–9)*. Further, findings suggest that localized signaling between stem cells and niche cells and cell adhesion between stem cells and niche cells are conserved mechanisms that are important for maintaining stem cell identity and adequate stem cell numbers in animals *(3)*. Improper regulation of stem cell behavior is known to lead to cancer formation, tissue degeneration, and premature aging. Understanding the molecular mechanisms governing stem cell self-renewal or differentiation in vivo is crucial to using stem cells for future regenerative medicine and gene therapy and will in turn help to better understand the mechanisms underlying cancer formation, aging, and degenerative diseases.

The *Drosophila* testis provides an excellent in vivo system for studying the relationships between niches and stem cells at cellular and molecular levels *(10–20)*. The adult *Drosophila* male has two testes, each with a long coiled tube that is closed at the apical end and opens to the seminal vesicle at the basal end (*see* **Fig. 3.1**). At the apical tip of the *Drosophila* testis is a germinal proliferation center

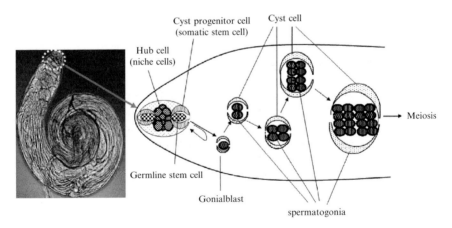

Fig. 3.1 Wild–type *Drosophila* testis visualized by phase-contrast microscopy. The schematic diagram highlighted from the tip of the testes shows five to nine germline stem cells (GSCs; only two GSCs are shown in this scheme) surrounded by approx. twice as many cyst progenitor cells (somatic cyst cells, SSCs), and both are anchored around somatic hub cells at the tip of the *Drosophila* testis. The testis proliferation center is composed of hub, germline, and somatic stem cells; gonialblasts; and 2- to 16-cell spermatogonia

composed of a cluster of 12 quiescent somatic cells called the hub; anchored around the hub are five to nine germline stem cells (GSCs) and twice as many somatic stem cell (SSCs, also called cyst progenitor cells) (*see* **Fig. 3.1**) that maintain spermatogenesis *(11,21,22)*.

Spermatogenesis is a complex, continuous, and highly organized event composed of several steps of cell proliferation or differentiation in which daughter cells of GSCs develop into mature sperm. Beginning in the testes tip, spermatogensis occurs through an asymmetric division of a GSC that produces two daughter cells; the one adjacent to the hub retains a stem cell identity *(21)*, and the other one, displaced away from the hub, becomes the gonialblast and begins to differentiate *(23)*. Similarly, the division of an SSC produces two daughter cells; the one closer to the hub remains an SSC, and the one farther away differentiates *(21)*. The gonialblast then undergoes four rounds of synchronous mitotic division with incomplete cytokinesis to generate 16 interconnected spermatogonia; the somatic cyst cells will grow without further division and enclose the spermatogonial clusters *(15–17)*. Next, the spermatogonial cells undergo the premeiotic S-phase, becoming spermatocytes, and enter a prolonged G2 period followed by meiosis and spermatid differentiation.

In the *Drosophila* male GSCs, adhesion between niche cells and stem cells controls their self-renewal. GSCs begin to differentiate when they lose contact with the niche (hub). Tight anchorage of GSCs to their niche is correlated with high levels of Drosophila E-cadhrein (DE-cadherin) and β-catenin at the interface between the hub and GSCs *(23)*, which in turn depend on Gef26, a guanine nucleotide exchange factor for the Rap GTPase (guanine triphosphatase) *(18)*. Loss of Gef26 results in the loss of both GSCs and SSCs because of impaired adherens junctions between the GSCs and hub cells *(18)*. Further, DE-cadherin and Armadillo at the hub–GSC interface interact with Apc2, a *Drosophila* homolog of the mammalian tumor-suppressor gene adenomatous polyposis coli protein (APC), which anchors astral microtubules to orient centrosomes and the spindle perpendicular to the hub for asymmetric stem cell division *(23)*. Further loss of APC2 and CNN (centrosomin) function result in increased male GSCs, misaligned spindle, and misoriented centrosomes *(23)*. It has been shown that in the *Drosophila* male germline, stereotype spindle orientation and asymmetric outcome of stem cell divisions are developmentally programmed asymmetric behavior and inheritance of mother and daughter centrosome in which the mother centrosome remains in contact with the niche and retained by the GSC through many GSC generations, and the daughter centrosome moves to the opposite side of the cell before spindle formation *(24)*.

In addition to the physical attachment, signaling between stem cells and their niches is required to control self-renewal or differentiation of GSCs/SSCs in *Drosophila* testes. Several signaling pathways, such as that of Janus kinase–signal transducer and activator of transcription (JAK/STAT), transforming growth factor-β (TGF-β), DPP, bone morphogenetic protein (BMP), epidermal growth factor receptor (EGFR), mitogen-activated protein kinase (MAPK), are known to regulate stem cell self-renewal or differentiation in *Drosophila* testes. The JAK/STAT pathway is the major signaling pathway *(25,26)* involved in stem cell maintenance in *Drosophila* testes *(27)*. The hub cells, which function as a niche for stem cells, express a growth factor, upd (unpaired), which acts as a short-range signal to activate the JAK/STAT pathway in GSCs *(28,29)*. The *upd* gene instructs GSCs to undergo self-renewal through the Mom (master of marelle)/Hop (hopscotch)/Stat92E (signal-trasducer and activator of transcription protein at 92E) pathway; Stat92E then enters the nucleus to activate expression of genes that instruct the self-renewal of GSCs and SSCs. Overexpression of the Hop/Stat92E pathway caused by overexpressed *upd* leads to unrestricted stem cell division, and loss-of-function mutations in the Hop/Stat92E pathway lead to loss of GSCs and early germ cells, suggesting that signaling maintains stem cell fate or viability *(28,29)*. Furthermore, GSCs null for *stat92E* can produce differentiating daughter cells but cannot maintain stem cell fate. These data suggest that Hop/Stat92E signaling instructs stem cell fate rather than maintaining cell viability *(29)*. Moreover, early spermatogonia with limited mitotic divisions can repopulate the niche and be reverted to stem cell identity by conditionally manipulating Hop/Stat92E signaling *(30)*. Overexpressing *upd* leads to an accumulation of somatic cells in testes, suggesting that *upd* also controls the maintenance of SSCs *(28,29)*.

BMP (TGF-β) is the other signaling pathway required to maintain GSCs/SSCs in the *Drosophila* male testes *(31,32)*. The members of the BMP pathway, such as *gbb* (glass bottom boat) and *dpp* (decapentaplegic), are expressed in the hub and somatic stem cells and regulate GSC self-renewal *(33,34)*. Other members of the BMP pathway, *punt* and *shn* (schnurri), are expressed in the somatic cyst cells and regulate the proliferation of spermatogonial cells *(32)*. Loss of the BMP pathway components, including Tkv (thick veins), Put (punt), med (medea), Mad (mothers against Dpp) and Shn (schnurri), or misexpression of the negative regulator Dad results in the loss of GSCs. However, ectopic expression of *dpp* in the GSCs leads to slight expansion of GSCs *(32–34)*. *Bam* and *Bgcn* pathways, which are the target of BMP (TGF-β) signaling, act autonomously in the GSCs to restrict the proliferation of amplifying germ cells *(35)*. Loss of *Bam* (bag of marble) *or Bgcn* (benign gonial cell neoplasm) activity leads to the overproliferation of undifferentiated GSCs *(35)*. We showed that a *Drosophila* homolog of the human tumor suppressor gene Birt–Hogg–Dubé syndrome (DBHD) interacts with JAK/STAT and Dpp signal transduction pathways and is required for male GSC maintenance in the fly testes by suppressing the GSC overproliferation phenotype associated with overexpression of either *upd* or *dpp* *(19)*.

The EGFR/MAPK pathway also regulates the fate of GSCs by promoting their differentiation in *Drosophila* testes *(36,37)*. The *Drosophila* EGFR (DER) functions in SSCs through the Raf/MAPK pathway. The DER/Raf/MAPK pathway mediates an SSC-to-GSC signal that restricts self-renewal and promotes differentiation of the GSCs. If the cyst cells do not send the signal to GSCs through the DER/Raf/MAPK pathway, both daughters of the GSCs will retain stem cell identity. Loss of function of *der* or *raf* in SSCs, resulting in too many GSCs and gonialblasts, suggests that SSCs play a guardian role, ensuring the balance between the self-renewal and differentiation of GSCs *(36,37)*.

Table 3.1 Primary antibody for labeling male germline stem cell and their progeny

Antibody	Cell type	Species raised in	Suggested dilution	Source
Hts 1B1	Fusome and spectrosome	Mouse	1:4	DSHB
FAS III	Hub	Mouse	1:10	DSHB
Armadillo	Hub	Mouse	1:50	DSHB
DE-Cadherin	Hub	Rat	1:20	DSHB
Vasa	GSCs and all germ cells	Rabbit	1:2,000	R. Lehmann
STAT92E	GSCs	Rabbit	1:500	S. Hou
p-Mad	GSCs and gonialblasts	Rabbit	1:500	P. Ten Dijke
Bam-C	Spermatogonia	Rat	1:500	D. McKearin
Traffic Jam (TJ)	SSCs and early cyst cells	Guinea pig	1:5,000	D. Godt
Eyes absent (Eya)	Late cyst cells	Mouse	1:100	DSHB
pMapk	Cyst cells neighboring GSCs and spermatocytes	Mouse	1:100	Sigma
GEF26	Interface between hub and GSCs	Rabbit	1:2,000	S. Hou
APC-2	Interface between hub and GSCs	Rabbit	1:10,000	M. Bienz
Notch	GSCs, gonialblasts, and early spermatogonia	Mouse	1:1,000	DSHB
CNN	Centrosome	Rabbit	1:1,000	T. Kaufmann
α-Tubulin	Microtubules and spindle	Mouse	1:100	Sigma
γ-Tubulin	Centrosome	Rabbit	1:1,000	Sigma
Phosphorylated histone 3	M-phase	Rabbit	1:3,000	Upstate Biotechnology
β-Galactosidase	For M5-4 lacZ enhancer trap line to label GSCs, hubs, and gonialblasts; for 842 marker to mark hubs, SSCs, and early cyst cells	Rabbit	1:1,000	Cappel
Anti-GFP	For GFP-fusion protein lines	Mouse/Rabbit	1:100/1:200	Molecular Probes

GFP, green fluorescent protein; GSC, germline stem cell; SSC, somatic cyst cell. Hts-1b1, Huli-tai shao; FasIII, Fascilin III; p-mad, phosphorylated Mad

Drosophila is used to study stem cells in the testes because it has several useful characteristics, including the vast availability of cell-specific molecular markers, which enables stem cells to be easily identified and compared to their neighboring cells (*see* **Table 3.1**). In addition, the mitotic activity of stem cells and their early progeny facilitates cell lineage labeling. Further, the gene function in these stem cells and their niches can be manipulated using sophisticated genetic tools such as the flipase/flipase recombination target (FLP/FRT) technique *(38)*, GAL4-GAL80 (mosaic analysis with a repressible cell marker, MARCM) *(39)*, and positively marked mosaic lineage (PMML) technique *(40)*. Efficient genetic screens can be used to identify genes (mutations) that are important for stem cell self-renewal or differentiation. Finally, a functional genomics/proteomics approach can be used to identify intrinsic and extrinsic factors that regulate stem cell self-renewal or differentiation.

Immunohistological staining is a versatile technique that uses antibodies to detect the presence or absence of a protein, its tissue distribution, and subcellular localization. Antibodies are used to detect the protein product of a specific gene and its expression patterns in development that facilitate the analysis of both transcriptional and posttranscriptional gene regulation. In addition, staining with multiple antibodies will correlate the expression patterns of different proteins in different tissues at the same time. In *Drosophila*, the available molecular markers, lineage labeling, available phenotypes, and use of versatile genetic tools and immunohistological staining methods will help to better understand the male GSC system at both cellular and molecular levels. The immunohistological staining methods presented in this chapter work well in different laboratory conditions and with a variety of antibodies (monoclonal or polyclonal; expressed on cell surface, in cytoplasm, or in nuclei), as well as with enhancer trap-lacZ lines and green fluorescent protein (GFP)-fusion lines expressed in GSCs, SSCs, and their progeny (*see* **Table 3.1**).

3.2   Materials

1. *Drosophila* adult males (0–3 d old).
2. Standard CO_2 source for anesthetizing the flies.
3. Fine paintbrushes for sorting out males.
4. A pair of fine tweezers (Roboz) for dissection.
5. Glass microslides (25 × 75 mm) for dissection and mounting the tissue and with one end frosted for labeling the genotypes.
6. Glass microcoverslips (22 × 60 mm).
7. Nonstick microcentrifuge tubes (1.5 mL) in white and amber.
8. Pipets and pipet tips.
9. Falcon tubes/conical tubes (15 and 50 mL).
10. Glass bottles (100 mL, 500 mL, and 1 L).
11. Timer.
12. Plastic dropper.
13. Kimwipes.
14. Parafilm for sealing the Eppendorf tubes.
15. Minivortex (VWR Scientific Products).
16. Tube rocker or shaker (Labquake).
17. Aluminum foil.
18. Gloves (Kimberly-Clark).
19. Microcentrifuge tube rack (Fisher Scientific).
20. Dissecting microscope with attached light source.
21. Distilled water.
22. Dissecting solution (*Drosophila* ringer's solution): 130 mM NaCl, 4.7 mM KCl, 1.9 mM $CaCl_2$, and 10 mM HEPES (4-(2-hydroxyethyl)-1-piperazineethanesulfonic acid), pH 6.9. Dissolve 7.5 g NaCl, 0.35 g KCl, 0.21 g $CaCl_2$, and 2.38 g

HEPES in approx. 1 L H_2O and stir to dissolve. Adjust to pH 7.2 with 1*N* HCl and bring to a final volume of 1 L with water. Store in a glass bottle at 4°C.

23. Phosphate-buffered solution (10X phosphate-buffered saline [PBS]): Dissolve 76.0 g NaCl (1.3*M*), 3.6 g NaH_2PO_4 (0.03*M*), and 9.94 g $Na_2HPO_4.H_2O$ (0.07*M*) in 1000 mL distilled water. To make 1X PBS, dilute 10X PBS at a 1:9 ratio in distilled water. Adjust to pH 7.4 with HCl. Store in glass bottles at room temperature. Alternatively, use 10X PBS, pH 7.4 (Gibco), and dilute to use 1X PBS.

24. Triton X-100 (Sigma).

25. Bovine serum albumin (BSA; Sigma). Use 0.5%.

26. PBX solution: 1X PBS plus 0.1% Triton X-100, 0.5% BSA. Store in a glass bottle at room temperature and aliquot at 4°C.

27. Formaldehyde (formalin) (37%) aqueous solution (Sigma).

28. Fixation solution: 4% formaldehyde in 1X PBX. Make 50 mL solution by mixing 5.4 mL 37% formaldehyde solution in 1X PBX solution. Mix well in vortex. This solution should be prepared fresh every time but may be stored at 4°C for 1 wk.

29. Normal goat serum (NGS; Vector laboratories). Store at 4°C.

30. Blocking solution (2% NGS): To make 50 mL of solution, add 1 mL NGS to 49.0 mL of 1X PBX. Mix well in vortex and store in Falcon tube at 4°C.

31. Primary antibodies: The primary antibodies available to study male GSCs in *Drosophila* are listed in **Table 3.1**, which shows specific cell-type expression, species in which it was raised, the suggested dilution, and the sources. We use anti-β-gal for lacZ reporter lines and anti-GFP for GFP-fusion protein lines (*see* **Table 3.1**). Store the antibodies at 4°C with 0.02% sodium azide. Keep an aliquot of each in the freezer at −20°C with 50% glycerol and at −80°C for long-term storage. Avoid frequent freezing and thawing.

32. Secondary antibodies: Several secondary antibodies are available from different companies. We use secondary antibodies of goat antimouse, goat antirat, goat antirabbit, and goat antiguinea pig immunoglobulin G (IgG) conjugated to Alexa 488 or Alexa 594 from Molecular Probes (Invitrogen). Store all secondary antibodies in a dark place at 4°C. We used a 1:400 dilution of all secondary antibodies in 1X PBX containing 0.5% BSA.

33. 4,6-Diamidino-2-phenyldole dihydrochloride (DAPI; Molecular Probes) for staining DNA. Dissolve in 1X PBS for counterstaining. Store in the dark at 4°C.

34. Ribonuclease (RNase) A, 10 mg/mL stock (store at 4°C). This component is required only if using propidium iodide.

35. Glycerol ultrapure (Sigma).

36. Mounting medium: 50% glycerol in 1X PBS (pH 7.4) containing 1% antifade (1,4-diazabicyclo [2,2,2] octane [DABCO]; Sigma) compounds. Store at room temperature in a glass bottle or Falcon tube.

37. Waterproof permanent marker/pencil for labeling the slide.

38. Clear or white nail polish for sealing the slides.

39. Microslide plastic folder (VWR Scientific Products).

40. Fluorescence microscope.

41. Zeiss LSM510 confocal microscope for imaging.

42. Computer and appropriate software for image processing.

3.3   Methods

This protocol is summarized in **Fig. 3.2**. The protocol for immunostaining is described in detail next.

3.3.1   Testes Dissection

The adult *Drosophila* male internal reproductive system consists of paired testes, seminal vesicles, vasa deferentia, accessory glands, and an unpaired ejaculatory duct with an appended ejaculatory bulb *(41)*. Normally, paired testes lie within the abdomen in an asymmetrically bilateral position. The coiled portion of the testes lies ventrolaterally on one side of the body between abdominal segments 3 to 5; the free end extending to the opposite side reaches the anterior border of segment 2, or the tips may lie as far back as segment 3. However, the exact position of the testes is not fixed

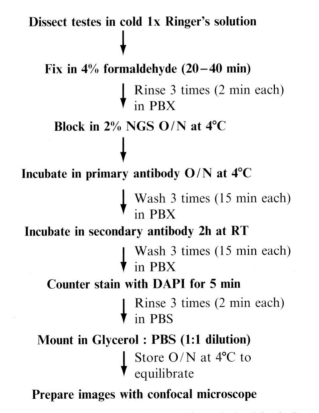

Dissect testes in cold 1x Ringer's solution

↓

Fix in 4% formaldehyde (20–40 min)

↓ Rinse 3 times (2 min each) in PBX

Block in 2% NGS O/N at 4°C

↓

Incubate in primary antibody O/N at 4°C

↓ Wash 3 times (15 min each) in PBX

Incubate in secondary antibody 2h at RT

↓ Wash 3 times (15 min each) in PBX

Counter stain with DAPI for 5 min

↓ Rinse 3 times (2 min each) in PBS

Mount in Glycerol : PBS (1:1 dilution)

↓ Store O/N at 4°C to equilibrate

Prepare images with confocal microscope

Fig. 3.2 Schematic overview of the steps for immunohistological staining in *Drosophila* testes. DAPI, 4,6-diamidino-2-phenylindole; NGS, normal goat serum; O/N, overnight; PBX, 1X phosphate-buffered saline (PBS) plus 0.1% Triton X-100, 0.5% bovine serum albumin and varies according to the age of the flies and nutritive condition *(41)*. These organs are colorless at the time of eclosion, but testes, seminal vesicles, and vasa deferentia become a bright butter yellow with age. The color of these organs fades after fixation *(41)*. The detailed protocol to dissect the *Drosophila* testes is as follows:

1. Anesthetize the flies on the CO_2 source surface and, using a fine paintbrush, sort out the young *Drosophila* males under the dissection microscope.
2. Put a clean glass microslide under the dissecting microscope and use a plastic dropper to place a few drops of Ringer's solution on the slide (*see* **Note 1**).
3. Using clean, fine tweezers, place a young male on a glass microslide containing dissecting solution (1X Ringer's solution).
4. Using two fine tweezers, dissect the testes from a young male in fresh Ringer's solution. Using one pair of tweezers, turn the body upside down, hold the top of the abdomen, and pull out the external genitalia with the other pair of tweezers. Remove the unwanted fly parts (*see* **Note 2**).
5. Transfer the dissected testes into a microcentrifuge tube containing 1 mL cold Ringer's solution.
6. Place dissected testes from 10 to 15 males in each tube. Wait until the testes settle and label the tube with marker or pencil with the specific genotype.
7. Pipet off the dissecting solution from the microcentrifuge tube and retain the testes in the tube.

3.3.2 Fixation

For immunohistological analysis, it is essential to preserve the tissue architecture and cell morphology efficiently and adequately because inappropriate and prolonged fixation can lead to poor antibody binding to the antigen sites. Fixation is required to prevent artifactual diffusion of soluble tissue components, to avoid decomposition of the structure, to arrest enzymatic activity, and to protect the tissue against the deleterious effects that occur at various stages of the immunohistological process. The best fixative differs from tissue to tissue and antigen to antigen. There are several fixatives available to fix the testes tissues. However, here we use formaldehyde (formalin) to fix the testes tissues; it works very well and does not affect the accessibility of the antigens.

1. Prepare 4% formaldehyde (fixing solution) in 1X PBX.
2. Add 500 μL 4% formaldehyde in the microcentrifuge tube containing dissected testes tissues.
3. Fix the testes by incubating the tube on a rocker/shaker at room temperature for 20–40 min (*see* **Note 3**).
4. Stop the rocker/shaker and place the microcentrifuge tube in the tube rack to allow the testes to settle in the tube.
5. Remove all the fixative solution and rinse the testes three times, for 2 min each, in 1X PBX.

3.3.3 Blocking and Staining

Normal serum is the most popular blocking agent for immunohistological staining. We use normal goat serum because our secondary antibodies are raised in goat. The proper working dilution of the antibodies is essential for the best results because the same dilution will not work for every antibody. First, optimize the primary antibody dilution that gives the strongest specific antigen staining with the lowest nonspecific background; then, optimize the dilution of the secondary antibody. The detailed protocol for blocking and staining of the testes tissues is as follows:

1. Prepare the 2% NGS (blocking solution) in 1X PBX in a conical tube.
2. Add 1 mL blocking solution in the microcentrifuge tube and block the testes by incubating on a rocker/shaker overnight at 4°C or 30 min at room temperature (*see* **Note 4**).
3. Stop the rocker/shaker and place the microcentrifuge tube in the tube rack to allow the testes to settle in the tube.
4. Pipet off the blocking solution.
5. Prepare primary antibody to working dilution in 1X PBX containing 0.5% BSA. Mix well by hand or with a vortex (*see* **Note 5**).
6. Add 200–400 μL of diluted primary antibody in microcentrifuge tubes containing the testes tissues (*see* **Note 6**).
7. Incubate the testes with the primary antibody on a rocker/shaker overnight at 4°C or at room temperature for 2 h (*see* **Note 7**).
8. Stop the shaker and place the microcentrifuge tube in the tube rack to allow the testes to settle in the tube.
9. Pipet off the primary antibodies and save at 4°C to reuse (*see* **Note 8**).
10. Rinse the testes three times with 1X PBX and wash three times for 15 min each with 1-mL changes of PBX on a rocker/shaker at low speed and room temperature.
11. Prepare secondary antibody to desired concentration (*see* **Note 9**).
12. Add 200–400 μL of diluted secondary antibody to the microcentrifuge tube containing the testes tissues (*see* **Note 10**).
13. Incubate the testes with secondary antibodies on a rocker/shaker at room temperature for 2 h or overnight at 4°C.
14. Stop the shaker and place the microcentrifuge tube in the tube rack to allow the testes to settle in the tube.
15. Remove secondary antibody.
16. Rinse the testes three times with 1X PBX and wash three times for 15 min each with 1-mL changes of PBX on a rocker/shaker at room temperature.
17. Rinse three times in 1X PBS.
18. To counterstain with DNA, add 1 μg/mL DAPI (from 1 mg/mL stock) in 1X PBS to the microcentrifuge tube containing the testes tissues and incubate for 5 min at room temperature on a rocker/shaker (*see* **Note 11**).
19. Remove the DAPI solution and store in a dark place at 4°C to reuse later (*see* **Note 12**).
20. Rinse the testes with 1X PBS three times for 2 min each.

3.3.4 Mounting and Microscopy

1. Prepare mounting medium in 1X PBS (PBS:glycerol ratio of 1:1) containing 1% DABCO.
2. Add 100 μL of mounting medium to the microcentrifuge tube containing the testes tissues.
3. Incubate the testes tissues in a dark place overnight at 4°C to allow tissues to equilibrate.
4. Using a pipet, transfer the testes with mounting medium to a glass microslide frosted at one end.

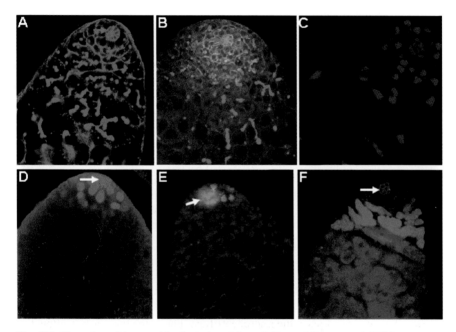

Fig. 3.3 Examples of immunohistological staining of *Drosophila* testes. **A** Wild-type testes immunostained with anti-Arm (*green, yellow arrow*) to label the hub cells and with 1B1 to mark the fusome (*green*). **B** Wild-type testes immunostained with anti-Arm (*green, yellow arrow*) to label the hub cells, anti-vasa antibody (*red*) to label the germ cells, 1B1 (*green*) to label the fusome, and DAPI (*blue*) to label DNA. **C** Wild-type testes immunostained with anti-Tj (*red*) to label the SSCs and early cyst cells. **D** M5-4 marker testes immunostained with anti-Arm (*red, white arrow*) and anti-β-galactosidase (*green*) to label the hub cells, GSCs, and nearby gonial-blasts. **E** Wild-type testes immunostained with antiphosphorylated histone 3 (*red*) to label the mitotic active cells such as GSCs, gonialblasts, and spermatogonia; anti-Arm (*green, white arrow*) to label the hub cells; and DAPI (*blue*) to label DNA. (**F**) The bam-GFP line testes immunostained with anti-arm (*red, white arrow*) to label the hub and anti-GFP (*green*) to label the 2- to 16-cell spermatogonia. DAPI, 4′,6-diamidino-2-phenylindole; GFP, green fluorescent protein; GSC, germline stem cell

5. Arrange the testes under the dissecting microscope using low levels of light.
6. Cover the testes with a glass microcoverslip (22 × 60 mm). Remove excess mounting medium using Kimwipes (*see* **Note 13**).
7. Seal the edges of the microcoverslip with nail polish.
8. Use permanent marker or pencil to label the slide for genotype.
9. Store the slides in the dark at 4°C until observation (*see* **Note 14**).
10. Confirm staining by examining the slides using fluorescence microscopy (*see* **Note 15**).
11. For highest-quality images, use confocal microscopy. We use a Zeiss LSM 510 confocal laser scanning microscope attached to a Pentium personal computer to capture the images and LSM 5 Image Browser to download the images.
12. Process the images using Adobe Photoshop.

Some of the examples of staining using these methods are presented in **Fig. 3.3**. However, for testing new antibodies or other markers and obtaining optimal conditions, it is advisable to use appropriate fixative conditions and antibody dilutions (*see* **Note 15**). We also used these protocols to label the female GSCs and other tissues in *Drosophila*.

3.4   Notes

1. Clean the entire dissecting area, including slides, tweezers, and other tools, with Kimwipes and make sure all equipment and materials are dust and lint free.
2. Use 0- to 3-d-old males because they show best morphology. Dissect the testes carefully, avoiding puncture with tweezers, which can destroy the morphology of the germ cell. To prevent evaporation, make sure to have enough dissecting solution on the slide.
3. Formaldehyde (formalin) is highly toxic and carcinogenic. It is readily absorbed through the skin and is irritating or destructive to the skin, eyes, mucous membranes, and upper respiratory tract. Avoid breathing the vapors and wear appropriate gloves during handling. Formaldehyde (4%) preserves cellular structure and minimizes background fluorescence. For best results, limit the fixation time to 25–40 min because prolonged fixation can cause poor antibody penetration and abolish staining.
4. Blocking is used to discourage nonspecific binding and reduce background staining. Block the testes using serum made from the same species as the secondary antibodies. We use NGS because our secondary antibodies are raised in goat.
5. Testes can be double/triple stained with primary antibodies that originated in two different species. To avoid weak or background staining, test the optimal dilutions for each primary antibody, which can differ from each other. Avoid freeze-and-thaw cycles, which can inactivate the antibodies.
6. Wrap the microcentrifuge tube with Parafilm during incubation with primary antibodies to prevent leakage during rocking/shaking.
7. For best results, incubate the primary antibodies overnight at 4°C on slow speed in the rocker/shaker. For quick staining, a 2-h incubation period at room temperature also provides good results for some antibodies. Incubating primary antibodies for a long period of time can produce nonspecific staining and destroy the morphology of germ cells if the antibodies are diluted with 1X PBX. Use 1X PBS for longer incubation.
8. Saving the primary antibodies for reuse is possible, depending on the type of antibody (monoclonal or polyclonal) and its abundance. We reuse the primary antibodies four or five times

when they are stored at 4°C with 0.02% sodium azide. If the antibody is expensive and unavailable, make the aliquots and store at −20°C.

9. Keep secondary antibodies in the dark at 4°C. Avoid freeze-and-thaw cycles. We use a dilution of 1:400 for all secondary antibodies in 1X PBX containing 0.5% BSA. For optimal staining, replace secondary antibodies that are more than a year old with new ones.

10. For best results, use amber-color microcentrifuge tubes for secondary antibody incubation. If using the white microcentrifuge tubes, wrap them with aluminum foil to protect the tissues from light. Incubation is more effective in the dark because exposure to light can inhibit the fluorescence signal.

11. DAPI is a carcinogen and irritant. It may be harmful by inhalation, ingestion, or skin absorption. Wear appropriate gloves and avoid breathing the dust and vapors. Dispose all tips and microcentrifuge tubes in a special beaker exposed to DAPI.

12. DAPI diluted in 1X PBS (without Triton) can be reused multiple times if kept in the dark at 4°C. Avoid diluting DAPI in 1X PBX (with Triton), which can reduce DAPI staining signal.

13. Carefully place the coverslip onto tissues to avoid trapping air bubbles. Do not allow the coverslip to move during the removal of the excess mounting medium; movement can cause damage to the testes tissues.

14. Place prepared slides in microslide plastic folder or slide box. Wrap the folder/box with aluminum foil and store in the dark at 4°C. For optimal imaging, slides should be stored in the dark at 4°C and images collected within 2–3 d. In addition, prepared slides may be stored at −20°C for up to 2 wk with similar results.

15. If no staining or weak staining occurs, this could be because reagents were applied in the wrong order or steps were omitted; insufficient time was allowed for incubation of antibodies; tissues were overfixed; or incompatible primary and secondary antibodies were used. If overstaining or high background occurs, this could be caused by a high concentration of antibodies and long incubation time, insufficient blocking time, or insufficient washing time. For optimal staining, repeat the experiment carefully, using the correct time for fixation, blocking, and washing as well as the correct concentration of antibodies.

References

1. Reya, T., Morrison, S. J., Clarke, M. F., and Weissman, I. L. (2001) Stem cells, cancer, and cancer stem cells. *Nature.* **414,** 105–111.
2. Spradling, A., Drummond-Barbosa, D., and Kai, T. (2001) Stem cells find their niche. *Nature.* **414,** 98–104.
3. Fuchs, E., Tumbar, T., and Guasch, G. (2004) Socializing with the neighbors: stem cells and their niche. *Cell.* **116,** 769–778.
4. Ohlstein, B., Kai, T., Decotto, E., and Spradling, A. (2004) The stem cell niche: theme and variations. *Curr. Opin. Cell Biol.* **16,** 693–699.
5. Li, L., and Xie, T. (2005) Stem cell niche: structure and function. *Annu. Rev. Cell Dev. Biol.* **21,** 605–631.
6. Xi, R., and Xie, T. (2005) Stem cell self-renewal controlled by chromatin remodeling factors. *Science.* **310,** 1487–1489.
7. Scadden, D. T. (2006) The stem-cell niche as an entity of action. *Nature.* **441,** 1075–1079.
8. Li, L., and Neaves, W. B. (2006) Normal stem cells and cancer stem cells: the niche matters. *Cancer Res.* **66,** 4553–4557.
9. Le Bras, S., and Van Doren, M. (2006) Development of the male germline stem cell niche in *Drosophila. Dev. Biol.* **294,** 92–103.
10. Lin, H. (1997) The tao of stem cells in the germline. *Annu. Rev. Genet.* **31,** 455–491.
11. Fuller, M. T. (1998) Genetic control of cell proliferation and differentiation in *Drosophila* spermatogenesis. *Semin. Cell Dev. Biol.* **9,** 433–444.

12. Wong, M. D., Jin, Z., and Xie T. (2005) Molecular mechanisms of germline stem cell regulation. *Annu. Rev. Genet.* **39**, 173–195.

13. Xie, T., Kawase, E., Kirilly, D., and Wong M. D. (2005) Intimate relationships with their neighbors: tales of stem cells in *Drosophila* reproductive systems. *Dev. Dyn.* **232**, 775–790.

14. Terry, N.A, Tulina, N., Matunis, E., and DiNardo S. (2006) Novel regulators revealed by profiling *Drosophila* testis stem cells within their niche. *Dev. Biol.* **294**, 246–257.

15. Lin, H. (2002) The stem-cell niche theory: lessons from flies. *Nat. Rev. Genet.* **3**, 931–940.

16. Gilboa, L., and Lehmann, R. (2004) How different is Venus from Mars? The genetics of germline stem cells in *Drosophila* females and males. *Development.* **131**, 4895–4905.

17. Yamashita, Y. M., Fuller, M. T., and Jones, D. L. (2005) Signaling in stem cell niches: lessons from the *Drosophila* germline. *J. Cell Sci.* **118**, 665–672.

18. Wang, H., Singh, S. R., Zheng, Z., et al. (2006) Rap-GEF signaling controls stem cell anchoring to their niche through regulating DE-cadherin-mediated cell adhesion in the *Drosophila* testis. *Dev. Cell.* **10**, 117–126.

19. Singh, S. R., Zhen, W., Zheng, Z., et al. (2006) The *Drosophila* homolog of the human tumor suppressor gene BHD interacts with the JAK-STAT and Dpp signaling pathways in regulating male germline stem cell maintenance. *Oncogene.* **25**, 5933–5941.

20. Bunt, S. M., and Hime, G. R. (2004) Ectopic activation of Dpp signalling in the male *Drosophila* germline inhibits germ cell differentiation. *Genesis.* **39**, 84–93.

21. Hardy, R. W., Tokuyasu, K. T., Lindsley, D. L., and Garavito, M. (1979) The germinal proliferation center in the testis of *Drosophila melanogaster. J. Ultrastruct. Res.* **69**, 180–190.

22. Gonczy, P., and DiNardo, S. (1996) The germ line regulates somatic cyst cell proliferation and fate during *Drosophila* spermatogenesis. *Development.* **122**, 2437–2447.

23. Yamashita, Y. M., Jones, D. L., and Fuller, M. T. (2003) Orientation of asymmetric stem cell division by APC tumor suppressor and centrosome. *Science.* **301**, 1547–1550.

24. Yamashita, Y. M., Mahowald, A. P., Perlin, J. R., and Fuller, M. T. (2007) Asymmetric inheritance of mother vs daughter centosome in stem cell division. *Science.* **315**, 518–521.

25. Hou, S.X., Melnick, M. B., and Perrimon, N. (1996) *marelle* acts downstream of the *Drosophila* HOP/JAK kinase and encodes a protein similar to the mammalian STATs. *Cell.* **84**, 411–419.

26. Hou, X. S., and Perrimon, N. (1997) The JAK-STAT pathway in *Drosophila. Trend. Genet.* **13**, 105–110.

27. Hou, S. X., Zheng, Z., Chen, X., and Perrimon, N. (2002) The Jak/STAT pathway in model organisms: emerging roles in cell movement. *Dev. Cell.* **3**, 765–778.

28. Kiger, A. A., Jones, D. L., Schulz, C., Rogers, M. B., and Fuller, M. T. (2001) Stem cell self-renewal specified by JAK-STAT activation in response to a support cell cue. *Science.* **294**, 2542–2545.

29. Tulina, N., and Matunis, E. (2001) Control of stem cell self-renewal in *Drosophila* spermatogenesis by JAK-STAT signaling. *Science.* **294**, 2546–2549.

30. Brawley, C., and Matunis, E. (2004) Regeneration of male germline stem cells by spermatogonial dedifferentiation in vivo. *Science.* **304**, 1331–1334.

31. Schulz, C., Kiger, A. A., Tazuke, S. I., et al. (2004) A mis-expression screen reveals effects of bag-of-marbles and TGF-β class signaling on the *Drosophila* male germ-line stem cell lineage. *Genetics.* **167**, 707–723.

32. Matunis, E., Tran, J., Gonczy, P., Caldwell, K., and DiNardo, S. (1997) *punt* and *schnurri* regulate a somatically derived signal that restricts proliferation of committed progenitors in the germline. *Development.* **124**, 4383–4391.

33. Shivdasani, A. A., and Ingham, P. W. (2003) Regulation of stem cell maintenance and transit amplifying cell proliferation by TGF-β signaling in *Drosophila* spermatogenesis. *Curr. Biol.* **13**, 2065–2072.

34. Kawase, E., Wong, M. D., Ding, B. C., and Xie, T. (2003) Gbb/Bmp signaling is essential for maintaining germline stem cells and for repressing bam transcription in the *Drosophila* testis. *Development.* **131**, 1365–1375.

35. Gonczy, P., Matunis, E., and DiNardo, S. (1997) Bag-of-marbles and benign gonial cell neoplasm act in the germline to restrict proliferation during *Drosophila* spermatogenesis. *Development.* **124,** 4361–4371.
36. Kiger, A. A., White-Cooper, H., and Fuller, M.T. (2000) Somatic support cells restrict germline stem cell self-renewal and promote differentiation. *Nature.* **407,** 750–754.
37. Tran, J., Brenner, T. J., and DiNardo, S. (2000) Somatic control over the germline stem cell lineage during *Drosophila* spermatogenesis. *Nature.* **407,** 754–757.
38. Golic K. G., and Lindquist S. (1989) The FLP recombinase of yeast catalyzes site-specific recombination in the *Drosophila* genome. *Cell.* **59,** 499–509.
39. Lee T., and Luo, L. (1999) Mosaic analysis with a repressible cell marker for studies of gene function in neuronal morphogenesis. *Neuron.* **22,** 451–461.
40. Harrison, D., and Perrimon, N. (1993) Simple and efficient generation of marked clones in *Drosophila. Curr. Biol.* **3,** 424–433.
41. Millar, A. (1941). Position of adult testes in *Drosophila melanogaster* Meigen. *Proc. Natl. Acad. Sci. U. S. A.* **27,** 35–41.

Chapter 4
Genetic Tools Used for Cell Lineage Tracing and Gene Manipulation in *Drosophila* Germline Stem Cells

Wei Liu and Steven X. Hou

Contents

Summary The advancement of *Drosophila* germline stem cell research accompanies the development of powerful new tools for genetic analysis. These include the techniques of stem cell labeling, cell lineage tracing, mosaic mutant analysis, and gene manipulation in targeted cell populations, which together constitute the critical methodologies in stem cell research. We discuss four such techniques: the *tubulin-lacZ* positive-labeling system; the positively marked mosaic lineage (PMML) method; the flipase/flipase recombination target (FLP/FRT)-based mosaic mutant analysis; and the *GAL80*-based mosaic analysis with a repressible cell marker (MARCM) system.

Keywords *Drosophila* germline stem cell; FLP/FRT; GAL4/UAS; GAL80; MARCM; mosaic analysis; PMML; positive labeling.

4.1 Introduction

Stem cells are buried in large populations of differentiated cells. Thus, identifying the small number of stem cells found among the background cells becomes the first challenge in stem cell research. Next, genetic manipulations must be performed in targeted cells to investigate the functions of interesting genes. The study of *Drosophila* germline stem cells has been greatly enhanced by the availability of powerful tools for genetic analysis as well as the excellent genetic model system present in the species itself. The most useful technique for studying germline stem cells is to generate mosaic mutant cells in living animals.

In a mosaic animal, a genetically heterozygous organism is induced to produce homozygous mutant cell clones. Two factors are critical for the successful application of mosaic analysis. The first is the presence of a highly efficient, inducible recombination system. In *Drosophila*, this condition is reasonably satisfied by using the flipase/flipase recombination target (FLP/FRT) system, which has been demonstrated as more efficient than the X-ray-induced recombination *(1)*. The FLP/FRT system was originally derived from yeast. FLP is a recombinase that specifically recognizes the FRT sequence (FLP recombinase recognition target) and can catalyze the exchange of DNA downstream of two FRT sites *(2)*. If two FRT sites are located in the same genomic locus of two homologous chromosomes, recombination may occur with high efficiency at metaphase during mitosis. After cell division, recombination could result in the formation of two distinct daughter cells.

This process is normally controlled by an inducible *flp*, which can be expressed in a tissue- or time-specific manner. The heat-inducible *flp*, in which the *flp* gene is under the control of a fly *hsp70* gene promoter, is a commonly used FLP resource *(2)*. A simple heat shock at 37°C induces FLP, which catalyzes the FRT-mediated recombination.

The second factor that affects the mosaic analysis is the ability to differentiate the mosaic clones (genetically modified cells and their progeny). This can be achieved using either positively or negatively marked methods. In the positively marked method, only genetically modified cells and their progeny are highlighted; in the negatively marked method, the mosaic clones can be distinguished from the otherwise marked background cells. In both methods, an excellent labeling marker must be used. The ideal marker should be nontoxic, cell autonomous, and have low or no endogenous existence. In *Drosophila*, two proteins are commonly used as labeling markers: the bacteria β-galactosidase (LacZ) and the green fluorescent protein (GFP) from jellyfish.

Another important methodology in cell biology is modifying gene functions in targeted cells, which includes upregulation, downregulation, knock-in, and knock-out of candidate genes. In *Drosophila*, the GAL4/ upstream activation sequence (UAS) system is a very popular technique that was developed to deliver genes to specific places *(3)*. This system is also borrowed from yeast. GAL4 is a transcription activator and the UAS is the GAL4 recognition site. GAL4 binds to the UAS sequence and activates transcription of the gene nearby. In this application, the GAL4 gene is normally expressed under a controllable promoter; several UAS sequences are aligned in tandem before a basic promoter, followed by the open reading frame (ORF) of a gene to be overexpressed. The GAL4 and UAS transgenes are introduced into the genome separately to get different fly strains. The expression of a gene of interest will be switched on when they meet each other after appropriate crossed.

We discuss four techniques employing these methods. These techniques have been successfully used in *Drosophila* germline stem cell research. The applications are based on mitotic recombination. In adult flies, most cells are postmitotic; only stem cells and a few types of transient cells are able to divide. Therefore, the techniques based on mitotic recombination are biased to label stem cells in adult flies.

4.1.1 The tubulin-lacZ Positive-Labeling System

The *tubulin-lacZ* positive-labeling system positively labels mitotic stem cells and their progeny (*see* **ref. 1** and **Fig. 4.1A**). Two complementary transgenes were introduced in the same genomic locus as two homologous chromosomes. One bears

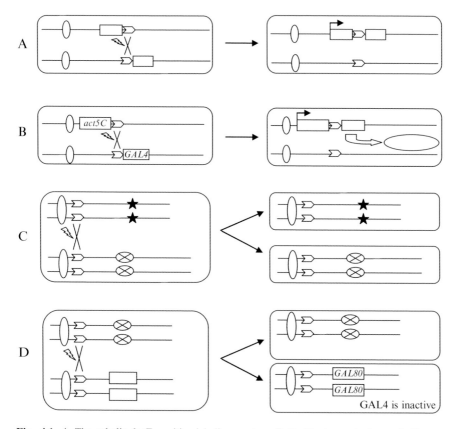

Fig. 4.1 **A** The tubulin–lacZ positive-labeling system. **B** Positively marked mosaic lineage (PMML). **C** Mosaic mutant analysis. **D** The GAL80-based mosaic analysis with a repressible cell marker (MARCM). $\boxed{\alpha\text{-tub}}$, *Drosophila* α-tubulin promoter; $\boxed{act5C}$, *Drosophila act5C* promoter; ★, dominant marker; ⟁ FRT sequence; ⚡, FLP induction; $\boxed{lacZ}$, *lacZ* open reading frame; $\boxed{GAL4}$, *GAL4* open reading frame; ⟨UAS-ORF⟩, upstream activation sequence promoter and open reading frame of candidate gene; ⊗ mutation; $\boxed{GAL80}$, *Drosophila* tubulin 1α promoter and *GAL80* open reading frame

a ubiquitously activated promoter, the *Drosophila* α1-tubulin promoter, followed by an FRT sequence. On the homologous chromosome, the ORF of a reporter gene, the β-*galactosidase* (*lacZ*), sits right after another FRT site. During mitotic division, when the chromosomes pair together at metaphase, an induced FLP can facilitate the homologous recombination between the two FRT sites. This leads to the construction of a functional gene cassette, which drives the *lacZ* gene expression in an extensively broad region. If this recombination happens in a stem cell, the stem cell and its progeny will be labeled with *lacZ*, and the clone will continue expanding and last a long time; if the recombination happens in a non-stem cell, a transient clone will be marked, and the clone will be cleared out after a short time. This technique is very sensitive because the marker gene is turned on immediately after recombination. However, the limitation of this method is that it cannot be used to manipulate the gene's activity in the clones.

4.1.2   Positively Marked Mosaic Lineage (PMML)

To generate positively marked cell clones and at the same time manipulate gene's activity in the clones, a positively marked mosaic lineage (PMML) technique has been developed (*see* **ref.** *4* and **Fig. 4.1B**) that combines the FLP/FRT and GAL4/UAS systems. In brief, two complementary transgenes are introduced in exactly the same genomic locus as two homologous chromosomes. One chromosome bears an actin5C promoter followed by an FRT sequence, and the other bears an FRT followed by the *GAL4* gene. Once the FLP is supplied, the homologous recombination could happen between the two FRT sites. This recombination will lead to the generation of a functional *actin5C-GAL4* gene, which can drive expression of any transgene under the UAS control. In the PMML system, a UAS-EGFP (enhanced green fluorescent protein) is introduced as the reporter.

More than one transgene under the UAS control can be introduced. For example, one transgene might be a UAS-EGFP reporter and the other a gene to be overexpressed. The gene's function can therefore be studied in the marked clones. If the transgene is a constitutively active form, we can study the influence of the gain function of that gene in the marked clones; if the transgene is a dominant negative form or expresses the RNA interference (RNAi) sequence of a gene, we can determine the consequences of turning down that gene's functions in the clones.

4.1.3   Mosaic Mutant Analysis

Diploid cells contain many essential recessive genes. No obvious phenotype can be detected if one copy of such a gene is lost because the second wild-type copy on the homologous chromosome will compensate for the functions of the lost copy. Studying such a gene's functions in adult animals is difficult because the homozygous mutant animal will die early, and the heterozygous animal may have no obvious phenotype. This problem can be overcome genetically by inducing mosaic cell clones that are homozygous for a mutation in an otherwise heterozygous genetic animal. Xu and Rubin developed such a technique in *Drosophila* using the FLP/FRT-mediated site-specific recombination system *(5)*.

The principle of this technique is explained in **Fig. 4.1C**. For a cell that is heterozygous for a mutation, two FRT sequences are introduced into the same genomic locus as two homologous chromosomes. Distal to one FRT site, a characterized mutation is on one chromosome, and a ubiquitously expressed dominant marker is on the complementary chromosome. During cell proliferation, the two chromosomes will be first duplicated and then paired at metaphase. The presence of FLP recombinase can promote the recombination between the two FRT sites. After the completion of cell division, there is a 25% possibility that one daughter cell is homozygous for that mutation. The homozygous mutant cell and its progeny will be identified by the absence of the marker. Then, we can study the gene's functions in the homozygous mutant clones.

In the mosaic mutant analysis system, the FLP can be introduced using a heat-inducible FLP (*hs-FLP*), and the markers (LacZ or GFP) are expressed ubiquitously. A complete set of modified fly chromosomes has been constructed to perform mosaic mutant analysis (*see* **Table 4.1**). Each chromosome carries a transgene with an FRT sequence close to a centromere. This transgene also includes a *mini-w+* or a heat-inducible *neo* cassette for introducing a mutation onto the FRT chromosome. Details about how to recombine a mutation onto an FRT chromosome can be found in **ref. 5**. Any mutation located distal to an existing FRT site can be placed after it. With the existing FRT chromosome collection, an estimated 95% of fly genes can be studied using mosaic mutant analysis *(5,6)*. However, this technique is not applicable for studying functions of genes on the X chromosome in male germlines because no paired X chromosomes exist in males.

4.1.4 *The GAL80-Based Mosaic Analysis with a Repressible Cell Marker*

Mosaic mutant analysis carries two significant drawbacks. First, the induced clones are negatively marked. Because the number of mutant clones and cells within each clone is relatively small compared to the marked background cells, it is not always easy to identify the mutant clones. Second, manipulating a gene's activity in marked clones is not feasible with this method. To overcome these drawbacks, a Gal80-based mosaic analysis with a repressible cell marker (MARCM) was developed (*see* **ref. 7** and **Fig. 4.1D**).

Gal80 is a Gal4-binding protein isolated from yeast that can interact with the transcription activation domain of GAL4 protein and block its activity. This repression was shown to work very well in fly with high specificity. In this system, a ubiquitously expressed *GAL80* transgene (*tubP-GAL80*) was placed on an FRT chromosome,

Table 4.1 Some useful transgenes and chromosomes[a]

Name	Purpose	Reference
hs-FLP[b]	Induces FLP recombinase by heat treatment (for all experiments)	(2)
UAS-EGFP[c]	Allows expression of EGFP under the control of GAL4	(4)
UAS-lacZ[c]	Allows expression of lacZ under the control of GAL4	(3)
X-15-33 (tub-FRT)	Provides the tubulin promoter after FLP/FRT-mediated homologous recombination (for the tubulin-lacZ positive-labeling system)	(1)
X-15-29 (FRT-LacZ)	Provides a functional lacZ gene after FLP/FRT-mediated homologous recombination (for the tubulin-lacZ positive labeling system)	(1)
FRT52B(w) (white-actin5C-FRT)	Provides the actin5C promoter after FLP/FRT-mediated homologous recombination (for PMML)	(4)
FRT52B(y) (yellow-FRT-GAL4)	Provides a functional GAL4 gene after FLP/FRT-mediated homologous recombination (for PMML)	(4)
arm-LacZ, FRT19A	Provides FRT on the X chromosome with a ubiquitously expressed lacZ (for mosaic mutant analysis)	(5) and flybase
arm-lacZ, FRT40A	Provides FRT on the left arm of the second chromosome with a ubiquitously expressed lacZ (for mosaic mutant analysis)	(5) and flybase
FRT42D, arm-lacZ	Provides FRT on the right arm of the second chromosome with a ubiquitously expressed lacZ (for mosaic mutant analysis)	(5) and flybase
arm-lacZ, FRT80B	Provides FRT on the left arm of the third chromosome with a ubiquitously expressed lacZ (for mosaic mutant analysis)	(5) and flybase
FRT82B, arm-LACZ	Provides FRT on the right arm of the third chromosome with a ubiquitously expressed lacZ (for mosaic mutant analysis)	(5) and flybase
FRT19A, tubP-GAL80	Provides FRT on the X chromosome with a ubiquitously expressed GAL80 (for MARCM)	(7)
FRTG13, tubP-GAL80	Provides FRT on the right arm of the second chromosome with a ubiquitously expressed GAL80 (for MARCM)	(7)
FRT82B, tubP-GAL80	Provides FRT on the right arm of the third chromosome with a ubiquitously expressed GAL80 (for MARCM)	(7)

FLP/FRT, flipase–flipase recombination target; MARCM, mosaic analysis with a repressible cell marker.

[a] For a complete description of selection markers and genotypes, please check the references and flybase (http://www.flybase.org/).

[b] Several different strains carry the hs-FLP transgene on different chromosomes for convenient crosses. See references and flybase (http://www.flybase.org/) for more information.

[c] Several reporter lines have been constructed for specific purposes, in which the reporter genes are fused to different subcellular localization signals, such as membrane-, microtubule-, cytoplasm-, or nuclear-targeted sequences. See references and flybase (http://www.flybase.org/) for more information.

a characterized mutation on the complementary FRT chromosome, and a tubP-GAL4 (or other tissue-specific GAL4s) and a UAS-GFP on the other chromosomes. If the FLP recombinase is present, it will promote the recombination between the two FRT sites. After the completion of cell division, there is a 25% possibility that one daughter cell is homozygous for that mutation without *tubP-GAL80*. In the mutant clones, the active GAL4 will actively drive UAS-GFP expression to mark the mutant clones. Other transgenes under the UAS control can also be included in these flies. Therefore, this technique can also be used to manipulate a gene's activity in the marked mutant clones.

A drawback of this technique is the persistence of the GAL80 protein. Because it is relatively stable, the daughter cell will inherit some functional GAL80 protein before it is eventually degraded. This would be a problem if you want to investigate the consequences immediately after the mutant cell is formed. Because the half-life of GAL80 protein is estimated to be 24 h in the central nervous system (CNS) and imaginal disks, strong labeling is generally not apparent until 48 h after clone induction. However, because *Drosophila* germline stem cells undergo division slowly, this waiting time may not be a major concern when this method is used to generate germline stem cell clones.

4.1.5 Comparison of the Four Techniques

Each of these techniques may be used to label stem cells and their progeny. In the mosaic mutant analysis method, the mutant clones are negatively marked; in the other three methods, the mutant clones are positively marked. In the *tubulin-lacZ* positive-labeling and the PMML systems, only FLP/FRT-mediated recombination is required to generate marked clones. Many *Drosophila* adult cells are polyploidy, having multiple copies of homologous chromosomes that pair together without cell division. A well-known example is the polytene chromosomes of *Drosophila* salivary gland cells. These cells can also be labeled by the *tubulin-lacZ* positive-labeling and the PMML methods (our unpublished observation). In the mosaic mutant analysis and MARCM systems, both FLP/FRT-mediated homologous recombination and chromosome segregation are required to generate marked clones. The PMML and MARCM methods can also be used to ectopically manipulate a gene's activity in the marked clones.

4.2 Materials

1. G418 (Geneticin; Invitrogen, cat. no. 11811).
2. G418-containing fly food: Punch a few holes in the standard fly food in vials. Add 0.2 mL 25 mg/mL freshly made G418 solution per 10 mL food. Let it air dry until there is no liquid on the surface.

3. Fly culture incubator.
4. Temperature-adjustable water bath tank (or 37°C bacteria incubator; *see* **Note 1**).
5. Some useful transgenes and chromosomes are shown in **Table 4.1**.

4.3 Methods

The methods introduced here are taken from publications and are used in female germlines. They can be also used for male germline clonal analysis.

The efficiency of clone induction is strongly influenced by different FRT chromosomes and an abundance of FLP. A simple way to control the FLP level is to adjust the heat shock conditions, including temperature and time. Most experiments require keeping a time-course record to follow the behavior of induced clones.

To get consistent and reliable results, a good control is critical. The control samples should be treated in exactly the same conditions as those of the experimental samples. We suggest setting the following controls: (1) without heat shock induction (for all methods); (2) without gene overexpression (for methods described in **Subheadings 4.3.3** and **4.3.5**); (3) without mutation (wild-type clones; for methods described in **Subheadings 4.3.4** and **4.3.5**).

4.3.1 Generate lacZ-Marked Clones in Adult Germlines

1. Produce female (or male) flies with the genotype *hs-FLP/+; X-15-33/X-15-29;* by standard crosses (*see* **Note 2**).
2. Immerse vials (or bottles) containing 1- to 3-d-old flies (progeny from the above cross) in a circulating water bath at 37°C (or 37°C bacteria incubator) for 1 h (*see* **Notes 3** and **4**).
3. Return flies to previous culture environment. Transfer them daily to fresh food with yeast to meet their optimal living requirements.
4. On the second day after heat shock and each day thereafter, flies can be dissected to detect the presence of lacZ (*see* **Note 5**).

4.3.2 Generate lacZ-Marked Clones in Developing Female Germlines (Adapted from Ref. 8)

1. Produce female flies with the genotype *hs-FLP/+; X-15-33/X-15-29;* by standard crosses.
2. Immerse vials (or bottles) containing third instar larvae (progeny from the above cross) in a circulating water bath at 37°C (or 37°C bacteria incubator) for 4 h.
3. Dissect 1- to 2-d-old females to inspect for lacZ expression.

4.3.3 Overexpress Transgene(s) in Positively Marked Adult Germline Clones (PMML)

For example, overexpress a transgene (*A*) in the GFP-labeled clones.

1. Produce flies with the genotype of *y, w, hs-FLP, UAS-GFP; FRT52B(w)/ FRT52B(y); UAS-geneA;* by standard crosses.
2. Use the procedures for heat shock treatment, culture conditions, and schedule of time-course records that are described in **Subheading 4.3.1, steps 2–4**; refer also to the same notes.

4.3.4 Generate Mosaic Germlines with Mutant Clones

4.3.4.1 Construct a Chromosome with Both an FRT Site and a Mutation

Take a mutation, *m-*, on the right arm of the second chromosome as an example.

1. Cross three or four female flies with the genotype *w; m-/ P[ry+]; hs-neo; FRT]42D, P[ry+; w+]47A* with two or three males of *w; Sco/CyO*.
2. Within 2 to 3 d, larvae should start to emerge from fertilized eggs. Transfer flies daily into a new vial with freshly prepared food containing G418.
3. Fertilized eggs should hatch 24 h after egg laying in 25°C incubators. During the first or second day after hatching, immerse the vials containing the young larvae in a water bath at 37°C (or in a 37°C bacteria incubator) for 1 h to induce Neo production.
4. After eclosion, most of the flies should be *w+*; only a small number of flies should have white eyes (*w-*). Select each *w-* virgin from these vials and expand it by crossing it with a balancer stock (such as w; *Sco/CyO*). Check for the presence of the mutation and make a balanced stock. The correct strain should have a genotype as *w-; P[ry+]; hs-neo; FRT]42D, m-/CyO*.

4.3.4.2 Generate Mutant Clones (m-) in Adult Germline Stem Cells

1. Cross the above-generated flies with *w, hs-FLP; FRT42D, arm-lacZ* flies to get offspring with the genotype of *w, hs-FLP/+ (or Y); FRT42D, m-/FRT42D, arm-lacZ*.
2. Perform heat shock treatment on these offspring at 1–3 d of age. Administer heat shock six times in 3 d consecutively by immersing in a water bath at 37°C for 1 h (or placing in a bacteria incubator). Treat once in the morning and once in the late afternoon, with an 8-h interval between treatments (*see* **Note 6**).
3. Return flies to previous culture environment. Transfer them daily to fresh food with yeast to meet their optimal living requirements.
4. On the second day after the last heat shock, flies can be dissected for study (*see* **Note 5**).

4.3.5 Overexpress a Candidate Gene (B) in GFP-marked Mutant (n−) Female Germline Clones

1. Produce flies with the genotype of *y, w, hs-FLP; actin-gal4, UAS-EGFP/UAS-geneB; FRT82B, tub-GAL80/FRT82B, n-;* by standard crosses.
2. Use the procedures for heat shock treatment, culture conditions, and schedule of time-course records that are described in **Subheading 4.3.4.2, steps 2–4**; refer also to the same notes.

4.4   Notes

1. If a circulating water bath is not available, a 37°C bacteria incubator also works well.
2. Because hormones provided by males during copulation are important for maintaining a constant egg-laying speed among the females, keeping males with the females is helpful to allow copulation.
3. For methods described in **Subheadings 4.3.1** and **4.3.2**, a single heat shock at 37°C for 1 h is generally sufficient to produce enough marked clones because of the high sensitivity of these techniques (see **Subheading 4.1.5**). The heat shock time can be varied depending on different hs-flp resources.
4. After heat shock, dew may form on the inner wall of vials. To prevent flies from attaching to the wall, transfer them to dry vials.
5. Because of the slow growth speed of germline cells, time-course records of the marked clones may be made at 2-d intervals. However, flies may also be checked daily.
6. Because of the small number of germline stem cells within each fly and the slow division time (estimated at 24–48 h), a harsh heat shock treatment for elevated FLP level is recommended for the less-sensitive methods described in **Subheadings 4.3.3 and 4.3.4**, although mild treatments were also successfully applied in different labs (see the listed references).

References

1. Harrison, D. A., and Perrimon, N. (1993) A simple and efficient generation of marked clones in *Drosophila. Curr. Biol.* **3**, 424–433.
2. Golic, K. G., and Lindquist, S. (1989) The FLP recombinase of yeast catalyzes site-specific recombination in the *Drosophila* genome. *Cell.* **59**, 499–509.
3. Brand, A. H., and Perrimon, N. (1993) Targeted gene expression as a means of altering cell fates and generating dominant phenotypes. *Development.* **118**, 401–415.
4. Kirilly, D., Spana, E. P., Perrimon, N., Padgett, R. W., and Xie, T. (2005) BMP signaling is required for controlling somatic stem cell self-renewal in the *Drosophila* ovary. *Dev. Cell.* **9**, 651–662.
5. Xu, T., and Rubin, G. M. (1993) Analysis of genetic mosaics in developing and adult *Drosophila* tissues. *Development.* **117**, 1223–1237.
6. Margolis, J., and Spradling, A. (1995) Identification and behavior of epithelial stem cells in the *Drosophila* ovary. *Development.* **121**, 3797–3807.
7. Lee, T., and Luo, L. (1999) Mosaic analysis with a repressible neurotechnique cell marker for studies of gene function in neuronal morphogenesis. *Neuron.* **22**, 451–461.
8. Zhu, C. H., and Xie, T. (2003) Clonal expansion of ovarian germline stem cells during niche formation in *Drosophila. Development.* **130**, 2579–2588.

Chapter 5
Structural Polarity and Dynamics of Male Germline Stem Cells in an Insect (Milkweed Bug *Oncopeltus fasciatus*)

David C. Dorn and August Dorn

Contents

Summary Knowing the structure opens a door for a better understanding of function because there is no function without structure. Male germline stem cells (GSCs) of the milkweed bug (*Oncopeltus fasciatus*) exhibit a very extraordinary structure and a very special relationship with their niche, the apical cells. This structural relationship is strikingly different from that known in the fruit fly (*Drosophila melanogaster*)—the most successful model system, which allowed deep insights into the signaling interactions between GSCs and niche. The complex structural polarity of male GSCs in the milkweed bug combined with their astonishing dynamics suggest that cell morphology and dynamics are causally related with the most important regulatory processes that take place between GSCs and niche and ensure maintenance, proliferation, and differentiation of GSCs in accordance with the temporal need of mature sperm. The intricate structure of the GSCs of the milkweed bug (and probably of some other insects, i.e., moths) is only accessible by electron microscopy. But, studying singular sections through the apical complex (i.e., GSCs and apical cells) is not sufficient to obtain a full picture of the GSCs; especially, the segregation of projection terminals is not tangible. Only serial sections and their overlay can establish whether membrane ingrowths merely constrict projections or whether a projection terminal is completely cut off. To sequence the GSC dynamics, it is necessary to include juvenile stages, when the processes start and the GSCs occur in small numbers. The fine structural analysis of segregating projection terminals suggests that these terminals undergo autophagocytosis. Autophagosomes can be labeled by markers. We demonstrated acid phosphatase

From: *Methods in Molecular Biology, Vol. 450, Germline Stem Cells*
Edited by S. X. Hou and S. R. Singh © Humana Press, Totowa, NJ

and thiamine pyrophosphatase (TPPase). Both together are thought to identify autophagosomes. Using the appropriate substrate of the enzymes and cerium chloride, the precipitation of electron-dense cerium phosphate granules indicates the presence of enzymes and their location. Because the granules are very fine, they can be easily assigned to distinct cell organelles as the autophagosomes. Two methods, electron microscopy and immunocytochemistry, have pointed out a structural polarity and dynamics that are unprecedented for stem cells. We propose that these dynamics indicate a novel type of signal exchange and transduction between stem cells and their niche.

Keywords Electron microscopy; GSC dynamics; immunocytochemistry; male GSC; milkweed bug; structural polarity.

5.1 Introduction

It is a key feature of adult stem cells that they reside in a stem cell niche, which plays a central role in the regulation of stem cell maintenance, proliferation, and differentiation *(1)*. The stem cell–niche relationship has been the subject of intense research because of the possibility that its disturbance may cause serious human diseases. A more in-depth knowledge of their interactions may facilitate the application of adult stem cells in the repair of damaged tissues. But, stem cell–niche relationships have proven extremely difficult to explore in human and mammals because of their complexity. However, studies of *Drosophila* male and female germline stem cells (GSCs) and their niches have provided exciting insight over the last few years, and this species now represents an established model system (*see* **ref. 2** and **Chapters 1, 3,** and **4** of this volume). The success of the *Drosophila* model is based on the relatively simple anatomical structure of the GSC–niche complex in both sexes combined with the easy access to molecular genetic studies in this species.

Although models have proven extremely fruitful in uncovering principles in the control of basic biological processes common to vertebrates and invertebrates alike, the differences that may exist between animal phyla, within a phylum, or even between closely related species must not be underestimated. It is therefore important to extend the number of models to obtain a more detailed picture of the variability in the regulation of (seemingly) homologous processes among different organisms. This potentially could prevent prematurely transferring results gained from models to the human setting, with undesirable consequences in the medical application. On the other hand, it may emphasize the general validity and conservation of a process in evolution. To examine how "typical" the male GSC–niche complex of *Drosophila* is for insects in general, we performed comparative fine structural studies of the complex in the milkweed bug *(3–6)*.

5.1.1 Organization of Insect Testes

Testes of insects consist of one (cf. *Drosophila*) to more than 100 (cf. migratory locust) follicles of identical organization. The blind distal tips of the follicles—the follicular apices—include three types of cells: the GSCs, the cyst progenitor cells (CPCs), and the niche cells. The niche may be represented by only one large cell (cf. moths and locusts) or a small cluster of cells (cf. *Drosophila* and the milkweed bug) *(3,4,7)*. Because of their location, the niche cells were termed *apical cells*, which was about a century before their role as a niche was recognized *(8)*. Divergent from this general designation, the niche cells in *Drosophila* are called *hub cells*.

The CPCs and the niche cells are of somatic origin *(9,10)*. Niche cells normally stop dividing after embryogenesis. CPCs are associated with GSCs and display stem cell characteristics. In *Drosophila*, GSCs and associated CPCs (one GSC is embraced by two CPCs) undergo correlated divisions, leading to a gonioblast surrounded by two cyst cells that no longer divide during cyst maturation *(11)* and a daughter GSC that replaces the mother GSC and touches the niche *(12)*. In the adult *Drosophila*, all GSC divisions are asymmetrical, resulting in a gonioblast that is separated from the niche. In the milkweed bug, a gonioblast is ensheathed by only one cyst cell, which enlarges during spermatogenesis and becomes polyploid. Compared to the situation in *Drosophila*, gonioblast formation in the milkweed bug is at present not fully understood and seems far more complex than in the fruit fly *(5,6)*.

5.1.2 Organization of the Testis in the Milkweed Bug

Each of the paired testes of the milkweed bug contains seven identical testis follicles that are individually covered by a follicular envelope. In advanced developmental stages, the seven follicles together become enclosed in a sturdy testis envelope. Testis follicles of adult males consist of the apical complex and many cysts at different stages of spermatogenesis. **Figure 5.1** presents the light microscopic picture of the distal portions of two follicles (*see also* **Fig. 5.2** for a schematic representation of the most distal part of a follicle). The apical complex is comprised of the apical cells forming a sphere with a small lumen; the GSCs arranged around the apical cells (both together giving the impression of a rosette, hence the term *apical rosette*); and the CPCs, which occupy the most apical tip of a follicle in a single layer. Gonioblasts are formed primarily in the lateral and dorsolateral region of the apical rosette but not below the rosette, where dying germ cells and degenerating young cysts amass. Gonioblasts, which become enclosed by one cyst cell, enter the spermatogenetic pathway. All cysts border the follicular envelope. This contact is maintained during the entire spermatogenesis. The growing cysts move proximally, still surrounded by only one enlarging cyst cell with a nucleus that becomes polyploid *(5)*.

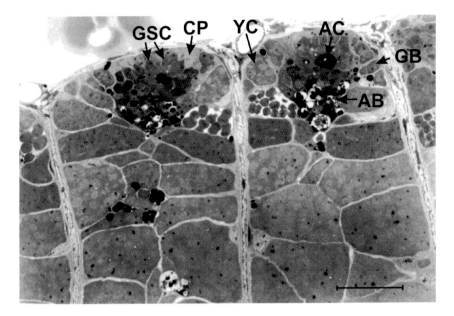

Fig. 5.1 Light micrograph. Apical regions of two testicular follicles in adult males, d 0. The longitudinal section shows the apical rosette, consisting of the germline stem cells (GSC) and the apical cells (AC), which represent the niche for the GSCs. Cyst progenitor cells (CP), gonioblasts (GB), and young cysts (YC) surround the apical rosette. Apoptotic bodies (AB) of dying gonio-blasts and young cysts amass below the apical rosette. Bar = 67.8 μm. (From **ref. 5** with kind permission of Balaban, Philadelphia/Rehovot.)

The schematic representation of a follicular apex, depicted in **Fig. 5.2**, shows a more detailed picture of the GSCs. These cells express a strong structural polarity as can only be recognized in electron microscopic studies *(4–6)*. In median sections of the apical complex (*see* **Fig. 5.3**), electron micrographs reveal an almost neuron-like structure of the GSCs consisting of a perikaryon directed toward the periphery of the apical rosette and cellular projections, some of which are very long and directed toward the center of the rosette in which the apical cells (i.e., the niche of the GSCs) are located and which they often touch, sometimes in a bouton-like expression. Most intriguing, terminal segments of the projections are regularly seg-regated by transverse ingrowths of the cell membrane. The surface of the apical cells is largely covered by segregating or already segregated projection terminals (*see* **Fig. 5.4**). The terminals undergo a stereotyped degradation process character-istic of autophagocytosis (*see* **Subheadings 5.1.4** and **5.1.5**). To demonstrate that the terminals become completely pinched off, serial sections have to be studied. Such series also allow the observation of the different stages of the separation process: Different degrees of transverse ingrowths of the cell membrane can be recognized, from slight indentations to an almost complete cutoff.

5.1.3 The Fine Structure of GSCs

The GSCs are the largest cells of the apical complex (**Figs. 5.1–5.8**). "Typical" GSCs are shown in **Figs. 5.7** and **5.8**. The nucleus includes many clumps of condensed chromatin and a large structured nucleolus. The perinuclear cytoplasm is characterized by many ribosomes arranged in small clusters. These clusters are randomly scattered in the perikaryon, with interspersed short cisternae of rough endoplasmic reticulum. Golgi complexes are rare and not prominent. Conspicuous are accumulations of mitochondria at the site where the projections arise (*see* **Fig. 5.8**). The organelle equipment of the cell projections differs between the basal region (where the projections sprout from the perikaryon) and the terminal region (next to the apical cells) and is dependent on the stage of terminal segregation process described below (*see* **Subheading 5.1.4**).

5.1.4 Dynamics of the GSCs

In the milkweed bug, the apical complex of adult males presents an extremely complex picture (*see* **Figs. 5.2** and **5.3**), and it is challenging to decipher the unique dynamics of GSCs in detail. However, it appears most likely that these processes

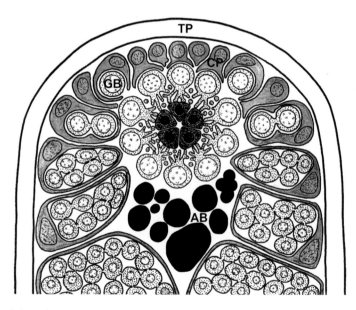

Fig. 5.2 Schematic representation of the apical region of a testicular follicle. The apical cells in the center of the apical rosette are densely stippled. The germline stem cells form projections toward the apical cells, which represent their niche. Cyst progenitor cells (CP) are located at the tip of the follicle and (at least some) embrace the GSCs. Cysts are formed laterally from the rosette GB, a gonioblast that is almost completely encapsulated by one cyst cell. During their development, cysts migrate proximally. AB, apoptotic bodies; TP, tunica propria. (From **ref. 5** with kind permission of Balaban, Philadelphia/Rehovot.)

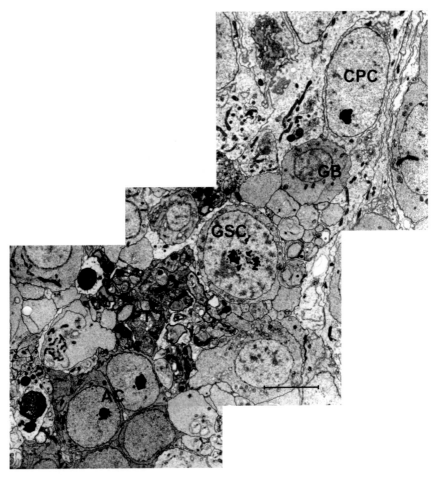

Fig. 5.3 Electron micrograph. Sector of the apical rosette depicting apical cells (AC), two germ-line stem cells (one marked with GSC), a (tangentially cut) gonioblast (GB), and a bordering cyst progenitor cell (CPC), which may enclose the GB and transform into a cyst cell. Note profiles of cell projections (or segregated projection terminals) below the perikaryon of GB; they are directed toward the apical cells. Profiles of projection terminals (PTs) that are touching the apical cells belong to the marked GSC. Some of the profiles represent completely segregated PTs, and some show signs of degeneration (*see* **Figs. 5.4** and **5.9**). Bar = 4.2 μm

are involved in GSC–niche interactions that concern the fate of GSCs and therefore would be of utmost importance also in respect to the established model *Drosophila*. The reasoning is that the structural relationships of GSCs and apical cells in the milkweed bug— gravely different from those of GSCs and *hub cells* in *Drosophila*— may indicate signaling interactions of GSCs and niche in the milkweed bug differing from those in *Drosophila*. Anticipating that apical complexes of larvae might present a less-elaborate picture, we examined the testes throughout all developmental stages, including larvae that had newly hatched from eggs.

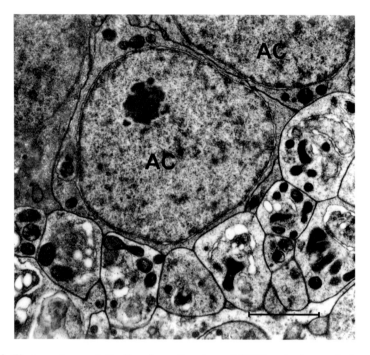

Fig. 5.4 Electron micrograph. Profiles of germline stem cell (GSC) projection terminals touching apical cells (AC). Some of the profiles presumably represent bodies that are completely segregated from the projections and show early signs of degeneration: abundance of mitochondria (depicted as electron-dense tubular organelles) and beginning of vesiculation. (Advanced stages of degeneration are shown in **Fig. 5.9**.) Note the scant edge of cytoplasm around the globular nucleus of apical cells. Cytoplasm and nucleus of apical cells are noticeably more electron dense than those of GSCs and cyst progenitor cells (CPCs) (*see* **Figs. 5.3** and **5.5**). Bar = 1.3 μm. (From **ref. 4** with kind permission of Balaban, Philadelphia/Rehovot.)

First instar larvae of the milkweed bug have much fewer GSCs than adult males. We serially sectioned two complete follicles of this early developmental stage and found four GSCs in each of these follicles (*see* **Fig. 5.5**). Intriguingly, already at this early stage GSCs depict the same principal structure as in adults, but their organization was much more easily unraveled than in adults because it was comprised of fewer cells. It was through the examination of the young larval stages that we were able to reconstruct a sequence of the GSC dynamics (*see* **Fig. 5.6**). The beginning of projection formation apparently starts during embryogenesis and was observed in newly hatched larvae (*see* **Fig. 5.7**). The GSC pictured in **Fig. 5.7** corresponds to the stage of early projection formation (*stage B/C* in **Fig. 5.6**).

Analyses of smaller series of ultrathin sections were carried out during all larval instars (the milkweed bug has five) and in mature males. They all confirmed the mode of projection formation, the segregation of projection terminals, and the degradation of segregated terminals in a stereotyped manner. An early step in the sculpturing of the projections is the inward growth of the cell membrane from the periphery of lobular

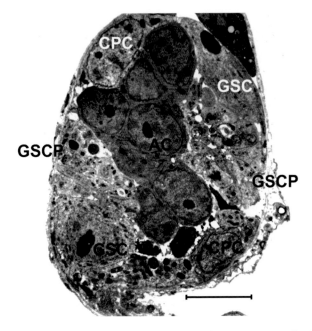

Fig. 5.5 Electron micrograph. A testicular follicle in a first instar larva immediately after hatching. The follicle includes three cell types: the germline stem cells (GSC), the apical cells (AC), and the cyst progenitor cells (CPC). GSCP denotes the area of GSC projections. Because of the small number of GSCs at this early developmental stage (only four in the pictured specimen), dynamics of GSCs can be studied more readily than at later stages of development. Bar = 5.7 μm. (From **ref.** 5 with kind permission of Balaban, Philadelphia/Rehovot.)

GSC protrusion (*see* **Figs. 5.6B,C** and **Fig. 5.7**). Some of the trabecular or septum-like ingrowths eventually cut off parts of the protrusions (*see* **Fig. 5.6D–F**). This process gives the impression of "carving out" and "modeling" of GSC projections. GSCs adopt a waist-like constriction between the perikaryon and the origin of the now finger-shaped projections (*see* **Figs. 5.6E** and **5.8**). At the same time, terminal parts of the projections (which are not all directed toward the apical cells) are segregated. Subsequently, the segregation process intensifies and may concern entire projections (*see* **Fig. 5.6F**). There are indications that new projections sprout from the perikaryon, but further investigations will have to be conducted to confirm detailed actions.

5.1.5   Analysis of the Degradation Process of Segregated GSC Projection Terminals

The mode of degradation of the segregated projection terminals was analyzed by immunohistological techniques on the fine structural level. During the segregation process of projection terminals, the number of mitochondria increases in these

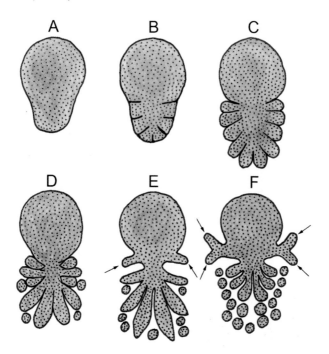

Fig. 5.6 Schematic representation of germline stem cell (GSC) dynamics deduced from observations during early larval stages. **A** Hypothetical precursor stage of GSCs (presumably found in embryonic testes) with lobular protrusion toward the apical cells. **B, C** Early stages of projection formation: Trabecular ingrowths of the cell membrane compartmentalize the cell protrusion. Such stages can be easily identified in testes of first and second instar larvae (compare with **Fig. 5.7**). **D** GSC projections in the process of being "carved out." GSCs show a waist-like constriction between the perikaryon and origin of projections. **E** Stage of GSC with finger-like projections and beginning of segregation of projection terminals. Probable sprouting of new projections at the parikaryon (*arrows*) (compare with **Fig. 5.8**). **F** Segregation process at projection terminals at an advanced stage. Sprouting of new projections seems enhanced (*arrows*). (From **ref. 6** with kind permission of Springer, Berlin/Heidelberg.)

segments; lysosomal bodies and autophagic vacuole-like vesicles become abundant (*see* **Figs. 5.4** and **5.9**). A first sign of degradation is the appearance of circular cisternae of rough endoplasmic reticulum, which are often concentrically arranged; these eventually lose their ribosomes and adopt a myelin-like structure (*see* **Fig. 5.9**). As the autotomy of the GSC projections increases, mitochondria become even more numerous in these fragmentation bodies and are larger and more electron dense than those in the perikaryon. Then, the fragmentation bodies undergo strong vesiculation. Finally, they are filled with myelin- and autophagosome-like bodies. Eventually, they seem to rupture and to release their contents. Debris is apparently taken up by adjacent GSCs and CPCs.

Although the electron microscopic studies gave strong indications that the degradation of GSC projection terminals occurs by autophagocytosis *(13,14)*, this had to be proven by the demonstration of digestive enzymes characteristic

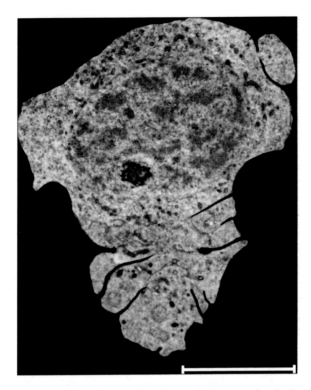

Fig. 5.7 Electron micrograph. Germline stem cell at an early stage of projection formation in a newly hatched larva (compare with **Fig. 5.6B**). Neighboring cells have been blackened to clarify the structure of the cell. Bar = 5.0 μm

for autophagic vacuoles. Such enzymes are acid phosphatase and TPPase (thiamine pyrophosphatase) *(15,16)*. The presence of both acid phosphatase and TPPase could be shown on the electron microscopic level in vacuoles that also ultrastructurally had the characteristics of autophagosomes (*see* **Figs. 5.10–5.12**).

Taking both ultrastructural and immunohistological studies into account indicates a unique relationship between stem cells and their niche. The GSCs of the milkweed bug present a cellular polarity that is extremely rare in stem cells. (Only the GSCs of moths show some similarities in this respect; *17*). Unprecedented for stem cells are the dynamics that have been observed in GSCs of the milkweed bug. We propose that the regression and probable regrowth of GSC projections that contact apical cells are involved in signal transduction between the stem cells and their niche. The increased lysosomal activity in projection terminals might indicate the endocytotic regulation of the signaling pathway *(18)*. The elucidation of the described unique stem cell dynamics may uncover a new mode of signaling *(6)*.

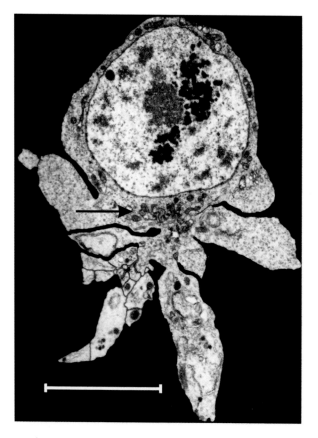

Fig. 5.8 Electron micrograph. Germline stem cell at an advanced stage of projection formation (compare with **Fig. 5.6E**). Note the waist-like constriction of the cell between perikaryon and origin of the finger-like projections. *Arrow* points to an aggregation of mitochondria in this cell region. Serial sections of this cell have shown that the ingrowths of the plasma membrane visible at some projections do not completely segregate the terminal parts. Completely segregated bodies are not shown but have been blackened as other neighboring cells. Bar = 4.6 μm

5.2 Materials

5.2.1 Breeding of the Milkweed Bug and Timing of the Developmental Stages

1. Incubator that can be adjusted to a constant temperature of 28°C ± 1°C.
2. Glass or plastic containers of different sizes for rearing bugs; fine gauze and rubber bands to cover the containers.
3. Filter paper; soft paper tissue; cotton-wool balls; tubes of cardboard.

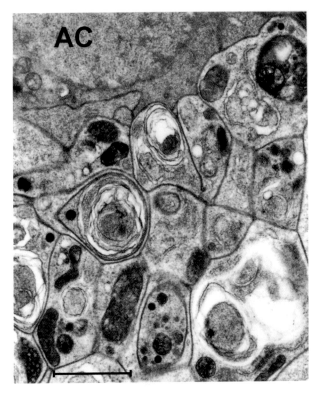

Fig. 5.9 Electron micrograph. Profiles of projection terminals and cell bodies of segregated terminals of germline stem cells. Some of the projection terminals and segregated bodies show signs of ongoing autophagocytosis. Note autophagic vacuoles with myelin-like figures (*center*) and a complex autophagic vacuole (*upper right*). AC, part of apical cell. Bar = 1.5 µm

4. Petri dishes of different sizes.
5. Peeled sunflower seeds.
6. Soft tweezers and fine brush for handling bugs and bug eggs.

5.2.2 Preparation of Testes

1. Dissecting microscope.
2. Carbon dioxide (CO_2) gas.
3. Fine scissors (microscissors, as they are used for microsurgery); fine hard tweezers; fine needles with handles.
4. Insect Ringer solution: 10.4 g sodium chloride, 0.32 g potassium chloride, 0.48 g calcium chloride, and 0.32 g sodium hydrogen carbonate in 1 L distilled water; adjust to pH 7.0.

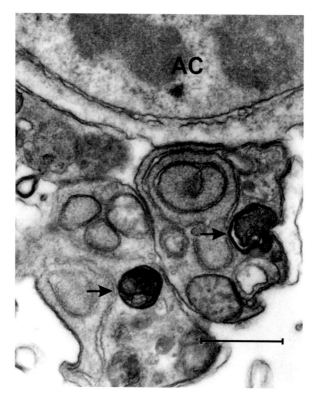

Fig. 5.10 Electron micrograph. Immunohistochemical demonstration of acid phosphatase activity in autophagic vacuoles of germline stem cell projections. The demonstration of acid phosphatase is indicated by immunocerium precipitation (*arrows*). Note that only autophagic vacuoles contain precipitations. AC, part of apical cell. Bar = 0.73 μm. (From **ref.** *4* with kind permission of Balaban, Philadelphia/Rehovot.)

5.2.3 Processing of Specimens for Electron Microscopy and Ultrastructural Analysis

Almost all the reagents used in processing of specimens for electron microscopy are potentially hazardous, such as chemicals used for fixation, washing, dehydration, embedding, and staining. They can be absorbed either by skin contact or by inhalation. A fume hood that ensures that fumes are sufficiently removed is essential in the laboratory. Hands should be protected by disposable gloves made of impermeable material. Eyes should be covered with gastight goggles *(19)*. Some solvents are highly flammable (i.e., ethanol, propylene oxide) and should not be used near an open flame. All chemicals should be of high purity.

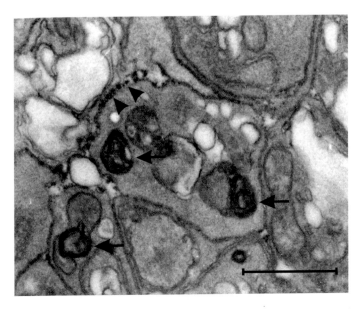

Fig. 5.11 Electron micrograph. Immunohistochemical demonstration of thiamine pyrophosphatase (TPPase) activity in autophagic vacuoles of germline stem cell projections. The demonstration of TPPase is indicated by immunocerium precipitation (*arrows*). Note precipitations in autophagic vacuoles. *Arrowheads* indicate nonspecific precipitation of cerium at plasma membrane. Bar = 0.77 μm. (From **ref. 4** with kind permission of Balaban, Philadelphia/Rehovot.)

5.2.3.1 Chemicals

1. Glutaraldehyde (electron microscopy purity grade).
2. Cacodylate 0.2*M* buffer stock solution: Solution A: 42.8 g sodium cacodylate in 1000 mL distilled water; solution B: 10 mL concentrated HCl (36–38%) in 603 mL distilled water for 0.2*M* HCl. To obtain pH 7.2, add 4.2 mL solution B to 50 mL solution A and dilute with distilled water to total volume of 200 mL.
3. Sucrose.
4. Osmium tetroxide.
5. Agar.
6. Ethanol (should not be denatured).
7. Propylene oxide.
8. Araldite.
9. Uranyl acetate.
10. Lead citrate.
11. Sodium hydroxide.
12. Methylene blue stain stock solution and staining: Add 1% methylene blue and 1% azure B to 1% sodium borate. For staining of semithin section: Dilute

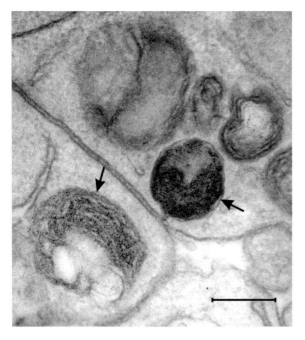

Fig. 5.12 Electron micrograph. Immunohistochemical demonstration of thiamine pyrophosphatase (TPPase) activity in autophagic vacuoles of germline stem cell projections (*arrows*). This high-resolution micrograph shows that the immunocerium precipitation (i.e., cerium phosphate) consists of fine, electron-dense granules. Their association with distinct cellular structures can be clearly recognized. In the phagosome to the *lower left*, the granules mark enclosed cisterns of endoplasmic reticulum. Bar = 0.26 μm

stock solution 1:10 with distilled water and cover dried sections using a pipet. Place slide on heat bench at 70°C for 5 to 10 min. Wash off with tap water and let slide dry.

5.2.3.2   Equipment and Tools

1. Embedding molds for electron microscopy. Choose small flat molds that allow orientation of specimens.
2. Heating bench.
3. Incubator that can be adjusted to 62°C ± 1°C.
4. Ultramicrotome.
5. Glass knives for cutting semithin and ultrathin sections. Diamond knifes produce nicer ultrathin sections than glass knives, but their handling needs prior experience.
6. Copper grids.
7. Light microscope.
8. Electron microscope.

5.2.4   Immunocytochemistry

All chemicals, equipment, and tools necessary for processing of specimens for electron microscopy and ultrastructural analysis (*see* **Subheading 5.2.3**) are also needed for immunocytochemistry. The following list includes only the additional chemicals and equipment needed.

1. Sodium-β-glycerophosphate.
2. Thiamine pyrophosphate.
3. Cerium chloride.
4. Maleate (tris-hydroxymethyl-aminomethane maleate) buffer 0.2M stock solution: Solution A: Take 30.3 g tris-hydroxymethyl-aminomethane and 29.0 g maleic acid in a measuring cylinder and add distilled water to make 500 mL. Add 2 g charcoal, shake, let stand for 10 min, then filter. Solution B: 4% sodium hydroxide. To obtain pH 7.2, add 20 mL solution B to 40 mL solution A and dilute with distilled water to a total volume of 100 mL.
5. Triton X-100.
6. Manganese chloride.
7. Paraformaldehyde.
8. Potassium ferricyanide.
9. Calcium chloride.
10. Heatable water bath.

5.3   Methods

5.3.1   Breeding of the Milkweed Bug and Timing of the Developmental Stages

The (large) milkweed bug *Oncopeltus fasciatus* Dallas (Insecta, Heteroptera) is native to North America and feeds on milkweed seeds. Its migratory behavior makes it to one of the most interesting and thoroughly studied insects. In summer, the milkweed bug invades the northern temperate zone to take advantage of the large residential milkweed crop. In autumn, when the days get shorter, the bugs enter reproductive diapause and engage in a long period of migrating south, where they spend the winter *(20)*. High breeding temperatures (above 25°C) suppress the reproductive diapause and migratory behavior. The fact that, in captivity, they also feed on peeled sunflower seeds (which can be easily purchased in supermarkets) and other oily seeds and nuts instead of milkweed seeds (which are rather expensive and must be ordered from special seed shops) predestined the milkweed bug as an attractive laboratory animal for morphological, developmental, physiological, cytological, and genetic studies. Although

the milkweed bug is not a threatening pest, it is widely utilized to evaluate putative insecticidal and growth-regulating compounds because it was found that
the bug is highly responsive to such reagents. These features established the
milkweed bug *Oncopeltus fasciatus* as a broadly used model for a wide array
of applications.

Because the milkweed bug is frequently kept in laboratories (especially of agri-
cultural departments), starter colonies are easily obtained from these institutions.
Eggs as well as all larval stages, including adults, can be shipped without special
precautions. When raised at 28°C (the optimal temperature for the highest reproduction rate), the bugs do not enter reproductive diapause. Although in their natural
environment they show a distinct circadian behavior concerning feeding, copulation, and oviposition *(20)*, good breeding success is also achieved when kept in
constant darkness. A light/dark (L:D) regimen of L:D 12:12 or 16:8 h has the
advantage that most eggs are deposited around noon. Thus, a larger number of eggs
can be collected in a short period of time.

Timing of developmental stages is unproblematic because, when kept under
identical conditions, development is rather synchronous. Nevertheless, staging and
timing has to be tuned at each moult. The milkweed bug is a hemimetabolous insect
(i.e., has no pupal stage). It has five larval instars and moults directly to an adult
from the fifth larval stage. For timing, it is most convenient to start with eggs that
have been collected within 1 h after deposition. A timing accuracy of 1 d is sufficient for the described studies.

5.3.1.1 Breeding of the Milkweed Bug

1. Adjust incubator to a constant temperature of 28°C ± 1°C. Setting of a circadian
   rhythm (for example L:D 12:12 hrs) is advantageous for collecting large
   numbers of eggs (*see* **Subheading 5.3.1**) but not mandatory; bugs can be kept
   in constant darkness. Adjustment to a certain relative humidity is not required,
   but it should be kept below 70% (*see* **Note 1**).
2. Prepare glass or plastic containers of different sizes that can be tightly covered
   with fine gauze or a tight-fitting lid with a window covered with fine gauze (*see*
   Note 2).
3. Cover the floor of the containers with filter paper. In each container, put an open
   Petri dish with wet cotton-wool balls or wet filter paper (*see* **Note 3**), an open
   Petri dish with peeled sunflower seeds (*see* **Note 4**), and one or two tubes of
   cardboard (*see* **Note 5**).
4. To facilitate egg deposition, make "egg rolls" from soft paper tissue (about 1 cm
   in diameter and about 6 cm long) and put them in the container with adult bugs
   (*see* **Note 6**).
5. Place eggs, larvae, or adult bugs in the prepared containers. Put the containers in
   the incubator. Transfer the bugs about every week (depending on how densely a
   container is populated) into new clean containers.

5.3.1.2   Timing of Developmental Stages

1. To obtain eggs that are 0- to 1 h old, take a container with adult, mature bugs, clear it from all previously laid eggs, and place new egg rolls in it, preferably around noon. Collect these egg rolls 60 min later, harvest the eggs with a fine brush, put the eggs in a newly prepared container, and return the container into the incubator (*see* **Subheading 5.3.1.1**). Egg development takes about 120 h (5 d) at 28°C.

2. For timing of the first larval instar, examine eggs after 4 d of incubation and remove all prematurely hatched larvae. Reexamine the eggs the next day; about 70–80% will now have hatched. Remove the unhatched eggs and discard them. First instar larvae (L1) can now be collected at daily intervals. The designation is L1d0 for first instar larvae 0 d old (i.e., 0–24 h old), L1d1 for first instar larvae 1 d old (i.e., 24–48 h old), and so on. Under the given conditions, L1 lasts 3 d, L2 lasts 3 d, L3 lasts 4 d, L4 lasts 6 d, and L5 lasts 7 d. Copulation behavior of males starts 6 d after adult moult (i.e., Ad5). Spermatogenesis occurs throughout adult life until senescence (examined until Ad46).

3. For timing of advanced larval instars and adults, observe the development of the bugs in a container until about 20% of them have reached the desired larval stage. Now, remove those larvae and all bugs that are older. Put the container back into the incubator until the following day. Then, select the newly molted bugs and place them into a new container. These bugs are designated as d0 of the larval instar you selected (i.e., L2d0, L3d0, etc.) or adult d0 (Ad0), respectively.

5.3.2   *Preparation of Testes*

1. Determination of sex: At the adult stage, the male milkweed bug is usually smaller than the female, and a rounded genital capsule closes the posterior end, whereas the female has an ovipositor that is partly visible, although it is retracted. In addition, the female has a median process on the posterior edge of the fourth abdominal sternum, which is lacking in the male *(21)*. In fifth instar larvae, the sexes differ—as in adults—in the median process of the fourth abdominal sternum. In addition, the posterior edge of the fifth abdominal tergum has lateral lobes. These lobes are black in the female but not in the male larvae *(21)*. In fourth instar larvae, the sexes differ only slightly but markedly. The sternum of the seventh abdominal segment has a median black spot. This spot is small in males and considerably larger in females, in which it extends into the posterior part of the sixth segment. Larvae of the first, second, and third instars are difficult to sex. Males have to be identified by the histology of their gonads. This is uncomplicated because, already at the time of hatching, the apical testicular regions show their typical organization (*see* **Fig. 5.5**).

2. Narcotize specimens with carbon dioxide (CO_2) (*see* **Note 7**).
3. Separation of the abdomen: Transect the specimen behind the hind legs. Abdomens of first, second, and third instar larvae are immediately submerged in fixative (*see* **Note 8**). Abdomens of fourth and fifth instar larvae as well as abdomens of adults are placed in a small Petri dish with insect Ringer for dissection of the testes.
4. To remove the dorsal integument of the abdomen, cut at each side along the pleural fold and remove the dorsum beginning where the abdomen was cut off from the thorax. Now, you have a dorsal view of the abdominal body cavity.
5. Separation of the testes: The testes are located at each side of the midgut. They can be recognized by their pale orange pigmentation (all other tissues are milky white or cream color). The testes are rather sturdy because of their strong envelope and therefore can be easily dissected with the help of two fine tweezers or a fine tweezers and microscissors. Avoid touching the apical parts of the testicular follicles. The preparation of the testes should be carried out as quickly as possible because the Ringer solution may cause tissue alterations. It is advisable to practice testis preparation prior to fixation. Transfer the isolated testes immediately into the fixative for prefixation (*see* **Subheading 5.3.3**).

5.3.3 Processing of Specimens for Electron Microscopy and Ultrastructural Analysis (19)

During dissection of the testes and preparation for electron microscopy, the specimens should never be allowed to dry out and should be submerged in each fluid during processing. Although this is usually unproblematic with isolated testes, whole abdomens (concerning first, second, and third instar larvae) tend to float on the surface of hydrophilic fluids. This is caused by the lipoid surface layers of the cuticle. Should it be the case that abdomens do not submerge by themselves, cover the tissue with stiff gauze and press them gently below the surface (*see* **Note 9**).

1. Fixation is carried out in two steps: prefixation and postfixation. (a) The fixative for prefixation contains 2.5% glutaraldehyde in $0.08M$ cacodylate buffer, pH 7.4, with the addition of 4% sucrose. The osmolarity of the fixative should be about 270 mOsm, which corresponds to the osmolarity of the milkweed bug hemolymph. Prefixation needs 1 h at room temperature. After prefixation, wash the specimens in cacodylate buffer three times for 20 min and put them into postfixative. (b) The fixative for postfixation contains 2% osmium tetroxide in $0.08M$ cacodylate buffer, pH 7.4. Postfixation requires an additional 1 h at room temperature.
2. Wash specimens in distilled water three times for 10 min.
3. Orientation of specimens (*see* **Note 10**): For this purpose, liquefy 1.5% agar in a beaker in a hot water bath. Warm embedding molds (as they are used for embedding of specimens in araldite for electron microscopy) on a heating bench

(60°C–70°C) and fill them with the liquid agar. Transfer the testes from the distilled water into the agar and orient them with a heated needle in a manner that later allows you to carry out exact cross sections or (of another testis) exact longitudinal sections of the follicles. Take the forms with the oriented specimens from the heating bench and let the agar solidify. Take the small agar blocks out of the forms and put them in 30% ethanol.

4. Dehydrate specimens via increasing ethanol concentrations of 50%, 70%, 96%, and 100% (twice for 10 min for each concentration).

5. Embedding in araldite: Transfer the specimens from 100% ethanol to propylene oxide (twice for 10 min), then to a mixture (1:1) of propylene oxide and araldite (overnight), and finally to pure araldite. Change araldite after 24 h. After an additional 24 h, embed specimens in embedding molds. If necessary, trim agar blocks; they have to be well enclosed by araldite in the molds. Orient the agar blocks with the enclosed specimens in a way that allows precise cross and longitudinal sections of the testis follicles (*see* **step 3**). Put molds with embedded specimens in an incubator adjusted to 62°C for 48 h.

6. Cutting of semithin and ultrathin sections with an ultramicrotome: Semithin sections (1 μm) are placed on glass slides and stained with toluidine blue for light microscopy. These sections are important for orientation (*see* **Note 11**). Ultrathin sections (50–70 nm) (*see* **Note 12** for the production of serial sections) are placed on copper grids and double stained with uranyl acetate (2% uranyl acetate in 50% ethanol) for 2 min and lead citrate (1 g lead citrate in 42 mL distilled water and 8 mL $1N$ sodium hydroxide) for 10 min. Between the first and second staining, rinse three times in distilled water. After the second staining, rinse grids in $0.2M$ sodium hydroxide, rinse twice in distilled water, and let them dry.

7. Observe the sections by transmission electron microscope.

8. Serial sections: To elucidate the intricate structure of GSCs and their relationships to apical cells and CPCs, complete series of serial sections were made from two testis follicles of L1d0 (**Fig. 5.5** shows a section from such a series). At advanced developmental stages, serial sections were made from individual GSCs (**Figs. 5.7** and **5.8** show GSCs from such a series) (*see* **Note 13**).

9. Reconstruction of testis follicles (L1d0) and individual GSCs: Photos are taken of all sections of a series. Most convenient are digital micrographs, which can be overlaid using image-processing computer programs (*see* **Note 14**). The analysis of a series ensures, for example, which profile of a GSC projection belongs to which perikaryon and which profile represents a completely segregated projection terminal. Such reconstructions in combination with developmental studies allow the schematic representations shown in **Figs. 5.2** and **5.6**.

5.3.4 Immunocytochemistry

Immunological demonstration of acid phosphatase and TPPase has been performed at the electron microscopic level. This allows better assignment of the label to a certain cell organelle but is much more difficult than light microscopic immunological demonstrations. The procedure for the demonstrations of the two antigens follows largely the same protocol. Differences are specifically pointed out; otherwise, follow the in-common described course.

1. Preparation of testes is carried out as described in Subheading 5.3.2. The protocol outlined next has been optimized for young adult males (Ad5).

2. Fixation comprises three steps. Prefixation (first step of fixation) is carried out in 1% glutaraldehyde in 0.1M cacodylate buffer, pH 7.2, for 15 min at 4°C.

3. Washing in cacodylate buffer (five times for 10 min): Before incubation in the respective substrate of either one of the antigens, cut off the apical third of the testes and use only these parts (which include the apical complexes) for further procedures. The cutting edge allows better penetration of the substrate and cerium chloride (next **step**).

4. For the demonstration of acid phosphatase, incubate the testes (i.e., apical parts) in the following substrate: 0.1M sodium-β-glycerophosphate with 2 mM cerium chloride and 5% sucrose in 0.1M maleate buffer, pH 5.0. For improved penetration of the substrate and cerium chloride, add 0.002% Triton X-100 to the incubation medium. Incubate specimens for 1–2 h at 37°C (22).

5. Controls: Use the same incubation medium but without substrate (i.e., without sodium-β-glycerophosphate) and incubate as above.

6. Specimens incubated in substrate and controls are then washed three times for 10 min in cacodylate buffer and subjected to the second step of fixation. (Continue with step 10.)

7. For demonstration of TPPase, prepare specimens as described in steps 1–3. Incubate the apical parts of the testes in the following substrate: Thiamine pyrophosphate (TPP) with 5 mM manganese chloride, 4 mM cerium chloride, and 5% sucrose in 0.1M maleate buffer, pH 7.2. Add 0.002% Triton X-100 to the incubation medium. Incubate specimens for 1–2 h at 37°C (23).

8. Controls: Use the same incubation medium but without substrate (i.e., TPP) and incubate as in **step 4**.

9. Specimens incubated in substrate and the controls are then washed three times for 10 min in cacodylate buffer and subjected to the second step of fixation.

10. Second step of fixation: This step of fixation, as well as the following procedure, is the same for the demonstration of both acid phosphatase and TPPase, respectively. The fixative contains 4% paraformaldehyde and 2% glutaraldehyde in 0.1M cacodylate buffer, pH 7.2. Incubate for 3 h at 4°C.

11. Wash three times for 10 min in cacodylate buffer and subject specimens to the third step of fixation (postfixation).

12. Postfixation is carried out in a solution of 1% osmium tetroxide, 0.8% potassium ferricyanide, and 5 mM calcium chloride in 0.1M cacodylate buffer for 1 h at room temperature.

13. After washing in distilled water, specimens are dehydrated and embedded in araldite as described above (*see* **Subheading 5.3.3, steps 4** and **5**).

14. Semithin sections (1 μm) are placed on glass slides and stained with toluidine blue for light microscopy (*see* **Subheading 5.2.3.1, step 12**). These sections are important for orientation. Ultrathin sections (50–70 nm) are *not* stained.

The potassium ferricyanide added to the postfixative provides sufficient contrast of the tissue to enable the recognition of the different cell types.

15. Analysis of the immunolabeling of acid phosphatase and TPPase: Lysosomes and autophagosomes will be labeled in treated tissue but not in controls. At low and medium magnification (5,000- to 20,000-fold), the labeled cell organelles will appear dark, whereas unlabeled organelles and membranes appear faint. This is demonstrated in **Fig. 5.10** for acid phosphatase and in **Fig. 5.11** for TPPase. At higher resolution (>30,000-fold), it shows that the dark label is caused by the precipitation of fine black (i.e., electron-dense) amorphous granules (*see* **Fig. 5.12**). The granules consist of water-insoluble cerium phosphate. Because of their fineness, their location can be clearly assigned to distinct cellular structures (*see* **Fig. 5.12**). It is of utmost importance to compare sections from tissue that was incubated with substrate with sections from tissue that was not incubated with substrate but otherwise treated equally (i.e., controls). Such a comparison delineates the degree of nonspecific cerium precipitation. This "background" labeling, which can rarely be completely avoided (*see* **Fig. 5.11**) has to be subtracted from the labeling in experimental specimens.

5.4 Notes

1. It is not required to adjust the relative humidity in the incubator to a certain level, but it should not exceed 70%. If the humidity is too high, sunflower seeds will mold, which may lead to epidemic diseases of the milkweed bug.
2. Milkweed bugs do not mind crowding, provided they are kept clean. For example, a container of about 30 × 20 × 20 cm (length × width × height) may hold 200–300 adult bugs.
3. It is important that the bugs always have access to water (i.e., wet cotton-wool balls or wet filter paper).
4. Milkweed bugs suck the oily components of the sunflower seeds. Seeds have to be replaced when they start to shrivel or when they are contaminated with excrement. Usually, this is necessary once to twice a week.
5. The cardboard tubes allow the bugs to hide and hang on during the moulting process.
6. In their habitats, females try to deposit egg clutches where they are protected. Egg rolls are accepted as oviposition sites. This also makes the collection of eggs easy.
7. Experimental specimens are placed into an Erlenmeyer flask. They are exposed to carbon dioxide (adjust the manometer of the CO_2 bottle to about 2 bar) until they no longer move (after about 5 min).
8. For better penetration of the fixative, cut off the posterior tip of the abdomen (*see also* **Note 9**).
9. Abdomens tend to float until they reach higher concentrations of ethanol. When pressing them gently below the surface of fluids, use chemically inert material (special gauze), particularly during postfixation with osmium tetroxide. Floating may also be prevented by fixation in vacuum or treatment with solutions of common detergents prior to fixation.
10. Testis follicles are polar organs with extremely polarized GSCs (in the case of the milkweed bug) (*see* **Figs. 5.1–5.5**). Analysis of such tissue under the electron microscope is extremely difficult if the exact orientation of the follicles is not clear.
11. It would be very laborious to approach the location of the apical complex by ultrathin sections, even after careful orientation of the specimen in the block. It is therefore time saving to start a block by cutting semithin sections. Stain each fifth section with toluidine blue and examine it under the light microscope. After identification of the apical complex, continue with ultrathin sections. The most difficult step is the recognition of the apical complex. It is therefore advisable

to make a complete series of semithin sections through an apical complex and to study it carefully by light microscopy before performing studies with the electron microscope.

12. Ribbons of about five ultrathin sections are placed on a copper grid. Blocks should be trimmed in a way that it can be determined which is the anterior and which is the posterior section. Grids must be placed in consecutive order. A complete series through a testis follicle of a first instar larva (L1d0) comprises about 200 sections at 70 nm (and such follicles contained only four GSCs).

13. To obtain a series of sections comprising a complete GSC of an advanced developmental stage, a very large number of sections are still needed. It is best to make semithin sections up to the middle of the apical complex and to then start with serial ultrathin sections.

14. Depending on the complexity of the structures, there are several methods for three-dimensional reconstruction. Most are painstaking and time consuming. Those especially interested in this field should refer to the Hayat's excellent book *(19)* and the references given therein.

References

1. Lin, H. (2002) The stem-cell niche theory: lessons from flies. Nat. Rev. Genet. 3, 931–940.
2. Spradling, A., Drummond-Barbosa, D., and Kai, T. (2001) Stem cells find their niche. Nature. 414, 98–104.
3. Hardy, R.W., Tokuyasu, K.T., Lindsley, D.L., and Garavito, M. (1979) The germinal proliferation center in the testis of Drosophila melanogaster. J. Ultrastruct. Res. 69, 180–190.
4. Schmidt, E.D., Papanikolaou, A., and Dorn, A. (2001) The relationship between germline cells and apical complex in the testes of the milkweed bug throughout postembryonic development. Invert. Reprod. Dev. 39, 109–126.
5. Schmidt, E.D., Sehn, E., and Dorn, A. (2002) Differentiation and ultrastructure of the spermatogonial cyst cells in the milkweed bug, Oncopeltus fasciatus. Invert. Reprod. Dev. 42, 163–178.
6. Schmidt, E.D., and Dorn, A. (2004) Structural polarity and dynamics of male germline stem cells in the milkweed bug (Oncopeltus fasciatus). Cell Tissue Res. 318, 383–394.
7. Szöllösi, A. (1982) Relationships between germ and somatic cells in the testes of locusts and moths, in *Insect Ultrastructure* (King, R.C., and Akai, H., eds.), Plenum, New York, pp. 32–60.
8. Grünberg, K. (1903) Untersuchungen über die Keim- und Nährzellen in den Hoden und Ovarien der Lepidopteren. Z. Wiss. Zool. **74,** 327–395.
9. Gönczy, P., Viswanathan, S., and DiNardo, S. (1992) Probing spermatogenesis in *Drosophila* with P-element enhancer detectors. *Development.* **114,** 89–98.
10. Heming, B.S., and Huebner, E. (1994) Development of the germ cells and reproductive primordial in male and female embryos of *Rhodnius prolixus* Stål (Hemiptera: Reduviidae). Can. J. Zool. **72,** 1100–1119.
11. Gönczy, P., and DiNardo, S. (1996) The germ line regulates somatic cyst cell proliferation and fate during *Drosophila* spermatogenesis. *Development.* **122,** 2437–2447.
12. Yamashita, Y.M., Jones, D.L., and Fuller, M.T. (2003) Orientation of asymmetric stem cell division by the APC tumor suppressor and centrosome. *Science.* **301,** 1547–1550.
13. Locke, M., and Sykes, A.K. (1975) The role of the Golgi complex in the isolation and digestion of organelles. *Tissue Cell.* **7,** 143–158.
14. Dunn, W.A. (1990) Studies on the mechanisms of autophagy: formation of the autophagic vacuole. *J. Cell Biol.* **110,** 1923–1933.
15. Cheng, H.-W., and Chiang, A.-S. (1995) Autophagy and acid phosphatase activity in the corpora allata of adult mated females of *Diploptera punctata. Cell Tissue Res.* **281,** 109–117.
16. Yang, D.-M., and Chiang, A.-S. (1997) Formation of whorl-like autophagosome by Golgi apparatus engulfing a ribosome-containing vacuole in corpora allata of the cockroach *Diploptera punctata. Cell Tissue Res.* **287,** 385–391.
17. Klein, C. (2000) Die postembryonale Gonadenentwicklung und Gametogenese bei *Lymantria dispar* L. (Lepidoptera) mit besonderer Berücksichtigung des Gonadensomas. Licht- und

elektronenmikroskopische Untersuchungen. Thesis, Universität des Saarlandes, Saarbrücken, Germany.

18. Seto, E.S., Bellen, H.J., and Lloyd, T.E. (2002) When cell biology meets development: endocytic regulation of signaling pathways. *Genes Dev.* **16,** 1314–1336.

19. Hayat, M.A. (2000) *Principles and techniques of electron microscopy. Biological applications,* 4th ed., Cambridge University Press, Cambridge, UK.

20. Dingle, H. (1996) *Migration. The biology of life on the move,* Oxford University Press, New York, Oxford.

21. Bonhag, P.F., and Wick, J.R. (1953) The functional anatomy of the male and female reproductive systems of the milkweed bug, *Oncopeltus fasciatus* (Dallas) (Heteroptera: Lygaeidae). *J. Morphol.* **92,** 177–283.

22. Robinson, J.M., and Karnovsky, M.J. (1983) Ultrastructural localization of several phosphatases with cerium. *J. Histochem. Cytochem.* **31,** 1197–1208.

23. Angermüller, S., and Fahimi, A. (1984) A new cerium-based method for cytochemical localisation of thiamine pyrophosphatase in the Golgi apparatus of rat hepatocytes. *Histochemistry.* **80,** 107–111.

Chapter 6
High-Resolution Light Microscopic Characterization of Spermatogonia

Hélio Chiarini-Garcia and Marvin L. Meistrich

Contents

Summary It is possible to distinguish the morphological features of the spermatogonial nuclei and nucleoli and to further identify their distinct generations using an appropriate method to fix whole testes via vascular perfusion with glutaraldehyde, postfixation by immersion in reduced osmium, embedding in araldite, and staining of semithin tissue sections. A well-trained individual can distinguish each of the spermatogonial types in rodents ($A_{undiferentiated}$, A_1, A_2, A_3, A_4, In, and B) using this tissue preparation technique based on their morphological details and without correlation with the stages of the epithelium cycle or other parameters. The possibility of distinguishing each spermatogonial type by their morphological characteristics allows a more accurate evaluation of their kinetics during the spermatogenic cycle. Moreover, the understanding of spermatogonial behavior is a means to elucidate the functional control of the spermatogenesis, which consequently allows the determination of their effects on the fertility of humans and other animals.

Keywords High-resolution light microscopy; morphology; perfusion; spermatogonia; stem cell.

6.1 Introduction

The morphology of spermatogonia has been extensively studied by light microscopy since the 1960s; such researchers as Clermont *(1)*, Huckins *(2)*, Oakberg *(3)*, de Rooij *(4)*, Meistrich *(5),* and their coworkers have described many properties of these cells during spermatogenesis in diverse species. They have mainly applied two methods for spermatogonial studies: cross sections of testes tubules (after paraffin embedding) or in toto preparations of whole mounts of tubules. In the method of cross-sectioning tubules, fixation by Zenker-formal or Bouin's fixative and standard paraffin embedding were most commonly used. Sections 5 μm thick were stained with hematoxylin and eosin or PAS (periodic acid Schiff)–hematoxylin *(6)*. In the whole-mount method, isolated and fixed seminiferous tubules were partially flattened by a coverslip on a slide, and the cells on the base of the seminiferous tubules, including spermatogonia, were identified by adjusting the plane of focus of the microscope lens onto the basal lamina.

Various important characteristics of spermatogonial biology were described in different species under normal and various experimental conditions by using paraffin embedding and whole-mount methods. Specifically in laboratory rodents (mouse and rats), there are numerous generations of spermatogonia. The first cell in the spermatogenic process, the stem cell or the stem spermatogonia, is an A single (A_s) spermatogonium *(2,3)*. The A_s spermatogonia are classified as single cells because they have no intercellular bridge connections with other spermatogonia. After the A_s differentiation division, all subsequent generations of spermatogonia remain connected by intercellular bridges *(7)*. The result of the first stem cell differentiating division is a pair of spermatogonia or A_{pr} cells. Subsequent divisions lead to chains of 4, 8, and 16 cells, designated as A-aligned (A_{al}) spermatogonia. Quantitative studies of the development of undifferentiated spermatogonia used mainly the whole-mount method by which different spermatogonial types were identified by the number of cells in the chain. The subsequent cells following these primitive spermatogonial types, in mice and rats, are the differentiated type A spermatogonia (A_1, A_2, A_3, and A_4) and type In and B spermatogonia *(8)*. The cytological features of these different stages of spermatogonia are not very well detailed in testes prepared by standard paraffin embedding and the whole-mount method and do not allow confident recognition of all spermatogonial generations. In the past, the spermatogonial generations could only be classified based on their morphological characteristics as types A, In, and B *(8)*, and the quantitative evaluation of the different types of A spermatogonia in paraffin-embedded sections could only be performed based on the stage of the cycle of the seminiferous epithelium.

An alternative method of preparation of the testicular tissue, involving fixation of the testes by whole-body perfusion with glutaraldehyde, postfixation with reduced osmium, embedding in araldite resin, sectioning at 1-μm thickness, and toluidine blue–borate staining *(9,10)*, has been used for decades as a means for a more refined morphologic *(11,12)* and morphometric *(13,14)* evaluation of the testes. However, it was not employed to distinguish the different spermatogonial generations until 2001 *(15)*. Using this method, which provides a more accurate evaluation of the spermatogonial cytology, we could distinguish the spermatogonial subtypes in mice. The spermatogonial subtypes were distinguished mainly by their nuclear and nucleolar features. They were recognized as type A undifferentiated spermatogonia (A_s, A_{pr}, A_{al}; it was not possible to distinguish among them) and types A_1, A_2, A_3, A_4, In, and B spermatogonia. Later, we could successfully apply the same method to study spermatogonial generations in other species under normal and experimental conditions, like rat *(16,17)*, golden hamster *(18)*, donkey, marmoset, fish (data not published), and mutant *(19)* and knockout *(20)* mice.

The spermatogonial phase of the spermatogenic process, because of its exponential multiplication by mitoses, is very important for determining the number of spermatozoa produced by each particular species. Hence, the possibility of distinguishing the spermatogonial generations by their morphological features allows a more accurate evaluation of each spermatogonial type by stereological methods and permits the elucidation of their kinetics and topographical position in the seminiferous epithelium under physiological, experimental, and pathological conditions. Moreover, it allows a better understanding of the spermatogonial behavior and the mechanisms of altering spermatogonial production that could be related to the modulation of testis function.

6.2   Materials

1. Heparin (Liquemine, Roche).
2. Sodium thiopental (Thiopentax, Cristalia) intravenous bottle.
3. Three-way stopcock.
4. Catheter (Angiocath, BD).
5. Sodium chloride (0.9%).
6. Glutaraldehyde (biological grade).
7. Cacodylate buffer.
8. Perfusion and immersion fixation: 5% glutaraldehyde in $0.05M$ sodium cacodylate, pH 7.2–7.4.
9. Osmium tetroxide.
10. Potassium ferrocyanide.
11. Osmium-reduced postfixation: 1% osmium tetroxide and 1.5% potassium ferrocyanide in distilled water.
12. Propylene oxide (PO) or acetone.
13. Araldite 502 kit (EMS).
14. BDMA (benzyldimethylamine) resin (EMS).
15. Araldite resin (100 g): 54 g Araldite 502; 46 g dodenyl succinic anhydride (DDSA; Hardener); 2.5–3.0 g BDMA.
16. BEEM capsules.
17. Glass knife.
18. Toluidine blue O (Allied Chemical).
19. Sodium borate.

6.3 Methods

6.3.1 Fixation

The testes must be fixed by vascular perfusion for good morphological preservation *(6,9)*. This fixation method, when well performed, preserves the position of the testicular components, including the intertubular constituents (connective tissue cells, lymphatic spaces, and blood vessels) and the structure of seminiferous tubules (seminiferous epithelium and tunica propria). We describe a perfusion method for laboratory rodents or small animals in which whole-body perfusion is performed by a vascular route through the left ventricle. For large animals, the perfusion can be performed using the testicular artery. However, in some situations, the fixation can be performed by immersion (*see* **Note 1**).

1. To avoid intravascular clots during perfusion, 15 min before the perfusion fixation, heparin (125 IU/kg body weight) should be injected intraperitoneally (*see* **Note 2**).
2. The animals are then anesthetized (30 mg/kg body weight thiopental ip) and gently stretched on a restraining board (*see* **Note 3**).
3. Just after the abdominal cavity is opened up to the xiphoid process, the heart is exposed by cutting the thoracic cavity up to the axilla on both the right and left sides of the body.
4. The thoracic wall is grasped and fixed on the head side with hemostats to expose the heart *(10)*. At this moment, the animal is ready to be perfused (*see* **Fig. 6.1**).
5. Perfusion can be done using either a pump or a perfusion apparatus in which two intravenous bottles are connected to a single three-way stopcock. These bottles are positioned approx. 4 ft (about 120 cm) above the level of the heart (*see* **Fig. 6.1**).
6. One of the bottles contains saline and the other fixative solution (*see* **Note 4**). On the tip of the single tube that extends from the three-way stopcock, the needle that will be introduced into the left ventricle is attached (*see* **Notes 5** and **6**).
7. The needle is gently introduced into the left ventricle so that the solutions, when introduced, will pass through the aorta to the entire body. The right atrium is then sectioned with a forceps to allow release of the blood and perfusion solutions after passing through the body.
8. The valve of saline bottle is opened, and the saline starts slowly and progressively to push the blood through the body and out the right atrium. To obtain the proper flow rate and pressure of perfusion, the drops, which are seen in the small chamber in the tube just below the bottles, should be falling at approx. 3–4 drops/s (*see* **Fig. 6.1**).
9. The efflux of perfusion from the opened atrium will become clearer about 5 to 10 min after the beginning of the saline perfusion, which indicates that the blood has been washed from the animal (*see* **Note 7**).
10. At this moment, the saline valve is closed, and the valve of the fixative bottle is opened, allowing drops to fall with the same speed (3–4 drops/s). The perfusion time varies with the species; for example, adult mice are perfused for at least 20 min and adult rats for 30 min (*see* **Note 8**).

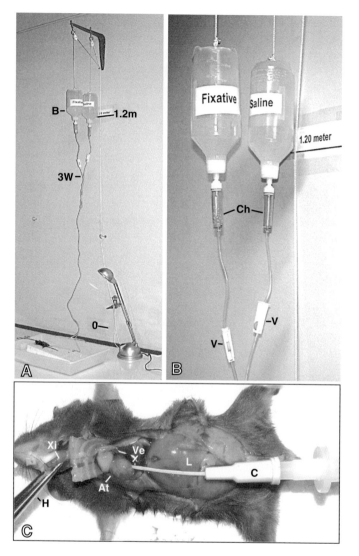

Fig. 6.1 **A** A perfusion apparatus consisting of two intravenous bottles (B) for fixative and saline connected to a single three-way stopcock (3W). **B** Chambers (Ch) where the solutions drop during perfusion and valves (V) that control the flow rate of the solutions. **C** Picture of a mouse that has been perfused through the heart. At, right atrium; C, catheter; H, hemostats; L, liver; Ve, left ventricle; Xi, xiphoid process

11. After the perfusion, the testes are removed, diced into slabs about 1 mm thick (*see* **Note 9**) and immersed in the same fixative for 12–24 h at 4°C. Afterward, the testicular fragments are washed in 0.05*M* cacodylate buffer, pH 7.2–7.4, twice for 20 min each. This completes the primary fixation of the testicular tissues; at this stage, the fragments can be kept for months in the cacodylate buffer at 4°C until embedding.

12. If the tissues will be studied by transmission electron microscopy (TEM) in addition to high-resolution light microscopy High-Resolution Light Microscopy (HRLM), the fragments should be subjected to a secondary fixation with reduced osmium (1% osmium tetroxide and 1.5% potassium ferrocyanide in distilled water) for 60–90 min just prior to embedding (*see* **Note 10**).

6.3.2 Embedding

The testes fragments should be progressively dehydrated, infiltrated with resin plus propylene oxide, and finally embedded in pure resin as described in this subheading (*see* **Note 11**) just after the secondary fixation with reduced osmium.

50% ethanol	10 min
70% ethanol	15 min
85% ethanol	15 min
95% ethanol	15 min (twice)
Absolute ethanol	20 min (twice)
Propylene oxide	20 min (twice)
1 PO : 1 araldite (closed vial)	60 min or more
1 PO : 3 araldite (closed vial)	Overnight or more
1 PO : 4 araldite (closed vial) (optional)	60 min or several hours
Pure araldite (closed vial)	60 min to several hours
Harden the araldite	
Oven at 45°C	Overnight or 12 h
Oven at 60°C	Overnight or more

The formula for the preparation of 100 g of araldite resin is shown in the Subheading 6.2 (*see* **Note 12**).

6.3.3 Microtomy

To identify the details of the spermatogonial generations, it is necessary to cut the testes fragments in semithin sections to avoid subcellular structure overlap. Use the following steps:

1. Specimen blocks have to be trimmed to expose the specimen and to form a trapezoidal cutting face. Usually, this is done with a razor blade or a handheld rotary grinding device (*9*) under a stereoscopic microscope.
2. Using a microtome and a glass knife with a small water trough attached, cut sections approx. 1 μm thick. Float the section on the water in the attached bath.

3. Pick up the section with a small loop (like the ones used for bacterial culture) and transfer the sections onto a droplet of distilled water on a glass slide.
4. Transfer the slide to a hot plate (70°C–90°C) until the water droplet evaporates.

6.3.4   Staining

To stain the semithin sections with good contrast, follow these steps:

1. Put several drops of 1% toluidine blue O dissolved in a 1% aqueous solution of sodium borate over the sections on a slide (*see* **Note 13**).
2. Heat on a hot plate (about 60°C) until the edges of the stain begin to dry and turn an iridescent green color.
3. Gently rinse the slide with distilled water from a wash bottle. If you wash too vigorously, the section may be released and lost.
4. Remove the excess water by pressing the slide, with the tissue facing down, gently over a piece of filter paper.
5. Allow the slide to dry completely at room temperature and mount with a cover-slip using any conventional medium.

6.4   Main Spermatogonial Features

The procedure described allows the identification of different generations of spermatogonia based mainly on the nuclear features in mice *(15)* and rats *(16)*. Depending on the spermatogonial generation, the chromatin appears heterogeneously dispersed throughout the nucleus, giving a mottled appearance and contrasting with small clear and dispersed spots of euchromatin. These aspects are more evident in undifferentiated type spermatogonia (A_s, A_{pr}, and A_{al}). On the other hand, in the most advanced spermatogonial generations—differentiated type A (A_1, A_2, A_3, A_4), types In and B—the mottling is finer than in the undifferentiated spermatogonia, giving a fine granular appearance (*see* **Fig. 6.2**).

Another important feature to distinguish the spermatogonial generations is the presence of the heterochromatin, which clearly starts to appear at the A_2 spermatogonial stage in rodents as flecks along the nuclear envelope. The amount of heterochromatin increases progressively from A_2 (in a rim covering about 10% of the nuclear envelope) to type In spermatogonia, in which it can reach 70% or even 100% of the internal nuclear envelope border (*see* **Fig. 6.2**).

The nucleoli clearly differ among spermatogonial types and are useful for distinguishing them. They are small, irregular, compact, and darkly stained in the more immature spermatogonia. However, in the A_1 spermatogonia they are well delineated and reticulated, and from A_2 to A_4 they become progressively less compact and irregular, with two to three nucleoli visualized per section and branches extending

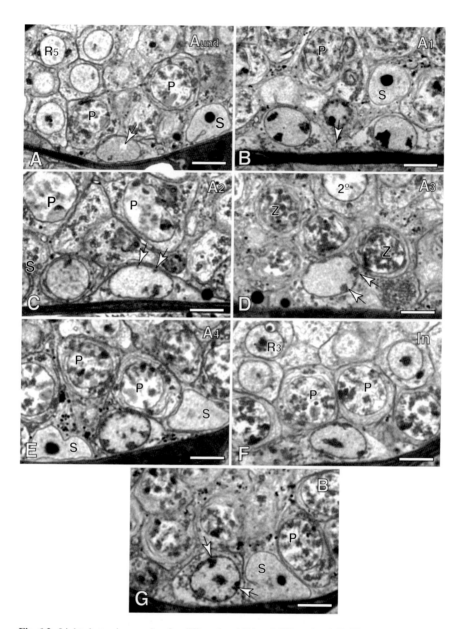

Fig. 6.2 Light photomicrographs of undifferentiated (**A**) and differentiated (**B–E**) type A, type In (**F**), and type B (**G**) spermatogonia of mouse (C57BL/6J strain) testes. Their main morphological features are shown that distinguish them from other generations by the HRLM method. **A** An A_{und} spermatogonia showing a mottled nucleus with an evident vacuole (*arrow*). **B** The arrow shows an intercellular bridge connecting two type A_1 spermatogonia. **C** An A_2 spermatogonium presents dispersed nucleoli and spots of heterochromatin adjacent to the nuclear envelope (*arrows*). **D** An A_3 spermatogonia with two nucleoli (*arrows*). **E** An A_4 spermatogonium with heterochromatin and dispersed nucleoli throughout most of the nucleus. **F** An In spermatogonium with a thick heterochromatin line surrounding the internal face of the nuclear envelope. **G** A B spermatogonium with large balls of heterochromatin close to some areas of the nuclear envelope (*arrows*). 2°, secondary spermatocyte. P, pachytene; R3, step 3 spermatid; R5, step 5 spermatid; S, Sertoli cell; Z, zygotene. Bars represent 12 μm

deeply into the nucleus. In the type In and B spermatogonia, the nucleoli become compact and darkly stained again.

The presence of vacuoles in the nucleus is only observed in the undifferentiated type A spermatogonia (A_s, A_{pr}, and A_{al}) of mice. Nuclear vacuoles are not found in differentiated type A spermatogonia in normal mice of various strains (B6, C3H, B6-C3H hybrids) and in any spermatogonia in rats (Sprague–Dawley and Wistar). However, in jsd (juvenile spermatogonial depletion) mutant mice, vacuoles are observed in nuclei of spermatogonia with morphological features of differentiated (A_1 to A_4) type A spermatogonia (19).

The cytoplasm rarely has clear features that can be used to distinguish the spermatogonial types. The only exception is that a very small number of cells (~1%) observed in rodents present a cytoplasm that has a flatter shape and is darker than other spermatogonia. Chiarini-Garcia and Russell (15) called these cells type A undifferentiated subtype 1 spermatogonia. These authors believed that they could be the A_s spermatogonia, the stem cell of the spermatogenic process. In the other spermatogonia under light microscopy, the cytoplasm appeared clear and homogeneous, showing small dark spots mainly close to the nucleus, corresponding to mitochondria. No other special details were observed in the spermatogonial cytoplasm under HRLM, although at the ultrastructural level some differences were noticed (21).

However, to distinguish spermatogonial types by HRLM, an individual has to be properly trained as certain spermatogonial cell types are very similar to others. Some uncertainty concerning the specific stage of spermatogonial identification can occur, but normally the errors are not greater than one stage away in development. Using another resin-embedding method (glycol methacrylate), nuclear details could not be well observed enough to distinguish the different spermatogonial generations.

In this way, studies concerning the numbers, kinetics, and niches of spermatogonia in normal, experimental, or pathological conditions can now be finely evaluated because of the possibility of distinguishing each spermatogonial generation when this procedure is properly applied.

6.5   Notes

1. Usually, fixation of testes by immersion leads to major histological artifacts. The dense connective tissue that surrounds the testes (tunica albuginea) does not allow adequate fixative penetration, resulting in irregular and poor preservation of the parenchyma; if fresh testes are cut, their parenchymas are so fragile that there is tubule dispersion. However, if the focus is on the study of the pool of spermatogonia, it is possible to study them if the seminiferous epithelium is well preserved. In this case, the testes can be cut in several large pieces and fixed by immersion, at 4°C, for 18–24 h in 5% glutaraldehyde in 0.05M cacodylate buffer at pH 7.2–7.4. After that, slabs of 1 mm thick are cut from the surface of the large slabs. As glutaraldehyde penetrates well up to 1 mm deep in the testes (9), it is expected that these fragments should be useful for proper study of the spermatogonial morphology. The remaining testes tissue from the interior of the large slabs should be discarded.
2. The use of heparin is important as the percentage of success of testes perfusion increased markedly when heparin was used before perfusion (6,10). If for some reason it is not possible to

inject heparin before perfusion (e.g., when blood collection without heparin is necessary), it is possible to dilute heparin into the saline solution (10 IU/mL) that will be perfused, and this will effectively prevent clots in the blood vessels.

3. When constraining the animal on the board, avoid stretching out its legs too much caudally. This position is not ergonomic for a quadruped and can compress organs and blood vessels, interfering with the quality of the perfusion.

4. The fixative solutions should be freshly prepared for best results. If it is necessary to prepare them beforehand, the diluted solution can be kept in the refrigerator (4°C) for several days or at −20°C for several weeks. Another important detail for the success of this method is the selection of the buffer. In studies with rats, it was shown that cacodylate buffer yielded excellent results, allowing recognition of fine details of the Sertoli cell nucleus. However, when phosphate buffer was used, the nucleoplasm of Sertoli cells became coarse, and they lost their smoothness, resulting in a nuclear morphology more similar to the nucleus of some spermatogonial generations (*see* **Ref. *17*** and **Fig. 6.3**). However, we have observed that phosphate buffer was adequate for spermatogonial recognition in marmoset and fish. Thus, it is advised to check the best buffer for each species to be studied.

5. Before the introduction of the needles into the left ventricle and the start of the perfusion, the tubes and the chambers of the three-way stopcock should be filled with the perfusion solutions. To accomplish this, first open the valve of the glutaraldehyde bottle and allow it to flow into a vial until all bubbles are flushed from the tube. After that, close this valve, open the saline valve, and turn the three-way stopcock to allow saline flow. A large amount of saline should be allowed to flow through the tube to remove all glutaraldehyde from the three-way stopcock and the tube leading to the needle or catheter (*see* **Fig. 6.1**). This procedure is important because the entry of even a small amount of fixative inside the circulatory system before the saline could be enough to block successful perfusion. The presence of fixative in the tube and the catheter is a common problem when multiple perfusions are done in succession. To avoid that, after one perfusion and before starting the next one, the tube that extends downward from the three-way stopcock and the needles has to be exhaustively washed with saline.

6. When the needle is introduced in the left ventricle, it is expected that the solutions flow through the aorta and gain access to the whole body, supplied by the general circulation, and reach the testes. However, if the needle is inserted too far, it can perforate the interventricular septum, and two things can happen: First, the solution will reach the right ventricle and circulate through the

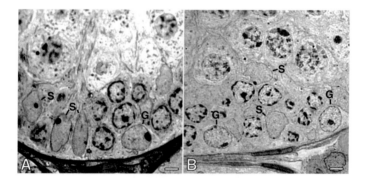

Fig. 6.3 Light photomicrographs of a section of rat (Wistar) testes showing the seminiferous epithelium perfusion fixed with glutaraldehyde in cacodylate (**A**) and in phosphate (**B**) buffer. Note that in phosphate buffer (**B**) the Sertoli cells' nuclei (S) are very granular and are not so clearly distinguished from the spermatogonial nuclei (G). P, pachytene; Pl, preleptotene. Bars represent 12 μm

pulmonary circulation or reflux though the vena cava; second, the remaining perfusion pressure in the left ventricle may not be enough to perfuse organs far from the heart, like the testes. Another important detail is the size of the needle. For the correct flow rate and pressure of perfusion in mice, a 20- or 22-gage needle should be used; for rats, an 18-gage needle is suggested. For the most reliable results, catheters that have needles with a plastic sheath that does not have a sharp tip should be used to avoid accidentally producing an additional hole in the heart that could compromise the perfusion. The needle of the catheter is inserted into the left ventricle just deep enough to make a hole in the muscular layer. After it is inserted, remove the metal part and introduce the plastic part deeper inside the ventricular chamber.

7. Washing the blood completely from the vascular system with saline will avoid the formation of clots, which may occur mainly in the organs far from the heart and in small veins and capillaries. A simple way to detect if the perfusion is running well is to check if the liver becomes progressively yellowish.

8. After 10–15 min of perfusion with the fixative, the body of the animal becomes very hard, showing that the perfusion process is running well. However, do not stop the perfusion because the testes are not normally well fixed at this time. The fixation time needs to be followed as defined, at least 20 min in mice and 30 min in rats. During the perfusion, it is not advised to open the scrotum or even start to move or change the positions of other viscera or put any pressure on the animal's body. Our experience has shown that these actions can compress or twist some of the blood vessels, reducing the flow of the solutions to the testes.

9. To obtain cross-sectioned seminiferous tubules, the testes have to be cut in a correct position. The cutting position depends on the rete testis location, which is the region where each seminiferous tubule starts and finishes. In mice and rats, for example, the testes are cut perpendicular to their long axis, considering the position of the rete testis and seminiferous tubules (*see* **Fig. 6.4**). As the rete testis can be located in different positions in the testes of different species, the best direction to cut the testes to obtain cross sections of seminiferous tubules has to be verified for each species. Avoid cutting tissues into cubes as it results in the loss of information on the proper orientation for embedding. The fragments should be obtained as slabs and embedded so that the sections will be cut parallel to the face with the largest area.

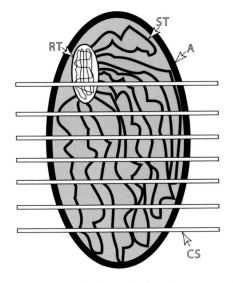

Fig. 6.4 Drawing showing the best position for cutting the rodent testes to obtain largest number of seminiferous tubules in cross sections (CS). A, albuginea; RT, rete testis; ST, seminiferous tubules

10. Osmium tetroxide is volatile, and its vapor is dangerous to mucosal membranes. Thus, it should be used under a proper hood. The mixture of osmium and ferrocyanide (reduced osmium), besides fixing the tissues, also increases the membrane contrast under TEM. However, for HRLM reduced osmium is not essential. Hence, by using reduced osmium as a secondary fixation, spermatogonial generation can be studied by HRLM and TEM. For best results, osmium and ferrocyanide solutions should be mixed just before use, and the testes fragments should be fixed for 60–90 min in the dark *(9)*.

11. Acetone could be used for infiltration instead of propylene oxide, apparently without adverse effects on spermatogonial morphology under light microscopy. During the infiltration procedure, resin penetration into the tissues is very difficult because of its viscosity. The penetration may be improved by placing the vial on a very slow rotary mixing plate *(9)*.

12. Araldite resin is used for HRLM, in preference to other epoxy resins, because of its capability to be cut into relatively large sections ($1–3 mm^2$), and because of its uniform hardening with little shrinkage, it does not crack in semithin sections cut at 1-μm thickness. We have used the Araldite 502 embedding kit (cat. no. 13900) from EMS (Hatfield, PA) based in Luft's formula. However, instead of using DMP-30, which is sold with the kit, we have used BDMA as the accelerator. Although the testes fragments are thin (about 1-mm thickness), they are large, so it is necessary to use BEEM capsules for embedding *(9)*.

13. The sodium borate is used to make the pH of the solution alkaline, which is necessary to optimize the staining of sections in epoxy resins with toluidine blue. After staining, do not dry the sections completely on the hot plate; this procedure stains the testis sections a homogeneous bluish color, without contrast. Instead, dry the sections with filter paper, preventing excess stain drying over the tissues. Note that the staining solution needs to be filtered before use to remove precipitates that may form.

Acknowledgments   The standardizing of this method is in part because of the efforts of Lonnie D. Russell (in memoriam), who applied it with a unique ability to show us under HRLM for the first time various important parameters of male reproductive biology. We thank Fernanda RCL Almeida and Deborah Amaral for their suggestions on the manuscript. This method was standardized during the development of different projects that were partially supported by Brazilian (CAPES, FAPEMIG, CNPq) and U.S. (National Institutes of Health, Latin American Fellowship) financial foundations.

References

1. Clermont, Y., and Bustos-Obregon, E. (1968) Re-examination of spermatogonial renewal in the rat by means of seminiferous tubules mounted in toto. *Am. J. Anat.* **122,** 237–248.
2. Huckins, C. (1971) The spermatogonial stem cell population in adult rats. I. Their morphology, proliferation and maturation. *Anat. Rec.* **169,** 533–557.
3. Oakberg, E. F. (1971) Spermatogonial stem-cell renewal in the mouse. *Anat. Rec.* **169,** 515–531.
4. de Rooij, D. G. (1973) Spermatogonial stem cell renewal in the mouse. I. Normal situation. *Cell Tissue Kinet.* **6,** 281–287.
5. Meistrich, M. L., and van Beek, M. E. A. B. (1993) Spermatogonial stem cells: assessing their survival and ability to produce differentiated cells. *Methods Toxicol.* **3A,** 106–123.
6. Russell, L. D., Ettlin, R. A., Sinha Hikim, A. P., and Clegg, E. D. (eds.) (1990). *Histological and histopathological evaluation of the testis,* Cache River Press, Clearwater, IL.
7. Weber. J. E., and Russell, L. D. (1987) A study of intercellular bridges during spermatogenesis in the rat. *Am. J. Anat.* **180,** 1–24.

8. de Rooij, D. J., and Russell, L. D. (2000) All you wanted to know about spermatogonia but were afraid to ask. *J., Androl.* **21,** 776–798.
9. Bozzola, J. J., and Russell, L. D. (1999) *Electron microscopy: principles and techniques for biologists,* 2nd ed., Jones and Bartlett, Sudbury, MA.
10. Spandro, R. L. (1990) Perfusion of the rat testis through the heart using heparin, in *Histological and histopathological evaluation of the testis* (Russell, L. D., Ettlin, R. A., Sinha Hikim, A. P., and Clegg, E. D., eds.), Cache River Press, Clearwater, IL, pp. 277–280.
11. Russell, L. D., and Clermont, Y. (1977) Degeneration of germ cells in normal, hypophysectomized and hormone treated hypophysectomized rats. *Anat. Rec.* **187,** 347–366.
12. Dym, M., and Fawcett, D. W. (1971) Further observations on the numbers of spermatogonia, spermatocytes, and spermatids connected by intercellular bridges in the mammalian testis. *Biol. Reprod.* **4,** 195–215.
13. Sinha Hikim, A. P., Bartke, A., and Russell, L. D. (1988) Morphometric studies on hamster testes in gonadally active and inactive states: light microscope findings. *Biol. Reprod.* **39,** 1225–1237.
14. Hess, R. A. (1990) Quantitative and qualitative characteristics of the stages and transitions in the cycle of the rat seminiferous epithelium: light microscopic observations of perfused-fixed and plastic-embedded testes. *Biol. Reprod.* **43,** 525–542.
15. Chiarini-Garcia, H., and Russell, L. D. (2001) High-resolution light microscopic characterization of mouse spermatogonia. *Biol. Reprod.* **65,** 1170–1178.
16. Chiarini-Garcia, H., Raymer, A. M., and Russell, L. D. (2003) Non-random distribution of spermatogonia in rats: evidence of niches in the seminiferous tubules. *Reproduction.* **126,** 669–680.
17. Avelar, G. F. (2004) Efeito da testosterone sobre o epitélio seminífero de ratos adultos após destruição seletiva das células de Leydig pelo etano dimetano sulfonato (EDS). Master's thesis, Universidade Federal de Minas Gerais, Instituto de Ciências Biológicas, Belo Horizonte, MG, Brazil.
18. Nascimento, H. F. (2004) Morfologia e distribuição topográfica dos diferentes tipos de espermatogônias de golden hamsters submetidos a curto e longo fotoperíodo. Master's thesis, Universidade Federal de Minas Gerais, Instituto de Ciências Biológicas, Belo Horizonte, MG, Brazil.
19. Bolden-Tiller, O. U., Chiarini-Garcia, H., Poirier, C., et al. (2007) Genetic factors contributing to abnormal spermatogonial differentiation in juvenile spermatogonial depletion (*jsd*) mice. *Biol. Reprod.* **77,** 237–246.
20. Russell, L. D., Chiarini-Garcia, H., Korsmeyer, S. J., and Knudson, C. M. Bax-dependent spermatogonia apoptosis is required for testicular development and spermatogenesis. *Biol. Reprod.* **66,** 950–958.
21. Chiarini-Garcia, H., and Russell, L. D. (2002) Characterization of mouse spermatogonial types by transmission electron microscopy. *Reproduction.* **123,** 567–577.

Chapter 7
Identification and Characterization of Spermatogonial Subtypes and Their Expansion in Whole Mounts and Tissue Sections from Primate Testes

Jens Ehmcke and Stefan Schlatt

Contents

Summary The different types of spermatogonia present in the testes of all mammalian species have a series of functions in the adult testis. Some cycle regularly to (1) maintain the spermatogonial population and (2) derive differentiating germ cells to maintain continuous spermatogenesis; other spermatogonia act as a functional reserve, proliferating only very rarely under healthy conditions but repopulating the depleted seminiferous tubules after gonadotoxic insult. The number, appearance, and function of different types of spermatogonia differ greatly between mammalian species, and therefore the precise number of mitotic steps and the number of identifiable stages in spermatogenesis, the spermatogenic efficiency, and the histological appearance of the seminiferous epithelium show remarkable variation. To characterize spermatogonial phenotypes and their respective functions and to understand the kinetics of spermatogenesis in any given species, a series of methods can be combined for best results. Conventional (hematoxylin or Periodic acid Schiff's reagent PAS/hematoxylin) staining on sections allows histological identification of the different types of spermatogonia and stages of spermatogenesis in the tissue. Immunohistochemical detection of the proliferation marker bromodeoxyuridine (BrdU) in sections and whole mounts of seminiferous tubules allows determination of which types of spermatogonia proliferate in which stage of spermatogenesis and determine the sizes of clones of proliferation spermatogonia in each stage. Combined, these methods allow the best possible characterization of spermatogenesis in any given mammalian species.

Keywords immunohistochemistry; primates; spermatogonial stem cells; whole mounts.

From: *Methods in Molecular Biology, Vol. 450, Germline Stem Cells*
Edited by S. X. Hou and S. R. Singh © Humana Press, Totowa, NJ

7.1   Introduction

In the testes of adult mammals, spermatogonia maintain their numbers by self-renewal and give rise to differentiating germ cells. Most of the diploid germ cells are differentiating spermatogonia undergoing several rounds of mitotic divisions before entering meiotic prophase. In primates, different types of spermatogonia have been identified *(1–6)*: the reserve stem cell A_{dark} spermatogonium, the renewing stem cell A_{pale} spermatogonium, and several generations of B spermatogonia (B_1, B_2, B_3, etc.). It has been unequivocally demonstrated that B spermatogonia are derived from A_{pale} spermatogonia *(1)*; the complex kinetics of the renewal and differentiation of A_{pale} spermatogonia, replenishing their own numbers and giving rise to B spermatogonia, have been described using the approach described here *(7,8)*.

Previous studies on spermatogenesis have focused on the cycle of the seminiferous epithelium and on the proliferation of spermatogonial stem cells in different species of nonhuman primates and other mammalian species *(1–5,9–16)*. Some of these studies examined whole-mount preparations of seminiferous tubules to obtain qualitative data on spermatogonial organization. Most previous studies applied morphometric approaches to (serial) cross sections of seminiferous tubules, producing valuable data on spermatogonial counts and labeling indices, but no previous approach has been able to present an in-depth evaluation of the clonal organization of spermatogonial stem cells at different stages of spermatogenesis *(1,15)*.

In the adult testis, proliferating spermatogonia and preleptotene spermatocytes in S-phase of meiosis can be detected by nuclear incorporation of bromodeoxyuridine (BrdU; *17*). Detecting BrdU incorporation in whole mounts of seminiferous tubules allows the study of the clonal organization of the proliferating spermatogonia. To enable the observer to assign each detected clone of proliferating spermatogonia to a specific stage of the cycle of the seminiferous epithelium, we employ immunofluorescent localization of acrosin, which allows accurate determination of the size and the shape of the acrosome and thus accurate determination of any stage of the cycle of the seminiferous epithelium. In the past, we have employed our novel staining method on sections and whole mounts of adult rhesus monkey testicular tissue. We analyzed and described the initiating divisions and the clonogenic expansion of spermatogonia, which allowed us to propose a well-defined model of premeiotic germ cell development in the rhesus monkey. Currently, we are employing our method to evaluate spermatogonial proliferation in the adult human testis.

In the following, detailed descriptions of the following procedures are given: (1) conventional embedding, sectioning, and staining of sections of testicular tissues (*see* **Fig. 7.1A**); (2) hematoxylin staining of whole mounts of seminiferous tubules (*see* **Fig. 7.1B**); (3) immunohistochemical staining of sections of testicular tissues (*see* **Fig. 7.2A**); and (4) immunohistochemical staining of whole seminiferous tubules (*see* **Fig. 7.2B,C**). Using a combination of these approaches, the stages of

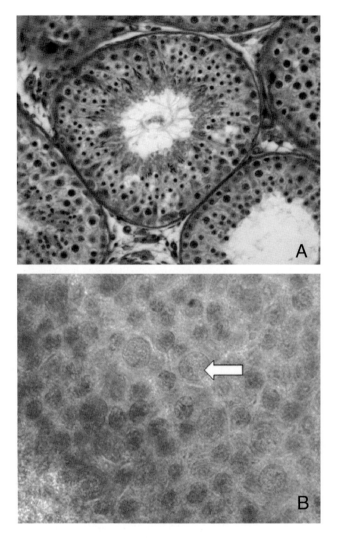

Fig. 7.1 A Conventional hematoxylin staining of adult rhesus monkey testicular tissue showing a central seminiferous tubule at stage II of the seminiferous epithelial cycle. **B** A_{pale} spermatogonia (large cells, *arrow*) and preleptotene spermatocytes (small cells) can be distinguished in specific planes of focus under high magnification when a whole mount of a seminiferous tubule of an adult rhesus monkey is conventionally stained with hematoxylin

the seminiferous epithelium and the status of spermatogenesis can be identified in testes from any primate species. In addition, the presence or absence of proliferating spermatogonia in any given stage of the cycle of the seminiferous epithelium can be determined, and the total number and clonal organization of proliferating spermatogonia can be described for all stages of the seminiferous epithelial cycle.

7.2   Materials

7.2.2   *BrdU Exposure in Vitro for Testicular Tissue Biopsies*

1. Cell culture medium: 100 mL high-glucose (4.5 g/mL) Dulbecco's modified Eagle's medium, 1% nonessential amino acid supplement, 100 IU penicillin/mL, 100 μg streptomycin/mL.
2. BrdU-labeling medium: Cell culture medium, 100 μ*M* 5-bromo-2'-deoxy-uridine.

7.2.2   *Tissue Fixation*

For tissue fixation, Bouin's fixative is used: 150 mL aqueous saturated picric acid solution, 50 mL formalin solution (37%), 10 mL glacial acetic acid (*see* **Note 1**).

7.2.3   *Whole-Mount and Section Conventional Staining*

1. Periodic acid solution: 1% periodic acid in distilled water (*see* **Note 2**).
2. Schiff's reagent can be purchased from a wide variety of commercial suppliers.
3. Hematoxylin solution: Mayer's hematoxylin solution can be commercially obtained from a variety of suppliers.
4. Aqueous mounting medium can be commercially obtained from a variety of suppliers.
5. Xylene-based mounting medium can be purchased from a wide variety of commercial suppliers.

---►

Fig. 7.2 **A** Section of the epithelium of a seminiferous tubule of an adult rhesus monkey in stage II of the spermatogenic cycle showing immunohistochemical staining for the identification of elongated spermatids (diamidino-2-phenylindole [DAPI], *blue fluorescence*), round spermatids (acrosin antibody, *red fluorescence*) and B$_2$ spermatogonia in S-phase of the cell cycle (BrdU antibody, *green fluorescence*). **B** A whole mount of a seminiferous tubule of a rhesus monkey showing immunohistochemical labeling of the proliferation marker BrdU (*green fluorescence*). The clones of A$_{pale}$ spermatogonia in the upper part of the micrograph and the groups of preleptotene spermatocytes in the lower part of this seminiferous tubule indicate that this section of the seminiferous tubule is in stage VII of the seminiferous epithelial cycle. **C** A whole mount of a seminiferous tubule of an adult rhesus monkey at stage VII of the seminiferous epithelial cycle showing immunohistochemical staining for BrdU (*green fluorescence*) and acrosin (*red fluorescence*). Note the small nuclei of the preleptotene spermatocytes, the big nuclei of the A$_{pale}$ spermatogonia, and the cap-shaped acrosomes of the round spermatids

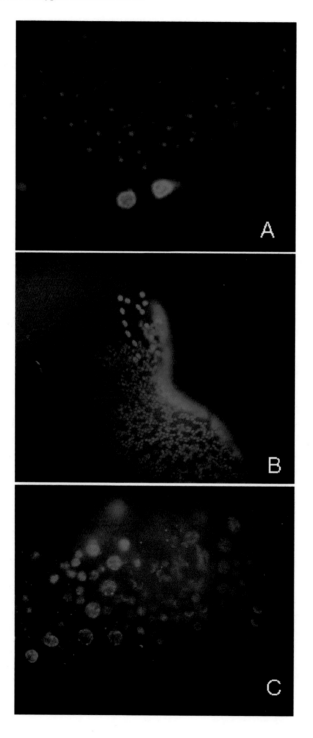

7.2.4   Whole-Mount and Section Immunostaining

1. HCl retrieval solution: $1M$ HCl.
2. Tris-buffered saline (TBS): $0.5M$ Tris-HCl, pH 7.6, $1.5M$ NaCl.
3. Trypsinizing solution: 0.1% trypsin in TBS.
4. Blocking solution: TBS, 0.1% bovine serum albumin (BSA), 5% goat serum.
5. Staining solution 1: monoclonal anti-BrdU antibody (Biomeda V3246), diluted 1:50 in TBS, 0.1% BSA.
6. Staining solution 2: Goat-anti-mouse antibody, biotin conjugated, diluted 1:100 in TBS, 0.1% BSA.
7. Staining solution 3: Monoclonal anti-acrosin antibody (Biosonda AMC-ACRO-C5F10-AS), diluted 1:100, and streptavidin-conjugated Alexa fluorochrome 488, diluted 1:100 in TBS, 0.1% BSA.
8. Staining solution 4: Goat-anti-mouse antibody conjugated with Alexa 546 fluorochrome, diluted 1:100 in TBS, 0.1% BSA.
9. Aqueous mounting medium with and without DAPI (diamidino-2-phenylindole, $1.5\,\mu g/mL$) can be commercially obtained from a variety of suppliers.

7.3   Methods

7.3.1   BrdU Labeling

Pieces of seminiferous tubules are isolated from fresh testicular tissue in ice-cold cell culture medium by careful separation with a pair of tweezers. This may not be possible in all cases (e.g., in prepubertal tissue made up from fragile and strongly intertwined tubules). In those cases, the tissue can be cut using fine scissors to obtain cuboidal tissue fragments with edges shorter than 2 mm. The separated tubules or tissue fragments are incubated in BrdU-labeling medium for 2 h in an incubator (5% CO_2, 35°C). The tissue is then washed twice for 5 min in cell culture medium to remove excess BrdU.

7.3.2   Fixation and Storage

The tissue is fixed in Bouin's fixative for 24 h at room temperature. The tissue is then washed in 70% ethanol until the yellow color of the picric acid has mostly disappeared and can be stored in 70% ethanol for prolonged periods of time.

7.3.3   Embedding and Sectioning

Tissue samples, stored in 70% ethanol, are dehydrated (transferred through 85% and 95% ethanol into 100% ethanol, 10 min in each concentration), routinely embedded in paraffin, and sectioned to $5\,\mu m$.

7.3.4 Conventional Staining

Sections are deparaffinized (xylene, twice for 15 min each), rehydrated (100% (twice), 95%, 85%, and 70% ethanol, 10 min each, followed by 3 min of distilled water), incubated for 15 min at room temperature in periodic acid solution, washed with distilled water (5 min), incubated for 15 min in Schiff's reagent, washed with distilled water (5 min), stained with hematoxylin solution (5 min), rinsed with distilled water, washed in tap water (10 min), and rinsed in distilled water. Sections are then dehydrated through rising ethanol concentrations (50%, 70%, 85%, 95%, twice at 100%, 10 min at each concentration) and transferred into xylene (twice, 15 min each) and are mounted with a permanent, xylene-based mounting medium.

Whole seminiferous tubules, stored in 70% ethanol, are rehydrated (50% and 30% ethanol, and distilled water, 10 min in each solution), incubated in TBS (three times at 10 min), incubated in trypsinizing solution (10 min; *see* **Note 3**), washed with distilled water (twice for 5 min), stained with hematoxylin solution (2 min), rinsed with distilled water, washed with tap water (10 min), rinsed with distilled water, and mounted with aqueous mounting medium (*see* **Note 4**).

7.3.5 Immunohistochemical Staining

Sections of testicular tissue are deparaffinized, rehydrated, incubated in HCl retrieval solution for 10 min at room temperature, washed in distilled water (3 min), incubated for 5 min at room temperature with trypsinizing solution, washed with distilled water (3 min) followed by TBS (three times for 5 min each), incubated for 30 min at room temperature with blocking solution, and then incubated overnight at 4°C with staining solution 1. The sections are then washed with TBS (three times for 5 min each), incubated for 1 h at room temperature with staining solution 2, washed with TBS (three times for 5 min each), incubated for 1 h at room temperature with staining solution 3, washed in TBS (three times for 5 min each), incubated for 1 h at room temperature with staining solution 4, washed with TBS (three times for 5 min each), and mounted using an aqueous mounting medium with DAPI (Vector, Burlingame, CA) containing DAPI (1.5 µg/mL; *see* **Note 5**).

Whole seminiferous tubules are temporarily transferred into 100% ethanol (1 h in 85%, 1 h in 95%, overnight in 100%, 1 h in 95%, 1 h in 85%, overnight in 70% ethanol) before staining. Note: this complete dehydration dissolves most of the membranes within the tissue, allowing easier diffusion of the staining solutions. Whole seminiferous tubules are stained using a similar protocol as described for sections, but the tubules are stained in suspension in 2-mL reaction vials using 200 µL of each staining solution. Also, the incubation times for the HCl retrieval solution and the trypsinizing solutions are increased to 15 min at room temperature each. Stained seminiferous tubules are mounted on microscope slides using VectaShield mounting medium without DAPI.

7.3.6   Tissue Analysis

Sections and whole mounts are analyzed using a fluorescent microscope with attached digital camera or a confocal scanning system. All images are acquired digitally.

7.3.7   Determination of Cycle Stages After Immunohistochemistry

After the tissue has been exposed to hydrochloric acid and trypsin to facilitate access of the antibody to BrdU incorporated into the nuclear DNA, the nuclear counterstaining with DAPI is intense in the heads of elongated spermatids from step 13 onward. In combination with the acrosin labeling, this allows distinguishing the following types of spermatids: round spermatids, DAPI-negative elongating spermatids of the earlier stages of the spermatogenic cycle, and DAPI-positive elongated spermatids of the later stages of the spermatogenic cycle.

The colocalization of BrdU as a proliferation marker, acrosin as a spermatid label, and DAPI for DNA staining of the most mature spermatids enables identification of proliferating cells and correlation of their presence with accurately determined stages of spermatogenesis depending on acrosomal structure and the presence of one or two types of spermatids *(7,8)*. In whole mounts of seminiferous tubules, DAPI staining is not employed because the presence of elongated spermatids in the seminiferous tubules can easily be determined using conventional phase-contrast microscopy.

7.3.8   Evaluation of Spermatogonial Clones

We defined cohorts of neighboring BrdU-positive spermatogonia as clones whenever the distance between two BrdU-positive nuclei was less than 1.5 times the diameter of the nucleus. Spermatogonia located at this distance are connected by cytoplasmic bridges. As an additional criterion, the clones had to show the same or a very similar intensity and pattern of BrdU labeling. These criteria were chosen because (1) a similar BrdU-staining pattern in various cells indicates that the timing of S-phase in these cells is synchronized, and (2) in hematoxylin-stained whole mounts, cytoplasmic bridges (indicating a clonal origin of several cells) between spermatogonia can only be detected in cells not farther apart than approx. 1.5 times their nuclear diameter. Using these criteria, the number of clones of BrdU-positive spermatogonia and the number of spermatogonia per clone can be determined for specific stages of the cycle of the seminiferous epithelium. The number of BrdU-labeled spermatogonia can be determined for each stage (a continuous part of the tubule showing the same acrosomal staining pattern).

7.4 Notes

1. Picric acid solution and formalin can be mixed to prepare a stock solution, but the glacial acetic acid is not to be added until right before use because the chemical reactions between the formaldehyde and the acetic acid cause the fixative to slowly deplete both the aldehyde and the acid, leading to a less-potent fixative.
2. Periodic acid solution has to be prepared fresh every time shortly before staining and has to be kept in the dark (drawer) during the staining procedure as the periodic acid solution is light sensitive and degrades rapidly when kept in light. The same is true for the Schiff's reagent in the subsequent staining step.
3. The trypsin digestion step is needed to clear most of the proteins from the tissue for better visualization. Without this step, the tissue remains opaque and difficult to observe. This trypsinizing step has to be adapted for each type of tissue used for this staining. Optimally, the tissue will be completely clear but still mechanically stable. Too long digestion times can cause the tissue to disintegrate during the later staining steps or at the time of mounting.
4. Hematoxylin-stained whole mounts can also be dehydrated (30%, 50%, 70%, 85%, and 95% ethanol for 10 min in each concentration and 10 min twice in 100% ethanol and xylenes twice for 15 min each) and can be mounted using a xylene-based mounting medium. Whereas the aqueous mounting provides a specimen with better integrity (no shrinkage), the dehydration and permanent mounting allows for long-term storage of the specimen and can also help to prevent mechanical damage to fragile whole mounts. Increased fragility can be caused by trypsin exposure for too long or can result from seminiferous tubules taken from challenged testicular tissue (cytotoxic insult, radiation, partial spermatogenic depletion caused by other causes).
5. When mounting the sections with mounting medium containing DAPI, only the heads of the elongated spermatids appear blue; all other nuclei are DAPI negative. This is because the DNA (which when intact captures the DAPI molecules for the nuclear labeling) in most cells has been denatured by the HCl treatment, which is a necessary precondition for the immunohistochemical staining of the proliferation marker BrdU. Only the modified, tightly packed DNA in the heads of the elongated spermatids is still able to retain DAPI molecules after this pretreatment. This allows for very selective labeling of elongated spermatid heads in this procedure.

References

1. Clermont, Y., and Leblond, C. P. (1959) Differentiation and renewal of spermatogonia in the monkey, *Macacus rhesus. Am. J. Anat.* **104**, 237–273.
2. Clermont, Y. (1969) Two classes of spermatogonial stem cells in the monkey (*Cercopithecus aethiops*). *Am. J. Anat.* **126**, 57–71.
3. Clermont, Y., and Antar, M. (1973) Duration of the cycle of the seminiferous epithelium and the spermatogonial renewal in the monkey *Macaca arctoides. Am. J Anat.* **136**, 153–165.
4. Kluin, P. M., Kramer, M. F., and de Rooij, D. G. (1983) Testicular development in Macaca *irus* after birth. *Int. J., Androl.* **6**, 25–43
5. Fouquet, J. P., and Dadoune, J. O. (1986) Renewal of spermatogonia in the monkey (*Macaca fascicularis*). *Biol. Reprod.* **35**, 199–207.
6. Zhengwei, Y., McLachlan, R. I., Bremner, W. J., and Wreford, N. G. (1997) Quantitative (stereological) study of the normal spermatogenesis in the adult monkey (*Macaca fascicularis*). *J. Androl.* **18**, 681–687.
7. Ehmcke, J., Luetjens, C. M., and Schlatt, S. (2005) Clonal organization of proliferating spermatogonial stem cells in adult males of two species of non-human primates, *Macaca mulatta* and *Callithrix jacchus. Biol. Reprod.* **72**, 293–300.

8. Ehmcke, J., Simorangkir, D. R., and Schlatt, S. (2005) Identification of the starting point for spermatogenesis and characterization of the testicular stem cell in adult male rhesus monkeys. *Hum. Reprod.* **20,** 1185–1193.

9. Dym, M., and Clermont, Y. (1970) Role of spermatogonia in the repair of the seminiferous epithelium following X-irradiation of the rat testis. *Am. J Anat.* **128,** 265–282.

10. Clermont, Y. (1972) Kinetics of spermatogenesis in mammals: seminiferous epithelium cycle and spermatogonial renewal. *Physiol. Rev.* **52,** 198–236.

11. Bellve, A. R., Cavicchia, J. C., Millette, C. F., O'Brian, D. A., Bhatnagar, Y. M., and Dym, M. (1977) Spermatogenic cells of the prepubertal mouse. Isolation and morphological characterization. *J. Cell Biol.* **74,** 68–85.

12. De Rooij, D. G., van Alphen, M. M., and van de Kant, H. J. 1986 Duration of the cycle of the seminiferous epithelium and its stages in the rhesus monkey (*Macaca mulatta*). *Biol. Reprod.* **35,** 587–591.

13. van Alphen, M. M., and de Rooij, D. G. (1986) Depletion of the seminiferous epithelium of the rhesus monkey, *Macaca mulatta*, after X-irradiation. *Br. J. Cancer Suppl.* **7,** 102–104.

14. van Alphen, M. M., van de Kant, H. J., and de Rooij, D. G. (1988) Depletion of the spermatogonia from the seminiferous epithelium of the rhesus monkey after X irradiation. *Radiat. Res.* **113,** 473–486.

15. van Alphen, M.M., van de Kant, H.J., and de Rooij, D.G. (1988) Repopulation of the seminiferous epithelium of the rhesus monkey after X irradiation. *Radiat. Res.* **113,** 487–500.

16. De Rooij, D. G., van Dissel-Emiliani, F. M., and van Pelt, A. M. (1989) Regulation of spermatogonial proliferation. *Ann. N. Y. Acad. Sci.* **564,** 140–153.

17. Rosiepen, G., Arslan, M., Clemen, G., Nieschlag, E., and Weinbauer, G. F. (1997) Estimation of the duration of the cycle of the seminiferous epithelium in the non-human primate *Macaca mulatta* using the 5-bromodeoxyuridine technique. *Cell Tissue Res.* **288,** 365–369.

Chapter 8
Epigenetic Control in Male Germ Cells

A Transillumination-Assisted Microdissection Method for the Analysis of Developmentally Regulated Events

Durga Prasad Mishra and Paolo Sassone-Corsi

Contents

Summary Germ cells constitute the vehicles of genetic information and thereby of inheritance through generations. The epigenetic control mechanisms that govern maintenance and reprogramming of the germline in diverse organisms have gained increasing interest as they reveal essential regulatory pathways implicated in health and disease. We describe a step-by-step transillumination-assisted microdissection method currently used in our laboratory to routinely characterize various spermatogenic cell types. This methodology is ideal to analyze nuclear chromatin condensation and allows further evaluation of epigenetic modifications in male germ cells.

Keywords Chromatin; epigenetic reprogramming; germ cells; meiosis; spermatogenesis.

8.1 Introduction

Epigenetic reprogramming in the germline is central to the adaptation of organisms to a changing environment without stable genetic alterations. During mammalian development, two periods are characterized by epigenetic reprogramming, gametogenesis and early embryogenesis. In the course of development of male germ cells or spermatogenesis, global sex-specific reprogramming of the epigenome occur as a wave of DNA demethylation followed by DNA methylation and chromatin

From: *Methods in Molecular Biology, Vol. 450, Germline Stem Cells*
Edited by S. X. Hou and S. R. Singh © Humana Press, Totowa, NJ

modifications *(1,2)*. This may contribute to a unique feature of the germline controlling gene function from one generation of an organism to the next. Maturation of male germ cells is characterized by an impressive degree of cellular restructuring and gene regulation that involves remarkable genomic reorganization *(2)*. These events are finely tuned but are also susceptible to the introduction of various types of errors.

The epigenetic control of gene function in germ cells follows a highly specialized program. During spermatogenesis, the male-specific epigenetic program is "reset," through a remodeling of the epigenome, to allow the fusion of the male and female gametes at the time of fertilization and to give rise to a zygote that is totipotent and thus able to give rise to any cell type. The fidelity of the resetting process is important for preventing aberrant epigenetic modifications that can be passed on to the next generation. Hence, the implications for human health are profound. Failure in epigenetic reprogramming may occur in male germ cells that have undergone assisted reproductive manipulations *(3)* or exposure to harmful environmental or chemical factors, leading to offspring with greater susceptibility to disease *(4)*.

Understanding how epigenetic patterning occurs in the germline may help in the prevention of heritable diseases, improvement of assisted reproductive technologies, and stem cell therapy *(5)*. Thus, techniques for the study of epigenetic reprogramming are essential to the evaluation of the molecular players and regulatory pathways involved in chromatin remodeling and gene regulation in male germ cells.

Spermatogenesis is characterized by a particularly spectacular chromatin remodeling central to the epigenetic reprogramming process, in which somatic linker histones are sequentially replaced by testis-specific variants, followed by the replacement of most histones with protamines *(5)*. The study of male germ cells is complicated by the exceptional organization of the seminiferous epithelium, and although the cells differ from each other in sedimentation properties and cell surface markers, isolation and study of epigenetic changes on the basis of these parameters are often inaccurate and cumbersome *(6–16)*. In this context, the technique described in the present chapter makes use of the unique light absorption pattern of the seminiferous tubule as seen under a light microscope correlating with a specific stage of the spermatogenic wave with every cross section of the tubule containing cell types in a specific combination *(17)*. Therefore, cells of particular stages can be isolated and characterized on the basis of their transillumination properties. The described technique of transillumination-assisted microdissection can be applied to characterize and evaluate the epigenetic reprogramming during chromatin remodeling, control of male germ cell differentiation, and stem cell biology.

8.2 Materials

1. Phosphate-buffered saline (PBS), pH 7.2–7.4 (Sigma Chemicals).
2. 90% ethanol.

3. Liquid nitrogen in a portable container.
4. Immersion oil (Leica).
5. 0.05 × 0.01-mm fine-tip forceps (two pairs) (World Precision Instruments).
6. 0.015 mm × 0.015-mm tip microdissection scissors and dissection scissors (Fine Science Tools).
7. Slides.
8. Coverslips.
9. Phase-contrast microscope and dissection microscope (Leica).

8.3   Methods

1. Sacrifice a sexually mature male mouse, remove the testes, and place them in a Petri dish containing PBS. Decapsulate the testes and transfer the seminiferous tubules to a new Petri dish containing PBS. All the procedures involving microdissection should ideally be completed within 2 h after sacrificing the mouse (*see* **Note 1**).
2. View the tubules on a transilluminating dissection microscope. A predictable light absorption pattern is produced when the seminiferous tubules are observed under a transilluminating dissection microscope. The greater the level of spermatid chromatin condensation, the greater is the amount of light absorbed, resulting in the differential appearance of tubule segments based on the stage of spermatogenesis. This pattern of light absorption can be used to isolate specific cell associations during sperm cell differentiation. Using fine forceps, gently pull apart the tubules, taking care not to cause damage by squeezing or shaking the tubules (*see* **Note 2).**
3. Determine the light absorption pattern and identify the weak spot (stages XII–I), the strong spot (stages II–VI), the dark zone (stages VII–VIII), and the pale zone (stages IX–XI) (*see* **Note 3**).
4. Cut a piece smaller than 0.5 mm of the stage of interest. Collect the piece of tubule in a 15-μL volume using a pipet. Transfer the tubule onto a glass slide. Carefully put a coverslip over the tubule. Avoid introducing air bubbles. The coverslip will squash the tubule, and the cells will flow out. This procedure may need to be modified to improve access of the antibodies to the antigen, particularly when studying nuclear proteins. For the study of nuclear proteins, cut the tubule (<0.5-mm piece) and subsequently add 15 μL 100 mM sucrose to it; the cells can be released by tweezing apart the tubule and pipeting the cells up and down to properly resuspend the cells in the sucrose solution. Cells should then be transferred onto a slide predipped in 1% paraformaldehyde and 0.15% Triton X-100. The slides can be dried in a humid chamber and then can be directly processed for immunocytochemistry or can be frozen for later use. For chromosome spreads, the tubules should be preincubated in hypotonic solution for 30 min before isolating the desired segment. For biochemical analyses (real-time polymerase chain reaction [PCR], methylation-specific PCR, chromatin immunoprecipitation analysis), tubules should be cut in sequential 2-cm sections.

When a sufficient amount of tubules is cut, they can be pooled in cryovials after removing excess PBS; freeze in liquid nitrogen and store at −80°C for later downstream applications. To study molecular and cellular events during a specific stage of germ cell differentiation, a more accurate staging method using a phase-contrast microscope is necessary (*see* **Note 4**).

5. Touch one edge of the coverslip with filter paper to remove extra fluid and to spread the cells evenly on the slide; the aim is to make a monolayer (*see* **Notes 4–6** and **Fig. 7.1**).

6. Examine the cells with a phase-contrast microscope using the ×40 objective. Identify the stage on the basis of the cell associations and morphological characteristics. For a more careful examination, use the ×100 objective with immersion oil (*see* **Note 7**).

7. After examination with a microscope, freeze the slide in liquid nitrogen for 20 s and remove the coverslip by flipping it off with a scalpel. Fix the cells in 90% ethanol for 2–5 min and then air dry. Store the slides at 15°C–25°C if they will be used within a few weeks or place at −80°C for long-term storage (*see* **Note 8**).

The methodology described here has been successfully used for over 30 years by groups around the world studying spermatogenesis. This method has been popularly

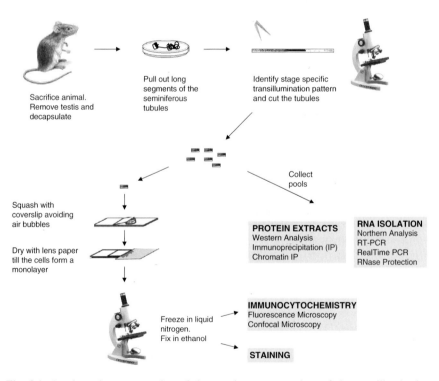

Fig. 8.1 A schematic representation of the step-by-step procedure of the transillumination-assisted microdissection method. Also see **ref.** *17*

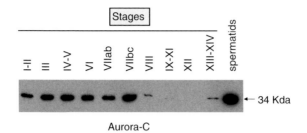

Fig. 8.2 Example of how the microdissection technique helps the identification of developmentally stage-specific expression of RNAs or proteins in male germ cells. Shown is a Western blot using a polyclonal antibody generated in our laboratory against the mouse Aurora-C kinase (N-terminus, amino acids 1–11). This antibody does not recognize the Aurora-A and Aurora-B kinases. Using staged preparations of rat seminiferous epithelium, it is evident that Aurora-C protein expression is restricted to spermatogenic stages when elongated spermatids are present, corresponding to periods when step 15–19 spermatids are present. Immunofluorescence studies of living cells obtained by the same microdissection technique revealed that the highest levels of protein are found in elongated speramatids in steps 13–15, and pachytene spermatocytes show weak punctate staining in the nucleus form stages VIII–XI. Additional information about Aurora-B and Aurora-C function in spermatogenesis is presented in **ref. 24**

used for the visualization of staged living spermatogenic cells in reproductive toxicology studies and for studying chromosome events and movements of organelles, such as the chromatid body *(18,19)*. In addition, we have pioneered new biochemical applications extending its use to studies of transcriptional regulation during spermatogenesis and to the characterization of spermatogenic gene-targeting mouse models *(20,21)*. At present, our laboratory is making use of this technique for evaluation of histone modifications in male germ cells as well identification of kinases that may be expressed in a developmentally regulated manner (*see* **Fig. 8.2**). The greatest challenge of this technique is the initial difficulty in learning to recognize the transillumination pattern of the seminiferous tubules and the spermatogenic cell types under phase-contrast microscopy *(22,23)*. Our goal here was to provide clear directions and accurate descriptions of the key features of the procedure. We have described its application for cellular and biochemical studies, and we believe that extension of this method will contribute to the study of epigenetic reprogramming, stem cell, and cancer biology.

8.4   Notes

1. It is necessary to use sexually mature mice (older than 60 d) to see a clear transillumination pattern of the tubule. In juvenile mice, there is no clear transillumination pattern before the chromatin of spermatids starts to condense. But, the wave organization of seminiferous epithelium is present

very early in development, and it is possible to prepare squash preparations and to study the morphology of the existing spermatogenic cell types in juvenile mice. All of the mouse and rat strains that we studied had the same transillumination pattern and predicted cell associations at each stage. The staging between mice and rats is slightly different (*6*).

2. The tubule and the light absorption patterns are damaged during separation and cutting. To avoid this, bend one tip of the forceps to create a small hook. Lift the tubule using the hook and cut with dissection scissors. In this way, you will avoid squeezing the tubule with the forceps.

3. From stages XII–VI, spermatids (steps 12–16) are arranged in bundles, resulting in the speckled appearance of the tubule. Stage XII is distinguished as the weak spot, a characteristic resulting from the increased chromatin condensation of step 12 spermatids and their organization into bundles, which are observed as individual spots. Spermatid bundles are most visible in the tubular region, termed the *strong spot*, comprised of stages II–VI. In the periphery of the tubule, the bundles have a striped configuration because they are seen sideways, whereas in the middle of the tubules the bundles have a spotty configuration because they are seen along the axis of the bundle. This pattern reflects the deep penetration of the Sertoli cells by the spermatid bundles. As spermatogenesis progresses, the bundling of the elongating spermatids stops, coincident with the movement of step 15 spermatids toward the tubule lumen. This produces the characteristically homogeneous dark zone corresponding to stages VII–VIII. Stage VIII marks the point of spermiation and corresponds to the release of elongated spermatids from the Sertoli cells into the lumen, where they begin their voyage to the epididymis. After the point of spermiation, the dark zone abruptly changes into the pale zone, representing stages IX–XI. For a full description, please see ref. *17*.

4. Aided by capillary diffusion that occurs when a piece of filter paper is placed at the edge of the coverslip, cells flow out of the tubule, yielding a live cell monolayer. Observe the spreading of the cells under a phase-contrast microscope and remove the filter paper when a monolayer has formed. Cells will have a slightly flattened appearance. Insufficient drying results in a squash preparation that is too thick, making it difficult to recognize the cell types. If the squash preparation has dried too much, cells look unhealthy and damaged. To avoid this after sacrificing the mouse, examine the squash preparations within a maximum time of 2 h. The appearance of pale spots in the nucleus is a typical feature of degenerating cells.

5. To control the cells moving under the microscope, use a syringe filled with immersion oil; seal the preparation by adding a drop of immersion oil along every edge of the coverslip. This will stop the cells from moving and will prevent the sample from drying, allowing longer observation of the preparation.

6. If it appears that the cells represent more than one stage, this is possible because not all tubules will have a uniform length of waves; some will have short waves, resulting in the 0.5-mm segment showing more than one stage.

7. The stages of spermatogenesis are determined on the basis of specific cell associations and key morphological criteria (*17*). The early spermatids have the most easily recognizable morphological features. The developmental state of the acrosome and nuclear shape and chromatin condensation of step 1–12 spermatids are the most commonly used hallmarks of specific stages. In addition, the formation of flagella, the appearance of particular types of meiotic spermatocytes or spermatogonia, and the possible organization of the spermatids into bundles can be used as criteria.

8. If there is a weak signal or too much background in immunocytochemistry, do not let slides defrost between freezing and fixation. Squash preparations can be stored at −80°C, but for some antibodies, storing at 15°C–25°C (for up to 1 wk) is better. Usually, the thinner the squash preparation is, the better the signal. Thus, try to cut as short a segment as possible and ensure that the cells are well spread.

Acknowledgments We thank Martti Parvinen, Noora Kotaja, Sarah Kimmins, and all the members of the Sassone–Corsi laboratory for critical reading of the manuscript, advice, and stimulating discussions.

References

1. Holliday, R. (1989) DNA methylation and epigenetic mechanisms. *Cell Biophys.* **15,** 15–20.
2. Sassone-Corsi, P. (2002) Unique chromatin remodeling and transcriptional regulation in spermatogenesis. *Science.* **296,** 2176–2178.
3. DeBaun, M.R., Niemitz, E.L., and Feinberg, A.P. (2003) Association of in vitro fertilization with Beckwith–Wiedemann syndrome and epigenetic alterations of LIT1 and H19. *Am. J. Hum. Genet.* **72,** 156–160.
4. Brinster, R.L. (2002) Germline stem cell transplantation and transgenesis. *Science.* **296,** 2174–2176.
5. Kimmins, S., and Sassone-Corsi, P. (2005) Chromatin remodelling and epigenetic features of germ cells. *Nature.* **31,** 583–589.
6. Russell, L.D., Ettlin, R.A., SinhaHikim, A.P., and Clegg, E.D. (1990) Mammalian spermatogenesis, in *Histological and histopathological evaluation of the testis,* Cache River Press, St. Louis, MO, pp. 1–40.
7. Perey, B., Clermont, Y., and Leblond, C.P. (1961) The wave of the seminiferous epithelium in the rat. *Am. J. Anat.* **108,** 47–77.
8. Parvinen, M., and Vanha-Perttula, T. (1972) Identification and enzyme quantitation of the stages of the seminiferous epithelial wave in the rat. *Anat. Rec.* **174,** 435–450.
9. Toppari, J., Eerola, E., and Parvinen, M. (1985) Flow cytometric DNA analysis of defined stages of rat seminiferous epithelial cycle during in vitro differentiation. *J. Androl.* **6,** 325–333.
10. Toppari, J., and Parvinen, M. (1985) In vitro differentiation of rat seminiferous tubular segments from defined stages of the epithelial cycle morphologic and immunolocalization analysis. *J. Androl.* **6,** 334–343.
11. Leblond, C.P., and Clermont, Y. (1952) Definition of the stages of the cycle of the seminiferous epithelium in the rat. *Ann. N. Y. Acad. Sci.* **55,** 548–573.
12. Parvinen, M. (1982) Regulation of the seminiferous epithelium. *Endocr. Rev.* **3,** 404–417.
13. Toppari, J., Bishop, P.C., Parker, J.W., and diZerega, G.S. (1988) DNA flow cytometric analysis of mouse seminiferous epithelium. *Cytometry.* **9,** 456–462.
14. Kangasniemi, M., Kaipia, A., Mali, P., Toppari, J., Huhtaniemi, I., and Parvinen, M. (1990) Modulation of basal and FSH-dependent cyclic AMP production in rat seminiferous tubules staged by an improved transillumination technique. *Anat. Rec.* **227,** 62–76.
15. Parvinen, M., Toppari, J., and Lähdetie J. (1993) Transillumination-phase contrast microscopic techniques for evaluation of male germ cell toxicity and mutagenicity, in *Methods in Reproductive Toxicology* (Chapin, R.E., and Heindel, J., eds.), Academic Press, Orlando, Fl, pp. 142–165.
16. Parvinen, M., and Hecht, N.B. (1981) Identification of living spermatogenic cells of the mouse by transillumination-phase contrast microscopic technique for "in situ" analyses of DNA polymerase activities. *Histochemistry.* **71,** 567–579.
17. Kotaja, N., Kimmins, S., Brancorsini, S., et al. (2004) Preparation, isolation and characterization of stage-specific spermatogenic cells for cellular and molecular analysis. *Nat. Methods.* **3,** 249–254.
18. Kotaja, N., Lin, H., Parvinen, M., and Sassone-Corsi, P.(2006) Interplay of PIWI/Argonaute protein MIWI and kinesin KIF17b in chromatoid bodies of male germ cells. *J. Cell Sci.* **119,** 2819–2825.
19. Kotaja, N., Bhattacharyya, S.N., Jaskiewicz, L., et al. (2006) The chromatoid body of male germ cells: similarity with processing bodies and presence of Dicer and microRNA pathway components. *Proc. Natl. Acad. Sci. U. S. A.* **103,** 2647–2652.
20. Martianov, I., Fimia, G.M., Dierich, A., Parvinen, M., Sassone-Corsi, P., and Davidson, I. (2001) Late arrest of spermiogenesis and germ cell apoptosis in mice lacking the TBP-like TLF/TRF2 gene. *Mol. Cell.* **7,** 509–515.

21. Martianov, I., Brancorsini, S., Gansmuller, A., Parvinen, M., Davidson, I., and Sassone-Corsi, P. (2002) Distinct functions of TBP and TLF/TRF2 during spermatogenesis: requirement of TLF for heterochromatic chromocenter formation in haploid round spermatids. *Development.* **129,** 945–955.

22. Henriksen, K., and Parvinen, M. (1998) Stage-specific apoptosis of male germ cells in the rat: mechanisms of cell death studied by supravital squash preparations. *Tissue Cell.* **30,** 692–701.

23. Ventelä, S., Toppari, J., and Parvinen, M. (2003) Intercellular organelle traffic through cytoplasmic bridges in early spermatids of the rat: mechanisms of haploid gene product sharing. *Mol. Biol. Cell.* **14,** 2768–2780.

24. Kimmins, S., Crosio, C., Kotaja, N., et al. (2007) Differential functions of the Aurora-B and Aurora-C kinases in mammalian spermatogenesis. *Mol. Endocrinol.* **21,** 726–739.

Chapter 9
GDNF Maintains Mouse Spermatogonial Stem Cells In Vivo and In Vitro

Hannu Sariola and Tiina Immonen

Contents

Summary   Spermatogonial stem cells (SSCs) produce sperm throughout the post-pubertal life of a male. Transgenic loss- and gain-of-function mouse models have shown that their self-renewal and differentiation are controlled in vivo by glial-cell-line-derived neurotrophic factor (GDNF) in a dose-dependent manner. After this in vivo observation, the culture conditions for mouse SSCs were rapidly developed. GDNF together with other growth factors, hormones, and vitamins maintain proliferation and self-renewal of SSCs for years in vitro. Both serum-supplemented and serum-free culture methods have been described. The cells are cultivated either on feeder layer or laminin-coated dishes. First reports from random and targeted mutagenesis of SSCs have been published. Some cells in the spermatogonial stem cell culture transform to embryonic stem cell-like cells and form teratomas in nude mice. In general, the spermatogonial stem cells maintain their germline identity in long-term culture. The mechanism for transformation to embryonic stem cell-like cells is not known, but the data suggest that germline and embryonic stem cells are closely related. We describe in detail the culture system of SSCs developed by Dr. Takashi Shinohara in 2003.

Keywords   GDNF family receptor α1; glial-cell-line-derived neurotrophic factor; mouse spermatogonial stem cells; Ret; spermatogonial transplantation.

From: *Methods in Molecular Biology, Vol. 450, Germline Stem Cells*
Edited by S. X. Hou and S. R. Singh © Humana Press, Totowa, NJ

9.1 Introduction

Spermatogonial stem cells (SSCs) self-renew and produce sperm cells or spermatozoa throughout the postpubertal life of a healthy male. The number of SSCs is
very low in a testis, in the mouse approx. 35,000 *(1)*. Therefore, the ratio of self-
renewal and differentiation must be strictly controlled. We have shown by transgenic loss- and gain-of-function mouse models that glial-cell-line-derived
neurotrophic factor (GDNF) critically controls the fate decision of SSCs *(2)*. This
observation has been of utmost significance to the recent rapid progression in the
characterization, cultivation, and genetic manipulation of SSCs.

GDNF is a distant member of the transforming growth factor-β family, in which,
with three other related molecules, it forms a subfamily called GDNF family ligands (GFLs). They all bind to and activate Ret receptor tyrosine kinase but require
GDNF family receptor α:s (GFRαs) for ligand binding and Ret activation. In neuronal cells, GFLs also use neural cell adhesion receptor as an alternative signal
transducing receptor *(3)*. The downstream targets of GDNF signaling in the SSCs
are poorly characterized, but at least a member of POZ (poxvirus and zincfinger)
family of transcriptional repressors, Bcl6b, is a verified target *(4)*.

SSCs reside in the periphery of the seminiferous tubules. Rodent SSCs form
intercalated chains of cells *(5,6)*. The true stem cell is called A$_s$ spermatogonia,
where s refers to single. It divides to two daughter cells that remain together with
an intercellular bridge. Through synchronized mitotic divisions, the intercalated
chain of SSCs elongates until it consists of 8–16 cells, sometimes even 32 cells,
when the cytoplasmic bridges are broken down. Singularity, characteristic microscopic morphology, location, and relationship with other cells in the seminiferous
tubules are useful characteristics of the SSC. It also displays a number of molecular
markers that have been used for identification and isolation of SSCs (*see* **Table 9.1**;
for comprehensive reviews, *see* **refs. 5–7**).

Table 9.1 Overview of markers used to identify spermatogonial stem cells and their progeny

Cell types	Markers
A$_s$ and A$_{pr}$	GFRα1
A$_s$, A$_{pr}$, and A$_{al}$	Plzf, Oct4, Ngn3,[a] Notch-1, Sox3, c-Ret
A spermatogonia	Rbm
Spermatogonia	EP-CAM
Premeiotic germ cells	Stra8, EE2
Cells on basal membrane and interstitium	CD9
Spermatogonia, spermatocytes, and round spermatids	GCNA1
Premeiotic spermatogonia and postmeiotic spermatids	Taf4B

Source: Modified from **ref. 5**.
[a]It was found by others that NGN3 is also expressed in some spermatocytes.
Abbreviations: GFLa1: GDNF family receptor a1, Plzf: Promyelocytic leuchaemia zinc finger, Oct4:
Octamer 4, Ngn3: Neurogenin 3, Sox3: SRY (sex determining region Y)-box 3, c-Ret: Rearranged
during transformation / GDNF receptor, Rbm: RNA binding motif protein, EP-CAM: Epithelial cellular adhesion molecule, Stra8: Stimulated by retinoic acid gene 8, EE2: EE2 114 kDa antigen,
GCNA1: Germ cell nuclear antigen 1, Taf4B: TATA box binding protein-associated factor 4B

Table 9.2 Markers that have been successfully used to isolate spermatogonial stem cell populations from the testis by either positive or negative selection

Selection method	Selection tool
Proteins with which an enriched population of spermatogonial stem cells is isolated by positive selection	CD9, integrin-α6, integrin-β, integrin-αV, THY-1, CD24
Proteins with which an enriched population of spermatogonial stem cells is isolated by negative selection	c-kit, MHC1, LY6A, CD34
Promoter (regions) used to direct expression of a marker to a subpopulation of germ cells, including spermatogonial stem cells	Stra8, OCT4

Source: From **ref. 5**.
Abbreviations: Thy-1: Thymocyte differentiation antigen 1 / CD90, c-kit: v-kit Hardy-Zuckerman 4 feline sarcoma viral oncogene homolog, MHC1: Major histocompatibility complex class I molecule, LY6A: Lymphocyte antigen 6 complex, locus A, Stra8: Stimulated by retinoic acid gene 8, Oct4: Octamer 4

SSCs have been isolated from testes by various means that are based on either positive or negative selection with markers listed in **Table 9.2**. Currently, the most commonly used antibodies for isolation are to Thymocyte differentiation antigen 1/ CD90 and GDNF family receptor α1 (GFRα1) *(7)*. SSC cultures have been initiated from both prepubertal and adult testes, but the neonatal mice seem to provide the most optimal source as they have the relatively highest ratio of germ cells *(8)*. In some culture systems and apparently stochastically, some SSCs are transformed in vitro to embryonic stem cell-like cells that form teratomas in nude mice *(9,10)*. Guan et al. *(10)* reported a variant culture method that efficiently (in 30% of the cultures) promoted embryonic stem cell-like cells, which contributed to embryogenesis, when transplanted in blastocysts. The basis of the high frequency of transformation of SSCs in this culture setup is currently unknown, but in other culture systems the SSCs are obviously much more resistant to transformation and epigenetic modifications during cultivation than other stem cells *(11)*. On the other hand, SSCs can be genetically modified in vitro, and after injection to infertile mouse testes, SSCs provide an alternative for embryonic stem cells in production of gene-targeted mice. The first reports from genetically modified mice by retroviral transfection and random or targeted insertion have been published *(12,13)*. According to these studies, the rate of transgene insertion and homologous recombination in cultured SSCs is at least as high as in embryonic stem cell lines.

If and when the cultivation of human SSCs becomes available, it will provide a tool to restore male fertility, for instance, after cancer therapies, which very often disrupt spermatogenesis. The SSCs may also serve as an ethically acceptable cellular resource for cell replacement therapies, at least if such culture conditions are developed that allow the cells to transform to embryonic stem cell-like cells. The particular advantage in the use of SSCs for cell therapies is that the cells can be taken from the patient, thereby evoking no immunological rejection. Furthermore, as compared to somatic stem cells, the growth potential of SSCs is much higher *(8,14)*. However, SSC cultures contain a heterogeneous group of cells as the stem cells also undergo partial differentiation in vitro. It has been estimated that only approx. 1% of cultured cells represent stem cells *(11)*.

Several methodological variants have been described for culture of SSCs. Basically, all currently used methods are based on supplementation of the culture medium with GDNF; other growth factors, such as fibroblast growth factor 2 (FGF2) and leukemia inhibitory factor (LIF); vitamins; and hormones (reviewed in **ref. 7**). Both serum and serum-free methods have been developed *(15–18)*. In some protocols, the cells are grown on feeder cells, either irradiated or mitomycin C-treated mouse embryonic fibroblasts or STO cells *(17,19)*; in others, the cells grow on laminin-coated dishes *(17)*. Even a suspension culture system has been reported *(20)*.

The ultimate technique used for verification of the stem cell nature of the in vitro cultivated SSCs is their transplantation to testes of infertile mice. The first transplantation method for mouse SSCs was published in 1994 *(21,22)*. The recipient mice are made infertile by irradiation or busulfan treatment, or they lack spermatogenesis because of genetic mutations (reviewed in **ref. 7**). The SSCs are injected directly in the seminiferous tubules or to rete testes, from where they colonize the seminiferous tubules and reinitiate sperm production. Also, xenotransplantations have been made. For instance, rat SSCs transplanted to mouse testes differentiate to sperm *(23)* that, after artificial insemination in rat oocytes, produce healthy rat progeny *(24)*.

We describe in detail the SSC culture technique originally developed by Dr. Takashi Shinohara and his colleagues at Kyoto University. Although there are several variants for the SSC culture conditions, this method allows efficient expansion of the germ stem cells, and this culture condition maintains the SSC characteristics, genetic stability, and growth potential even after 2 yr in culture. All studies in which this method is used should refer to the original article by Dr. Shinohara and his colleagues *(8)*.

9.2 Materials

9.2.1 SSC Isolation

1. SSC base: StemPro-34 SFM (Invitrogen 10639), 6 mg/mL D-(+)-glucose (Sigma G7021), 30 µg/mL pyruvic acid (Sigma P2256), 1 µL/mL DL-lactic acid (Sigma L4263), 5 µg/mL bovine serum albumin (BSA; ICN Biomedicals 810661), 2 mM L-glutamine, 50 mM 2-mercaptoethanol (Sigma M3148), 1X modified Eagle's medium (MEM) vitamin solution (Invitrogen 11120-052), 1X MEM nonessential amino acids solution (Invitrogen 11140-050), 100 µM ascorbic acid (Sigma A4544), 10 µg/mL d-biotin (Sigma B4501), 25 µg/mL insulin (Nakalaitesque 19251-24), 100 µg/mL transferrin (Sigma T1147), 60 µM

putrescine (Sigma P7505), 30 n*M* sodium selenite (Sigma S1382), 30 ng/mL
b-estradiol (Sigma E2758), 60 ng/mL progesterone (Sigma P8783), and 1% fetal
calf serum (FCS; JRH Biosciences JRH12003-78P). The SSC base can be used
at least for 3 wk after preparation. Transferrin, insulin, progesterone, estradiol,
sodium selenite, and putrescine should be prepared as aliquots and stored at
−20°C until use.

2. SSC medium: SSC base supplemented with StemPro Supplement (Invitrogen
10639), 20 ng/mL mouse epidermal growth factor (EGF; Becton Dickinson
40010), 10 ng/mL human basic fibroblast growth factor (bFGF; Becton
Dickinson 13256-029), 1000 U/mL ESGRO (murine leukemia inhibitory factor,
Chemicon ESG1107), 10 ng/mL recombinant rat GDNF (R&D Systems
512-GF).

3. The growth factors (EGF, bFGF, ESGRO [LIF], GDNF) and StemPro Supplement
should also be prepared as aliquots and stored at −20°C (*see* **Note 1**). The SSC
medium should always be prepared fresh from the SSC base by adding the supplements just before use.

4. Collagenase solution: 1 mg/mL collagenase (Sigma C5138) in Hank's balanced
salt solution (HBSS). Filtrate before use.

5. DNase (deoxyribonuclease) solution: 7 mg/mL DNase (Sigma DN25) in HBSS.
Filtrate before use.

9.2.2 Initiation of SSC Culture

1. 0.2% gelatin: 0.2% gelatin in phosphate-buffered saline (PBS), autoclaved.
Store at room temperature.

2. Trypsin stop solution: 5 mg/mL BSA in Iscove's modified Dulbecco's medium.

9.2.3 SSC Culture on Mouse Embryonic Fibroblast Feeders

1. Mouse embryonic fibroblast (MEF) growth medium (MEFGM): High-glucose
Dulbecco's modified Eagle's medium (DMEM), 10% FCS.

2. Mitomycin C stock solution: 0.4 mg/mL mitomycin C in sterile PBS. Be sure to
familiarize yourself with the Material Safety Data Sheet (MSDS) of mitomycin
C before opening the manufacturer's vial. To dissolve mitomycin C, add sterile
PBS directly to the original vial. Store at 4°C in the dark for a maximum of
2 wk.

3. 0.1% gelatin: 0.1% gelatin in sterile water.

4. Trypsin–EDTA (ethylenediaminetetraacetic acid): 0.05% trypsin, 0.53 m*M* EDTA.

9.3   Methods

9.3.1   SCC Isolation

1. Make gelatin-coated culture plate: 12-well tissue culture dishes should be coated with 0.2% gelatin for longer than 30 min at room temperature by adding enough gelatin solution to cover the bottom. After incubation, remove excess and let stand to dry.
2. Isolate testes from mice: Use P0–3 testes of DBA2 or ICR strains (*see* **Note 2**). Remove tunica albuginea with fine forceps in cold HBSS.
3. Wash two or three times with HBSS, transfer into 1–2 mL collagenase, and incubate at 37°C for 15 min (*see* **Note 3**).
4. Wash twice with HBSS and add 0.8 mL 0.25% trypsin plus 0.2 mL DNase solution; incubate at 37°C for 10 min.
5. Add 5 mL trypsin stop solution and repeat pipeting until the cells are dissociated (*see* **Note 2**).
6. Collect cells by centrifugation.

9.3.2   Negative Selection

1. Suspend the cells in fresh SSC medium and transfer into gelatin-coated culture plate at a density of 2×10^{-5} cells/well in 0.8 mL medium. Incubate overnight at 37°C.
2. Many cells attach to the plate after overnight incubation, but a significant number of germ cells, distinguished by their large size and characteristic pseudopod, remain floating. Passage the floating cells to a secondary culture plate after vigorous pipeting (use P1000 Pipetman, 10–15 times). The secondary culture plate must not be treated with gelatin. Very few germ cells are left on the original gelatin-coated plate, and cells transferred to secondary plates are relatively germ cell enriched.
3. Within 1 wk, the transferred cells proliferate and spread on the bottom of the well; round proliferating cells form colonies on top of the flat cell layer. Most of these primary colonies consist of a compact cluster of cells with unclear borders.

9.3.3   Passaging

Timing of the first passage is dependent on the growth of colonies. However, between 10 and 14 d after the initiation of culture (DIV; days in vitro) are recommended.

1. Wash the cells twice with PBS, add 0.25% trypsin, and incubate at 37°C for 4 min.
2. Add trypsin stop solution.

3. Replate at 1X dilution.
4. Colonies grow to the original size in about 10 d, and cells are again passaged (× 1/2 dilution).

9.3.4   Preparation of MEF Feeders (see Note 4)

1. Thaw one vial of cryopreserved primary MEFs, containing cells from one confluent 150-mm culture dish in MEF growth medium, 10% dimethyl sulfoxide (DMSO). Transfer the cells to 5 mL prewarmed (37°C) MEFGM.
2. Collect the cells by centrifugation, resuspend, and plate on a 150-mm culture dish in 25 mL MEFGM.
3. Let grow until the plate is confluent (this takes ~2–3 d).
4. Treat the cells with mitomycin C: Aspirate the medium and replace with fresh MEFGM, 10 µg/mL mitomycin C. Incubate for 2.5 h at 37°C.
5. During mitomycin C treatment, prepare gelatinated six-well culture dishes for MEFs. Cover the bottom with 0.1% gelatin and incubate for 15 min at room temperature. Remove excess and keep dry until needed (see Note 5).
6. After mitomycin C incubation, remove medium from MEFs, rinse three times with 10 mL PBS, add 10 mL trypsin–EDTA, and incubate at 37°C for 5 min.
7. Add 3 mL MEFGM and disperse the cells by pipeting up and down.
8. Collect the cells by centrifugation, resuspend the cell pellet in 10 mL MEFGM, and count the cells with a hemocytometer.
9. Plate the mitomycin C-treated MEFs onto gelatin-coated six-well plates (5×10^{-5} cells in 2 mL MEFGM/well).

9.3.5   Maintenance of SSCs on Feeder Culture

1. From the third passage, the cells should be transferred on mitomycin C-treated MEFs. Around 30 DIV, the growth of SSCs becomes stable. The established SSCs should be plated at a density of 3×10^{-5} cells/well in six-well culture plate (see Note 6).
2. Medium should be changed every 3 d (half medium change).
3. Culture should be passaged at a frequency of every 4–6 d depending on the proliferation. Established germ stem cells can continue to proliferate for more than 1 yr without losing stem cell activity.

9.4   Notes

1. Growth factors should be divided into small enough aliquots; repeated freeze–thaw cycles must be avoided. The stability of GDNF has been found to be limited at both −20°C and −80 °C even without intermittent thawing. The glycosylation state seems to affect the stability; the recombinant

rat GDNF recommended in the protocol is produced in insect cells and is likely to be more stable than recombinant GDNFs produced in *Escherichia coli* or yeast.

2. Use of neonatal/perinatal mice in isolation is recommended as the proportion of SSCs is higher than in older animals. There also seems to be a difference in the in vitro culture requirements of SCCs from different mouse strains *(16,17)*. SSCs especially from C57BL/6 and 129/Sv strains may be difficult to propagate with this protocol.

3. Gentle agitation can be applied during incubations, but the incubation times should not be exceeded. The disruption of tubules during collagenase IV treatment should be closely followed.

4. A thorough description of MEF isolation can be found in **ref. 25**, for instance. However, as MEFs can be cryopreserved analogously to cell lines, stocks can usually be obtained from laboratories that are routinely using MEF feeders.

5. The feeder plates can be used up to 1 wk after plating. However, use of feeders after an overnight incubation is highly recommended.

6. It is important to spread SSC cells uniformly on the MEF feeder layer. Otherwise, the cells will aggregate and differentiate. The cells that are starting to differentiate will detach from the feeder layer. In this situation, remove the medium and replace with fresh medium. In most cases, differentiating cells are removed, and the stem cells revive with this treatment.

Acknowledgment Special thanks are given to Dr. Takashi Shinohara, who was kindly ready to share his invaluable hands-on knowledge and the protocol details of this method.

References

1. Tegelenbosch, R. A. J., and de Rooij, D. G. (1993) A quantitative study of spermatogonial multiplication and stem cell renewal in the C3H/101 F1 hybrid mouse. *Mut. Res.* **290**, 193–200.

2. Meng, X., Lindahl, M., Hyvonen, M. E., et al. (2000) Regulation of cell fate decision of undifferentiated spermatogonia by GDNF. *Science.* **287**, 1489–1493.

3. Sariola, H., and Saarma, M. (2003) Novel functions and signalling pathways for GDNF. *J. Cell Sci.* **116**, 3855–3862.

4. Oatley, J.M., Avarbock, M.R., Telaranta, A.I., Fearon, D.T., and Brinster, R.L. (2006) Identifying genes important for spermatogonial stem cell self-renewal and survival. *Proc. Natl. Acad. Sci. U. S. A.* **103**, 9524–9529.

5. de Rooij, D. G., and Russell, L. D. (2000) All you wanted to know about spermatogonia but were afraid to ask. *J. Androl.* **21**, 776–798.

6. de Rooij, D. G. (2006) Rapid expansion of the spermatogonial stem cell toolbox. *Proc. Natl. Acad. Sci. U. S. A.* **103**, 7939–7940.

7. Aponte, P. M., van Bragt, M. P. A, de Rooij, D. G., and van Pelt, A. M. M. (2005) Spermatogonial stem cells: characteristics and experimental possibilities. *APMIS.* **113**, 727–742.

8. Kanatsu-Shinohara, M., Ogonuki, N., Inoue, K., et al. (2003) Long-term proliferation in culture and germline transmission of mouse male germline stem cells. *Biol. Reprod.* **69**, 612–616.

9. Kanatsu-Shinohara, M., Inoue, K., Lee, J., et al. (2004) Generation of pluripotent stem cells from neonatal mouse testis. *Cell.* **119**, 1001–1012.

10. Guan, K., Nayernia, K., Maier, L. S., et al. (2006) Pluripotency of spermatogonial stem cells from adult mouse testis. *Nature.* **440**, 1199–1203.

11. Kanatsu-Shinohara, M., Ogonuki, N., Iwano, T., et al. (2005) Genetic and epigenetic properties of mouse male germline stem cells during long-term culture. *Development.* **132**, 4155–4163.

12. Nagano, M., Brinster, C. J., Orwig, K. E., Ryu, B. Y., Avarbock, M. R., and Brinster, R. L. (2001) Transgenic ice produced by retroviral transduction of male germ-line stem cells. *Proc. Natl. Acad. Sci. U. S. A.* **98,** 13090–13095.
13. Kanatsu-Shinohara, M., Ikawa, M., Takehashi, M., et al. (2006) Production of knockout mice by random or targeted mutagenesis in spermatogonial stem cells. *Proc. Natl. Acad. Sci. U. S. A.* **103,** 8018–8023.
14. Jeong, D., McLean, D. J., and Griswold, M. D. (2003) Long-term culture and transplantation of murine testicular germ cells. *J. Androl.* **24,** 661–669.
15. Kubota, H., Avarbock, M. R., and Brinster, R. L. (2004) Culture conditions and single growth factors affect fate determination of mouse spermatogonial stem cells. *Biol. Reprod.* **71,** 722–731.
16. Kubota, H., Avarbock, M. R., and Brinster, R.L. (2004) Growth factors essential for self-renewal and expansion of mouse spermatogonial stem cells. *Proc. Natl. Acad. Sci. U. S. A.* **101,** 16489–16494.
17. Kanatsu-Shinohara, M., Miki, H., Inoue, K., et al. (2005) Long-term culture of mouse male germline stem cells under serum-or feeder-free conditions. *Biol. Reprod.* **72,** 985–991.
18. Hofmann M. C., Braydich-Stolle L., and Dym, M. (2005) Isolation of male germ-line stem cells; influence of GDNF. *Dev. Biol.* **279,** 114–124.
19. Creemers, L. B., den Ouden, K., van Pelt, A. M. M., and de Rooij, D. G. (2002) Maintenance of adult mouse type A spermatogonia in vitro: influence of serum and growth factors and comparison with prepubertal spermatogonial cell culture. *Reproduction.* **124,** 791–799.
20. Kanatsu-Shinohara, M., Inoue, K., Lee, J., et al. (2006) Anchorage-independent growth of mouse male germline stem cells in vitro. *Biol. Reprod.* **74,** 522–529.
21. Brinster, R. L., and Avarbock, M. R. (1994) Germline transmission of donor haplotype following spermatogonial transplantation. *Proc. Natl. Acad. Sci. U. S. A.* **91,** 11303–11307.
22. Brinster, R. L., and Zimmermann. J. W. (1994) Spermatogenesis following male germ-cell transplantation. *Proc. Natl. Acad. Sci. U. S. A.* **91,** 11298–11302.
23. Clouthier, D. E., Avarbock, M. R., Maika, S. D., Hammer, R. E., and Brinster, R. L. (1996) Rat spermatogenesis in mouse testis. *Nature.* **381,** 418–421.
24. Shinohara, T., Kato, M., Takehashi, M., et al. (2006) Rats produced by interspecies spermatogonial transplantation in mice and in vitro microinsemination. *Proc. Natl. Acad. Sci. U. S. A.* **103,** 13624–13628.
25. Spector, D. L., Goldman, R. D., and Leinwand, L. A. (eds.). (1998) *Cells: a laboratory manual,* Cold Spring Harbor Laboratory Press, Cold Spring Harbor, New York.

Part II
In Vitro Culture and Applications of Germline Stem Cells

Chapter 10
Ectopic Grafting of Mammalian Testis Tissue into Mouse Hosts

Ina Dobrinski and Rahul Rathi

Contents

Summary Mammalian spermatogenesis is a highly organized process of cell division and differentiation that requires intimate contact between germ cells and testicular somatic cells. Lack of a suitable in vitro system has caused many aspects of spermatogenesis, especially in nonrodent species, to remain elusive. We describe ectopic grafting of testis tissue from sexually immature males to immunodeficient mouse hosts as an in vivo culture system that allows recapitulation of complete spermatogenesis from diverse mammalian species with the production of fertilization-competent sperm in a mouse host. In this system, the donor species testicular environment is preserved allowing experimentation in a small rodent. The accessibility of the tissue in the mouse host makes it possible to manipulate spermatogenesis and steroidogenesis in a controlled manner that is often not feasible in the donor species. It also allows detailed analysis of the effects of toxins and compounds to enhance or suppress male fertility in an in vivo system without extensive experimentation in the target species. Finally, as it provides a source of male gametes even from immature gonads, grafting of fresh or preserved testis tissue offers an invaluable tool for the conservation of fertility in males if sperm cannot be obtained for cryopreservation.

Keywords Graft; mouse; spermatogenesis; testis.

From: *Methods in Molecular Biology, Vol. 450, Germline Stem Cells*
Edited by S. X. Hou and S. R. Singh © Humana Press, Totowa, NJ

10.1   Introduction

Spermatogenesis is a complex process that so far cannot be replicated in vitro. Transplantation of isolated male germ cells to the testis of a recipient animal is one approach to study and manipulate spermatogenesis *(1)*. However, this technique cannot be easily adapted between diverse species, and cross-species germ cell transplantation did not result in complete sperm production in species other than rodents *(2–5)*, likely because of an incompatibility between donor germ cells and recipient testicular environment.

Cotransplantation of the donor germ cells with their surrounding testicular tissue into a mouse host preserves the testicular environment and still allows experimentation in a small rodent. We therefore developed ectopic grafting of small pieces of testicular tissue under the back skin of immunodeficient mice as an alternate approach for the maintenance and propagation of male germ cells that can be more readily applied to different mammalian species *(6)*. Xenografting of testis tissue from newborn pigs and goats resulted, for the first time, in functional cross-species spermatogenesis from species other than rodents in a mouse host, and sperm could be obtained from neonatal donors. To date, grafting of testis tissue from sexually immature males to immunodeficient mice has resulted in germ cell differentiation and production of sperm from a variety of different mammalian species (*see* **Table 10.1**).

We also established that adult human testis tissue can survive when grafted into host mice. Although germ cell loss was noted in tissue with active spermatogenesis at the time of grafting, germ cells survived in tissue from patients with quiescent germinal epithelium *(7,8)*. Sperm recovered from allografts (mouse to mouse) and xenografts (monkey to mouse) supported embryo development when injected into oocytes *(9,10)*, and following embryo transfer, mouse sperm from allografts sired normal, fertile progeny *(9)*.

The onset of spermatogenesis in xenografted pig testis tissue occurred slightly earlier than in the donor species *(6)*, and testicular maturation and sperm production

Table 10.1   Testis tissue grafting: summary of results

Donor	Spermatogenesis	Comments	Reference
Mouse (allograft)	Complete	Offspring	*(9)*
Hamster	Complete		*(18)*
Rabbit (not ectopic)	Complete	Offspring	*(19)*
Pig	Complete	Embryos	*(6)*
Goat	Complete	Sperm recovery	*(6, 11)*
Cattle	Complete	Inefficient	*(13, 15)*
Cat	Complete	Sperm recovery	*(12)*
Horse	Complete	Inefficient	*(16)*
Rhesus monkey	Complete	Embryos	*(10)*
Banteng	Meiotic cells		*(20)*
Human (adult)	Germ cell survival		*(7, 8)*

in rhesus macaque testis tissue was significantly accelerated by exposure to the endocrine environment of the castrated adult mouse host *(10)*. This shortened time to sperm production is caused by accelerated maturation of the testicular somatic cells, whereas the length of the spermatogenic cycle remains unchanged *(11)*. In contrast, time to sperm production was comparable to that in the donor species or even slightly longer in xenografts from immature cats *(12)*, bulls *(13–15)*, and horses *(16)*. In these species, spermatogenic efficiency was low, with elongated spermatids present in only 5–15% of tubules of testis xenografts.

We also demonstrated that the global gene expression profile is very similar between testis tissue xenografts and testis tissue in situ *(17)*, further validating testis tissue xenografts as a representative model for mammalian spermatogenesis in the donor species.

10.2   Materials

10.2.1   Collection of Donor Testis Tissue

1. Disposable scalpel (no. 10 or higher).
2. Phosphate-buffered saline (PBS).
3. Ice bucket.

10.2.2   Preparation of Donor Testis Tissue

1. Phosphate-buffered saline.
2. Dulbecco's modified Eagle's medium (DMEM) (or other balanced cell culture medium).
3. Plastic tissue culture dishes (60 × 15 mm and 100 × 20 mm).
4. One pair iris forceps (~10 cm, 0.8-mm tips).
5. One pair small forceps (~11 cm, 0.1 × 0.6-mm tips).
6. One pair of curved iris forceps (~10 cm, 0.8-mm tips, curved).
7. One pair of dissecting scissors (~10 cm, straight).
8. Disposable scalpel (no. 10).

10.2.3   Grafting Procedure

For the grafting procedure, the recipient mice are immunodeficient (e.g., NCR nu/nu or SCID) mice, 6–8 wk old, male. Female mice can also be used, but castration is easier in males.

10.2.3.1 Anesthesia of Host Mouse

1. Reagents: β-Tribromoethyl alcohol, tertiary amyl alcohol.
2. Stock solution: 10 g tribromoethanol in 10 mL T-amyl alcohol. Dissolve the crystals by heating the tube in hot tap water. It is important that all of the crystals are dissolved. Store the solution in a foil-wrapped container (tribromoethanol is light sensitive) in the refrigerator.
3. Tuberculin syringes (1 mL).
4. Injection needles (26.5 gage).

10.2.3.2 Surgical Preparation

1. 70% ethanol.
2. Betadine solution.
3. Two pairs of iris forceps (~10 cm, 0.8-mm tips).
4. One pair of small forceps (~11 cm, 0.1 × 0.6-mm tips).
5. One pair of dissecting scissors (~10 cm, straight).
6. One needle holder (~12 cm).
7. Suture material with needle (6-0 silk braid with needle).
8. Wound clips (7.5 mm, e.g., Michel® clips).
9. Clip applying–removing forceps (12.7 cm, e.g., Michel).
10. Heat pad.

10.2.4 Graft Analysis

For graft analysis, use Bouin's solution (or any other fixative, depending on analysis method).

10.3 Methods

Testis tissue xenografting is a simple procedure in which tissue is obtained from the testis of a donor male, cut into small pieces, and inserted under the back skin of an immunodeficient, castrated mouse host, where it will undergo growth and differentiation (see **Fig. 10.1**). Depending on the hypothesis tested, treatments can be applied to the mouse host to affect aspects of donor spermatogenesis. Grafted tissue is then recovered from the mouse for analysis or harvesting of sperm.

10.3.1 Collection of Donor Testis Tissue

1. Obtain testis tissue by castration or biopsy from a donor male.
2. Maintain sterile conditions if at all possible.

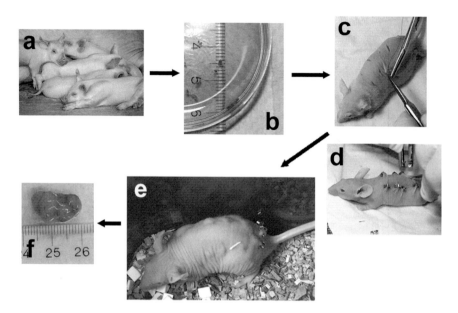

Fig. 10.1 Testis tissue xenografting illustrated for porcine testis tissue. **a** donor piglets; **b** testis tissue fragments prepared for grafting; **c** skin incisions have been made, and graft tissue is secured with suture material; **d** incisions are closed with wound clips; **e** grafted tissue has grown and is visible under the skin. Most wound clips have fallen off as the incisions healed; **f** tissue graft removed from mouse host. Note dramatic increase in size compared to **b**

3. If an entire testis is removed, take care not to cut into the tunica albuginea (*see* **Note 1**).
4. Immediately place the testis or biopsy into cold PBS on ice (*see* **Note 2**).

10.3.2 Preparation of Donor Testis Tissue

1. Wash the intact testis in PBS two to three times before transferring into a culture dish with PBS.
2. Remove any extra tissue (spermatic cord, epididymis, connective tissue), wash the testis again in cold PBS, and transfer it into a culture dish with DMEM.
3. Carefully remove the tunica surrounding the testis using a scalpel blade and a pair of scissors. If the testis is very small, the tunica can be removed by squeezing the testicular tissue out of the tunica with a small incision made on one end while holding the tunica with a pair of small forceps on the other end.
4. Discard the tunica.
5. Depending on the size of the testis, either the whole testis tissue can be cut into small pieces around 2 mm^3 using curved forceps and a scalpel blade or large

pieces of testis tissue can first be removed from the testis and then cut into smaller pieces. All this should be done in DMEM and under laminar airflow in a small culture dish (60 × 15 mm).

6. Keep the prepared tissue fragments in DMEM in small culture dishes on ice until grafting.

10.3.3   Grafting Procedure

10.3.3.1   Anesthesia of Host Mouse

1. Any suitable anesthetic can be used in accordance with animal care and use guidelines. In this chapter, the use of avertin is described (*see* **Note 3**).
2. Prepare a working solution: Dilute stock avertin 1:50 with warm distilled water or PBS (e.g., 200 µL stock in 10 mL water or PBS). If any crystals form during the dilution, discard and start again with warmer water. This solution can be used for 1 wk.
3. Dosage: Use about 0.6 mL per average size (~25 g) mouse. Adjust based on weight and on the strain. Inject intraperitoneally only.

10.3.3.2   Surgical Preparation

1. Keep the mouse warm during surgery (*see* **Note 4**).
2. Use immunodeficient mice to avoid tissue rejection by the mouse host.
3. Weigh the mouse and anesthetize with the appropriate dose (described in **Subheading 10.3.3.1**). Monitor anesthetic depth by absence of voluntary and reflex movement.
4. Position the mouse in dorsal recumbency for castration (described in **Subheading 10.3.3.4**).
5. Prepare sterile surgical field by clipping or plucking the hair (not necessary in nude mice) and wiping with 70% ethanol and Betadine solution (*see* **Note 5**).

10.3.3.3   Castration of Host Mouse

In most applications, the male host is castrated to provide high levels of follicle-stimulating hormone (FSH) and luteinizing hormone (LH) that will support initial growth of the grafted tissue.

1. Make a small, ventral midline skin incision to expose the peritoneum (*see* **Note 6**).
2. Lifting the peritoneum with a pair of small forceps to avoid accidentally injuring abdominal organs, make a small incision—about 0.5 cm long—in the peritoneum, exposing the peritoneal cavity.

3. Using one iris forceps to hold the peritoneum in an open position, use another pair of iris forceps to search for the fat pads attached to the epididymis and testis in the peritoneal cavity (*see* **Note 7**).
4. Gently pull the fat pad from one side out until the testis is on the exterior and the cauda epididymis is clearly visible.
5. Ligate the spermatic cord with a silk suture and remove the testis.
6. Replace the remaining spermatic cord into the abdomen.
7. Repeat the procedure for the second testis.
8. Suture the peritoneum and close the skin with suture clips.

10.3.3.4 Grafting

1. Position the mouse in ventral recumbency and prepare a sterile surgical field on its back as in **Subheading 10.3.3.2**.
2. Depending on how many grafts are to be inserted (generally four to eight per mouse) make 3-mm skin incisions on each side of the back of the castrated mouse.
3. Using the scissors, carefully open the incision and slowly make a pocket underneath the skin.
4. Holding the incised skin with one iris forceps, position a small piece of tissue deep into the pocket with another iris forceps (*see* **Note 8**). Close the wound with suture clips.

10.3.3.5 Postsurgical Care

1. Keep the mouse on a heating pad until it starts to recover from anesthesia.
2. Move the mouse to a cage with paper towels to provide additional insulation and cover. Place the cage on heated shelves until the mouse is fully recovered (*see* **Notes 9–11**).

10.3.4 Graft Removal

1. Sacrifice the host mouse according to animal care and use guidelines (*see* **Notes 12** and **13**).
2. Make a midline skin incision running from the tail to the neck and open the back skin. This exposes the grafts, which can be located either on the subcutaneous tissue or under the skin.
3. Carefully remove the grafts using forceps and a pair of scissors.
4. Record number of grafts recovered and size and weight of individual grafts.
5. Retrieve the seminal vesicles from the abdomen of the mouse and record their weight as an indication of testosterone production by the grafted tissue (*see* **Note 14**).

10.3.5   Graft Analysis

1. For recovery of sperm, gently mince graft tissue in warm media to release sperm.
2. Sperm release can be monitored under a dissecting microscope.
3. Remove debris from suspension by slow centrifugation (50g, 5 min); discard pellet. Sperm will be in the supernatant (*see* **Note 15**).
4. If not used immediately, sperm can be frozen in freezing extenders appropriate for the species.
5. For histology or immunocytochemical analysis, fix graft tissue. Choice of fixative depends on intended use (*see* **Note 16**).
6. For histological analysis of the tissue, best results are obtained by fixation in Bouin's solution overnight followed by washing at least three times in 70% ethanol, preferably at 24-h intervals.
7. For immunocytochemistry, fix tissue in 2–4% paraformaldehyde.

10.4   Notes

1. To avoid contamination and excessive tissue damage, it is best if a whole donor testis can be obtained. If a tissue fragment has to be used, wash extensively in DMEM containing antibiotics before processing for grafting.
2. If testes were obtained at a location different from where grafting will be performed, tissue can be shipped in PBS on wet ice. Tissue viability and graft survival comparable to fresh tissue can be expected for tissue stored up to 48 h at 4°C.
3. Avertin has a broad therapeutic width and is metabolized quickly. Therefore, additional drug can be administered if necessary during surgery.
4. A conventional small heating pad sterilized with 70% ethanol and covered with sterile paper towels works well. The area should be well lighted.
5. Take care not to soak the mouse in ethanol as this can easily lead to hypothermia caused by evaporative cooling.
6. The testicular fat pads are very obvious and can easily be located caudolaterally from the incision in the area of the inguinal canal.
7. Castration can also be performed through scrotal incisions.
8. Grafts can be secured in place with a small piece of suture material if desired.
9. Nude mice will easily get cold in mouse rooms/cages with high airflow. Provide additional bedding and shelter or keep cages on heated racks.
10. Care should be taken that mice from different lots are not placed in the same cage to avoid fighting.
11. Monitor mice for evidence of fighting. As grafts produce testosterone, recipient mice will show male-specific behavior.
12. Individual grafts can be removed surgically by anesthetizing the mouse and removing the graft through a skin incision. However, this may affect the endocrine balance, similar to the situation in hemicastration.
13. If indicated by the experiment, the mouse can be anesthetized prior to sacrifice, and blood can be collected for hormone measurements.
14. The seminal vesicles are androgen-dependent accessory sex glands. Their weight is an indicator of the levels of bioactive testosterone in the host mouse. In intact male mice, seminal vesicles weigh 100–300 mg; in castrated male mice, the weight is less than 10 mg.

15. If only very few sperm are present in the tissue, individual sperm for Intracytoplasmic sperm injection ICSI can be collected from the cell suspension using micromanipulators.
16. If the grafts are big (weighing over 100 mg), they should be cut into small pieces to ensure sufficient penetration of the fixative.

References

1. Brinster, R.L. (2002) Germline stem cell transplantation and transgenesis. *Science.* **296,** 2174–2176.
2. Dobrinski, I., Avarbock, M. R., and Brinster, R. L. (1999) Transplantation of germ cells from rabbits and dogs into mouse testes. *Biol. Reprod.* **61,** 1331–1339.
3. Dobrinski, I., Avarbock, M. R., and Brinster, R. L. (2000) Germ cell transplantation from large domestic animals into mouse testes. *Mol. Rep. Dev.* **57,** 270–279.
4. Nagano, M., McCarrey, J. R., and Brinster, R. L. (2001) Primate spermatogonial stem cells colonize mouse testes. *Biol. Reprod.* **64,** 1409–1416.
5. Nagano, M., Patrizio, P., and Brinster, R. L. (2002) Long-term survival of human spermatogonial stem cells in mouse testes. *Fertil. Steril.* **78,** 1225–1233.
6. Honaramooz, A., Snedaker, A., Boiani, M., Scholer, H. R., Dobrinski, I., and Schlatt, S. (2002) Sperm from neonatal mammalian testes grafted in mice. *Nature.* **418,** 778–781.
7. Schlatt, S., Honaramooz, A., Ehmcke, J., et al. (2006) Limited survival of adult human testicular tissue as ectopic xenograft. *Hum. Reprod.* **21,** 384–389.
8. Geens, M., De Block, G., Goossens, E., Frederickx, V., Van Steirteghem, A., and Tournaye, H. (2006) Spermatogonial survival after grafting human testicular tissue to immunodeficient mice. *Hum. Reprod.* **21,** 390–396.
9. Schlatt, S., Honaramooz, A., Boiani, M., Scholer, H. R., and Dobrinski, I. (2003) Progeny from sperm obtained after ectopic grafting of neonatal mouse testes. *Biol. Reprod.* **68,** 2331–2335.
10. Honaramooz, A., Li, M.-W., Penedo, M. C. T., Meyers, S. A., and Dobrinski, I. (2004) Accelerated maturation of primate testis by xenografting into mice. *Biol. Reprod.* **70,** 1500–1503.
11. Zeng, W., Avelar, G. F., Rathi, R., Franca, L. R., and Dobrinski, I. (2006) The length of the spermatogenic cycle is conserved in porcine and ovine testis xenografts. *J. Androl.* **27,** 527–533.
12. Snedaker, A. K., Honaramooz, A., and Dobrinski, I. (2004) A game of cat and mouse: xenografting of testis tissue from domestic kittens results in complete cat spermatogenesis in a mouse host. *J. Androl.* **25,** 926–930.
13. Oatley, J.M., de Avila, D. M., Reeves, J. J., and McLean, D. J. (2004) Spermatogenesis and germ cell transgene expression in xenografted bovine testicular tissue. *Biol. Reprod.* **71,** 494–501.
14. Oatley, J. M., Reeves, J. J., and McLean, D. J. (2005) Establishment of spermatogenesis in neonatal bovine testicular tissue following ectopic xenografting varies with donor age. *Biol. Reprod.* **72,** 358–364.
15. Rathi, R., Honaramooz, A., Zeng, W., Schlatt, S., and Dobrinski, I. (2005) Germ cell fate and seminiferous tubule development in bovine testis xenografts. *Reproduction.* **130,** 923–929.
16. Rathi, R., Honaramooz, A., Zeng, W., Turner, R., and Dobrinski, I. (2006). Germ cell development in equine testis tissue xenografted into mice. *Reproduction.* **131,** 1091–1098.
17. Zeng, W., Rathi, R., PAN, D., and Dobrinski, I. (2007) Comparison of global gene expression between porcine testis tissue xenografts and porcine testis in situ. *Mol. Reprod. Dev.* **74,** 674–679.
18. Schlatt, S., Kim, S. S., and Gosden, R. (2002) Spermatogenesis and steroidogenesis in mouse, hamster and monkey testicular tissue after cryopreservation and heterotopic grafting to castrated hosts. *Reproduction.* **124,** 339–346.

19. Shinohara, T., Inoue, K., Ogonuki, N., et al. (2002) Birth of offspring following transplantation of cryopreserved immature testicular pieces and in-vitro microinsemination. *Hum. Reprod.* **17,** 3039–3045.
20. Honaramooz, A., Zeng, W., Rathi, R., Koster, J., Ryder, O., and Dobrinski, I. (2005). Testis tissue xenografting to preserve germ cells from a cloned banteng calf. *Reprod. Fertil. Dev.* **17,** 247.

Chapter 11
Spermatogonial Stem Cell Transplantation, Testicular Function, and Restoration of Male Fertility in Mice

Derek J. McLean

Contents

Summary Mammalian spermatogonial stem cells, sometimes called male germline stem cells, are a small population of adult tissue-specific stem cells present in the testis. Formation of the spermatogonial stem cell population early in life and differentiation of spermatogonial stem cells in adults are responsible for continual production of sperm in the testis. Unfortunately, there are no specific biochemical or morphological markers for spermatogonial stem cells, so investigation of this cell type requires specific and consistent approaches to ensure valid data are obtained. Currently, the only assay for the presence of spermatogonial stem cells in a cell suspension is the spermatogonial stem cell transplantation technique. This requires the transfer of cells from a donor animal into the testis of a recipient animal, in which the spermatogonial stem cells will colonize and initiate donor-derived spermatogenesis. Although there is no specific marker for spermatogonial stem cells, several cell surface markers have been used to enrich for these cells prior to transplantation. Thus, selection and transplantation of spermatogonial stem cells can be used to investigate basic mechanisms regulating them. Successful transplantation and donor-derived spermatogenesis in recipient animals can lead to the restoration of fertility in infertile males. In combination with spermatogonial stem cell culture, this transplantation technique can also be used for the purpose of generating transgenic animals through the male germline. This chapter describes the methods for spermatogonial stem cell transplantation and how this approach is used to investigate testicular function.

Keywords Germline stem cells; spermatogonia; spermatogonial stem cells; spermatogenesis; spermatozoa; stem cell transplantation.

From: *Methods in Molecular Biology, Vol. 450, Germline Stem Cells*
Edited by S. X. Hou and S. R. Singh © Humana Press, Totowa, NJ

11.1   Introduction

Spermatogonial stem cell transplantation stands out as an approach to investigate germ cell biology because it affords researchers the opportunity to study the testis-specific male germline stem cell that is essential for the production of sperm. The continual production of sperm by sexually mature males requires an efficient and highly regulated process in the seminiferous tubules of the testis *(1)*. The somatic Sertoli cells support the differentiation of germ cells from diploid undifferentiated spermatogonia to mature, haploid spermatozoa that are morphologically distinct from all other cells in the testis. In most species, this is a remarkably productive process, and there are multiple examples of animals that produce over a billion sperm each day *(1)*. Thus, a constant supply of undifferentiated spermatogonia must be maintained so these cells can differentiate into sperm *(2)*.

The cell responsible for supplying undifferentiated spermatogonia is the spermatogonial stem cell. Similar to other tissue-specific stem cell populations, spermatogonial stem cells must produce cells that will differentiate to provide mature cells while maintaining a stem cell population for the production of future differentiating cells *(3)*. Extrinsic regulation of the spermatogonial stem cell self-renewal/differentiation process is thought to occur by the production of factors by somatic cells that contribute to the spermatogonial stem cell niche in the seminiferous tubule *(3,4)*. Characterization of the niche, which is poorly defined at this time, will provide valuable information about how cell fate is controlled for the spermatogonial stem cell. Investigation of the location of spermatogonial stem cells in the seminiferous tubule, regulation of spermatogonial stem cells, and characterization of the spermatogonial stem cell niche are active areas of research *(5)*. The development of the spermatogonial stem cell transplantation technique has accelerated the pace of research in this interesting area of germ cell biology *(6,7)*.

First reported in 1994 *(8)* in mice, the spermatogonial stem cell transplantation technique has been used by several research groups for multiple applications *(7)*. These applications include transplantation of male germ cells to determine the presence of spermatogonial stem cells in a cell population after positive or negative selection *(9–12)*, the production of transgenic mice *(13,14)*, and evaluation of in vitro approaches to selectively culture spermatogonial stem cells *(15,16)*. One of the more important applications of spermatogonial stem cell transplantation has been using the assay to determine if a gene mutation, either natural or created by targeted gene disruption, that results in incomplete spermatogenesis and infertility is caused by the inability of germ or somatic cells to function normally in the testis. This approach has provided useful information on several factors that are involved in spermatogonial stem cell differentiation, allowing more precise investigation of spermatogonial stem cell regulation *(17–19)*.

In addition to using spermatogonial stem cell transplantation with mice, the technique has been modified for application in several other species, including rats, pigs, and cattle *(13,20,21)*. However, the mouse is the predominant species for spermatogonial stem cell research for several reasons. First, transgenic mice with germ cells that express proteins such as β-galactosidase (LacZ) or green fluorescent protein (GFP) can be used as cell donors for straightforward identification following transplantation. At the time of analysis, establishment of donor-derived spermatogenesis from transgenic donor cells in recipients can be identified with the use of cell staining or fluorescent microscopy. Second, there are well-characterized mouse genetic lines that, because of a gene mutation, lack endogenous spermatogenesis *(22)*. If the mutation does not affect the hypogonadal–testicular axis or the somatic cells of the testis, then these mice can be used as germ cell recipients because exogenous germ cells will colonize and establish donor-derived spermatogenesis *(23)*. Similarly, approaches to eliminate the majority of differentiating germ cells in the testes of wild-type mice with drug treatments to generate stem cell recipients are straightforward and consistent *(24)*. Likewise, irradiation of mice can be used to eliminate differentiating germ cells and provide a suitable environment for transplantation *(25)*. Finally, immunocompromised mouse strains can be used as recipients for spermatogonial stem cells from other species *(26–28)*. Because of the inability of these mice to mount an immune response against foreign cells, exogenous spermatogonial stem cells will colonize the seminiferous tubules, and cell rejection will not occur. Complete differentiation of the donor-derived germ cells only occurs in closely related species such as rats and mice *(29)*. However, this approach can be used to evaluate the survival and proliferation of spermatogonial stem cells from other species. Analysis depends on the spermatogonial stem cells from the other species to colonize the testes of the immunodeficient mouse recipient.

Several applications of spermatogonial stem cell transplantation beyond basic research have been discussed *(16)*. For humans, treatment for cancer can lead to infertility, so obtaining and storage of spermatogonial stem cells from affected individuals would provide cells that could be transplanted back into the patient and restore fertility. A testicular biopsy would provide the spermatogonial stem cells for cryopreservation, storage, and transplantation. For the safety of the patient, the cell sample would need to be screened prior to transplantation to eliminate any cancerous cells that could be reintroduced to the patient during transplantation. This approach could be applied to adults or prepubertal patients after they have progressed through puberty. Similarly, spermatogonial stem cell transplantation has been suggested as a means to increase the amount of sperm produced from genetically superior livestock *(30)*. For example, spermatogonial stem cells from a superior bull sire could be transplanted into multiple recipients that would produce sperm with the donor bull genetic material. This approach, along with spermatogonial stem cell cryopreservation, could lead to continual production of donor bull sperm in recipients after the donor bull has died. Obviously, restoration of fertility for both applications is a key aspect of the success of this approach.

Research investigating restoration of fertility in mice following spermatogonial stem cell transplantation has been limited because of the focus on basic research with spermatogonial stem cells. It is known that younger mice are superior to older mice for the restoration of fertility following spermatogonial stem cell transplantation *(4)*. It is believed that the formation of stem cell niches is most active during prepubertal testis development, compared to other ages, and the introduction of donor spermatogonial stem cells into the testis prepubertally leads to greater colonization and more donor-derived spermatogenesis *(4,31)*. This results in higher fertility in recipient animals. Indeed, transplantation into the testis of the natural mutant W/W^v mouse, which has a very small number of endogenous germ cells, has the highest success in achieving restored fertility in recipient models investigated *(4,32)*. Restored fertility has also been reported in recipient goats transplanted in the testes with germ cells from transgenic goats *(33)*. Recipient goats were immature when the transplantation procedure was performed.

More research needs to be conducted to determine the best approach to achieve restoration of fertility following spermatogonial stem cell transplantation. In the case of humans, it is unlikely that patients would be sexually immature at the time of transplantation. Therefore, research aimed to define the conditions that lead to consistent restoration of fertility in mature animals following transplantation is needed. In livestock, it is more likely that sexually immature recipients could be used as recipients. However, these animals may serve as good models to investigate problems associated with the restoration of fertility in humans after transplantation.

A comprehensive understanding of the regulation and activity of spermatogonial stem cells and the niche that supports them in vivo will require application of new research technologies, such as global gene expression profiling and proteomics. Separation of spermatogonial stem cells from other testicular cells followed by culture in a serum-free environment has improved to the point at which novel approaches such as global gene expression analysis have been performed *(34)*. In addition, interest in stem cell biology in general motivates research investigating the plasticity of tissue-specific stem cells and the potential ability to dedifferentiate these cells so they can contribute to other tissues of an organism could have a significant impact on clinical applications. Similarly, the ability to stimulate embryonic stem cells to differentiate into germ cells in vitro has been reported *(35)*. It is critical that findings determined from in vitro experiments and observations are connected to in vivo studies. Thus, the application and use of spermatogonial stem cell transplantation is essential for all studies involving spermatogonial stem cells. In this chapter, the necessary tools and methods to conduct spermatogonial stem cell transplantation are covered in addition to how data from this technique is used to understand testicular function.

11.2 Materials

11.2.1 Mice

All mice can be purchased from Jackson Laboratory (Bar Harbor, ME). To maintain reproductive efficiency, mice should be housed and cared for according to the

National Research Council Guide for the Care and Use of Laboratory Animals and provided standard chow ad libitum.

1. Donor mice: Transgenic mice that express proteins markers such as LacZ or GFP in all germ cells are the ideal germ cell donors for spermatogonial stem cell transplantation if basic questions about spermatogonial stem cell biology are of interest. This is because the donor-derived testis colonization in recipient animals can be detected with straightforward means. Two transgenic mice lines that are useful donors for this purpose are (Jackson Laboratory strains) as follows:

    a. B6.129S7-*Gt(ROSA)26Sor*/J. These mice express LacZ in all cells in the testis and are often called Rosa26 mice.
    b. C57BL/6-Tg(ACTB-EGFP)1Osb/J. These mice express GFP in all cells in the testis.

There are naturally occurring mutant mice and many knockout mice lines that result in disrupted spermatogenesis. When the research question of interest involves determining the testicular cell type, either germ or Sertoli, responsible for the disruption of sperm production, then the knockout or natural mutant mouse must be used for the donor. One strategy for straightforward detection of colonization is to cross Rosa26 or GFP mice with the mouse line of interest so that donor germ cells can be detected in the recipient testis. It is critical that crossing mice lines does not increase the chance of immunological rejection of donor cells or impair reproductive efficiency (*see* **Note 1**). In the special circumstance that unique donor mice cannot be crossed with Rosa26 or GFP mice, then the choice of the recipient mouse is critical and should be carefully considered (*see* **Note 2** and **item 2** below).

2. Recipient mice: The selection of the best recipient mice depends on the application. For research projects designed to directly assay spermatogonial stem cell function in which Rosa26 or GFP mice are used as donors, then busulfan-treated wild-type mice are commonly used. The recipient strain must be immunologically compatible with the donor mouse to prevent rejection of the donor cells. When an immunologically compatible recipient mouse strain is not available, nude or SCID mice can be used as recipients.

Recipient mice immunologically compatible with Rosa26 (C57BL/6 background) and GFP (C57BL/6 background) are C57BL/6 background. Pure C57BL/6 mice are not usually used as recipients. We, and other laboratories, use the F1 hybrid cross of 129 SvCP X C57BL/6 males as recipients *(36)*.

As described when selecting proper donor mice for an experiment, when the research question focuses on characterizing a unique phenotype in which spermatogenesis is blocked in a particular strain, these mice can be used as recipients as well as donors. Again, careful attention should be paid to selection of donor mice, usually Rosa26 or GFP mice, to prevent immunological rejection of transplanted cells. W/W^v mice do not have endogenous spermatogenesis and are very useful recipients because if germ cell differentiation is observed after transplantation it has to be donor derived (*see* **Note 2**).

11.2.2   Media and Solutions

With a few exceptions noted, all chemicals and media can be purchased from Sigma.

1. Hank's balanced salt solution (HBSS): 5.33 mM potassium chloride; 0.44 mM potassium phosphate monobasic; 4 mM sodium bicarbonate, 137.93 mM sodium chloride; 0.3 mM sodium phosphate dibasic; 5.6 mM glucose; 0.01 g/L or 0.03 mM phenol red.

2. Phosphate-buffered saline (PBS; 1 L): 10 g NaCl, 0.25 g KCl, 1.44 g Na$_2$HPO$_4$, 0.25 g KH$_2$PO$_4$, pH 7.2.

3. Testis digestion medium: 0.5 mg/mL collagenase type IV, 0.25 mg/mL trypsin, 0.05 mg/ml deoxyribonuclease (DNase) I in HBSS. Prepare an additional stock of DNase I at 7 mg/mL for addition to digestion when needed.

4. Germ cell injection medium: Prepare with minimum essential medium α (MEMα) with 0.002% trypan blue. Prepare using trypan blue (0.2%; Invitrogen) as a stock solution.

5. Busulfan solution: Busulfan treatment is used to eliminate proliferating cells in the testes of recipient mice. Elimination of differentiating germ cells is believed to provide the necessary environment for donor spermatogonial stem cells to migrate to the basement membrane and establish a stem cell niche. Wild-type mice are treated with 40 mg/kg busulfan; immunodeficient mice are treated with a 33-mg/kg dose. Busulfan powder is dissolved in dimethyl sulfoxide (DMSO) prewarmed to 41°C–43°C. After the busulfan is dissolved, an equal volume of distilled water (41°C–43°C) is added to make the working solution. In our lab, we make a 4 mg/mL busulfan solution for mouse treatment.

6. 4% paraformaldehyde (PFA): We usually prepare 50 mL of 4% PFA. PFA should be used within 24 h of preparation. Heat half of the final volume of distilled water to 60°C and add PFA powder for final concentration of 4% (i.e., 2 g for 50 mL). Stir in fume hood, maintaining 60°C for 15 min. Add 1 drop 10N NaOH to clear the solution. More 10N NaOH can be added as needed, but it should not take more than 5–10 drops. Allow the solution to cool to room temperature and add remaining volume of 2X PBS. Adjust to pH 7.2 with HCl.

7. LacZ rinse buffer (1 L): 0.2M sodium phosphate, pH 7.3 (5.52 g monobasic monohydrate, 42.9 g dibasic heptahydrate, or 24.28 g dibasic anhydrous), 2 mM magnesium chloride (2 mL 1M stock solution), 0.02% NP40 (2 mL 10% stock), 0.01% sodium deoxycholate (1 mL 10% stock solution).

8. LacZ staining solution (10 mL): 9.6 mL LacZ rinse buffer, 5 mM potassium ferricyanide (16.5 mg), 5 mM potassium ferrocyanide (18.4 mg), 1 mg/mL 5-bromo-4 chloro 3 indolyl-beta-D galactopyranoside X-gal (0.4 mL 25 mg/mL solution of X-gal in dimethylformamide or DMSO) (see **Note 3**).

9. Anesthesia: We use a mixture of ketamine (0.1 mg/kg) and xylanzine (0.05 mg/kg) in 0.9% physiological saline to anesthetize mice for surgery. Alternatively, avertin can be used as a 2.5% solution prepared in PBS or physiological saline.

11.2.3 Equipment

1. Micropipet puller for the preparation of glass injection needles. Micropipet pullers can be purchased from several sources (e.g., Sutter Instruments, World Precision Instruments). Laboratories that generate transgenic mice usually maintain micropipet pullers (*see* **Note 4**).
2. Micropipet beveler is useful for smoothing the end of the glass micropipet so cells or debris do not stick to jagged edges created when the needle is prepared. This also provides a beveled point to the needle, which assists when inserting the needle into the mouse testis. Several companies sell micropipet bevelers.
3. A transjector is useful for injecting cells into the testis of the recipient mouse. An alternative strategy is to apply pressure through the needle by a mouth pipet or through a syringe. Polyethylene tubing is used to connect the needle to the pressure source.
4. A dissection microscope with ×20 to ×40 magnification that has a stage large enough for a mouse is needed because the efferent ductules that are used to guide the needle for cell injection cannot be visualized by the naked eye. The microscope is also used to visualize recipient mouse testes following colonization. Therefore, a digital camera documentation system is also useful for taking images of recipient testes. If GFP mice are used as donors, the microscope will require the equipment for fluorescent excitation to visualize colonization by donor GFP cells.

11.2.4 Additional Materials for Tissue Recovery, Cell Digestion, and Surgery

1. Instruments: Forceps (11–12 cm) and small scissors (10–12 cm) are needed for tissue removal, removing the tunica albuginea from the testis, and transplantation surgery.
2. An incubator or water bath maintained at 37°C is required for cell digestion.
3. Micropipets to make needles for cell injection: We use micropipets (25 µL) made from borosilicate glass that can be purchased from a variety of vendors.
4. Suture or skin staples are needed to close the skin after surgery. We use 6-0 absorbable chromic gut.
5. Optional surgical equipment could include a bead sterilizer, small cauterizer, and a heating pad to keep mice warm following surgery (*see* **Note 5**).

11.3 Methods

11.3.1 Testicular Cell Preparation for Donor Cell Injection

It is important to generate a single-cell suspension of cells in which greater than 80% of the cells are viable. Several approaches can be used to digest the testis to obtain a single-cell suspension, and several factors will affect the technique that is used. For example, the testes of neonatal mice can be digested with a single aliquot of digestion media containing trypsin and DNase I. In contrast, testes from adult mice require collagenase, trypsin, and DNase I and need at least one change of digestion media to generate a single-cell suspension. A general protocol is described, and researchers will need to refine the protocol for their needs.

1. Maintain aseptic technique as much as possible. All solutions should be filtered through a 0.2-μm filter prior to starting the preparation.
2. Following sacrifice, mice should be sprayed with 70% ethanol and the testes removed.
3. For adult mice, testes are transferred into digestion media and incubated for 15 min at 37°C to dissociate tubules. The tube should be inverted several times during the incubation, and the seminiferous tubules should come apart so individual tubules are observed.
4. The tube containing the suspension is transferred to ice and the tubules allowed to sediment for 5 min. This step will eliminate interstitial cells.
5. The supernatant is then removed and fresh digestion media added and incubated on shaker for 15 min at 37°C.
6. Additional DNase I can be added during the second digestion if needed. Following the second digestion, the cell suspension is gently pipeted for several minutes and checked to determine if there is a single-cell suspension.
7. The cell suspension is passed through a 40-μm cell strainer followed by washing the strainer with an equal volume of HBSS and centrifugation at 500g for 5 min at 4°C.
8. The cell pellet is suspended in HBSS containing 0.3 mg/mL soybean trypsin inhibitor, and the cell concentration is determined using a cell-counting chamber.
9. The cell suspension is then centrifuged at 500g for 4 min at 4°C, and the cell pellet is suspended in germ cell injection media at a concentration of 10^6 to 10^7 cells/mL (see **Note 6**).

Digestion of testes from neonatal or sexually immature mice does not require the collagenase or the second enzyme digestion step, although additional DNase I is often needed for complete digestion into a single-cell suspension. Cell viability can be tested by trypan blue exclusion. The cell suspension should be monitored throughout preparation by microscopy to determine if additional digestion is needed. Additional purification protocols to enrich for spermatogonial stem cells have been reported. These include brief incubation on laminin-coated plates (11), cell sorting (12), or magnetic bead sorting (10,36). Researchers are encouraged to explore the literature for the details of these additional steps.

11.3.2 Recipient Animal Preparation

As described, when mice with normal spermatogenesis or any degree of germ cell differentiation are used as recipients, the endogenous germ cells need to be eliminated. The method most used by researchers has been busulfan treatment.

1. Mice need to be 16–20 g body weight at the time of treatment. An intraperitoneal injection of a 40-mg/kg dose of busulfan is sufficient to eliminate endogenous spermatogenesis without causing harmful side effects.
2. Care should be taken to maintain the busulfan solution at 41°C–43°C. Treated mice are suitable as germ cell recipients 4–6 wk after busulfan treatment.

The use of W/W^v mice as recipients does not require busulfan treatment because these mice do not have endogenous spermatogenesis (*see* **Note 2**). Neonatal or sexually immature mice donors do not require busulfan treatment because donor germ cells will compete with endogenous germ cells to establish stem cell niches in the seminiferous tubules. Recipient mice can also be irradiated to eliminate endogenous germ cells in preparation for transplantation. However, it has been reported that irradiation damages the somatic cells, not germ cells, in the testis, and this damage leads to germ cell loss *(37)*. Therefore, the efficiency of spermatogonial stem cell colonization and donor-derived spermatogenesis in irradiated recipients may not be consistent with busulfan-treated mice.

11.3.3 Donor Cell Injection

Successful injection of the donor cell preparation requires patience and practice. The mouse should be anaesthetized and placed on its back (*see* **Note 7**).

1. A midline incision is made approx. 2 cm above the prepuce. The fat pad associated with the testis can usually be observed and used to pull the testis out of the mouse.
2. The testis should be positioned so good visualization of the arteries, veins, and ducts associated with the testis can be achieved.
3. There are three different routes in which germ cells can be injected into the testis.

   a. First, a small hole can be made with a 30-gage needle in the connective tissue surrounding the efferent bundle adjacent to the testis. This hole provides an opening to insert the glass needle containing the cell suspension. The needle can be inserted using the efferent tubules as a guide to the rete testis and the cells injected.
   b. Second, a small hole can be made in the tunica albuginea covering the rete testis. The needle can be injected through this hole and cells deposited in the rete.
   c. Third, the tunica albuginea on the surface of the testis is cut, and a seminiferous tubule is pulled through the hole. The needle can be inserted into the tubule and the cells injected into the testis.

4. In our experience, use of the efferent ductules as a guide to position the needle at the rete testis is the best method that causes the least damage to the testis.
5. Care should be taken to avoid damaging the efferent tubules or the rete testis. This is especially important if an endpoint of the project is donor-derived offspring from recipient mice. If using the efferent ductules as the route of injection is not successful, direct injection into the rete testis is still an option.

6. The cell suspension is injected into the testis through the needle using air supplied from a transjector, mouth pipet, or syringe. The volume of cells injected is usually 7–10 μL. If W/W^v mice are used as recipients, the injected volume is approx. 3–5 μL.
7. The trypan blue in the germ cell injection media allows visualization of filling of the seminiferous tubules. **Figure 11.1** shows the testis of a busulfan-treated mouse in which the seminiferous tubules are filled with germ cell injection media containing trypan blue.
8. Following injection, carefully return the testis to the body cavity in the same orientation it was removed. The incision and tissues should be regularly bathed with sterile physiological saline throughout the procedure.
9. Eye ointment should be placed on the mouse eyes to keep them moist. The incision can be sutured or closed using skin staples. Mice are allowed to recover and are monitored for at least 24 h after surgery (*see* **Note 8**).

The same basic procedure is followed for busulfan-treated mice or other types of sexually mature recipients. When neonatal or sexually immature mice are used to achieve a better opportunity for donor-derived offspring, the main difference is the method to anesthetize the mice. Hypothermia is an appropriate form of anesthesia for mice up to 3 wk of age (*see* **Note 9**). Testis transplantation in neonatal mice requires additional practice because of the smaller efferent ductules and testis.

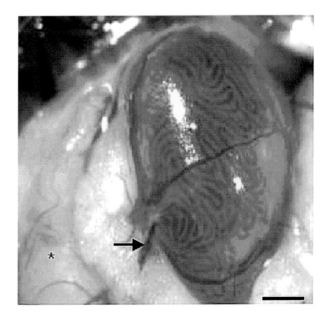

Fig. 11.1 Mouse testis following germ cell transplantation injection using the efferent ductules to guide the needle to the rete testis (*arrow*). Germ cells were suspended in germ cell injection media containing trypan blue. Seminiferous tubules containing the blue cell suspension can be observed. The epididymis is on the left side of the image (*). Scale bar = 0.5 mm

11.3.4   Recipient Testis Analysis

Testes in which germ cells have been injected and colonized can be analyzed at any point following transplantation. Individual spermatogonial stem cells colonize and establish a niche followed by cell differentiation. The colony will also expand later-ally along a seminiferous tubule. The number of donor-derived colonies can be counted and provides the most direct way of comparing the number of spermatogo-nial stem cells in two cell samples. Colony expansion can also be determined by measuring the length of each donor-derived colony. Both measurements should be conducted no longer than 6–8 wk after transplantation because donor-derived colo-nies will merge in the seminiferous tubules in the recipient testis. Counting colonies can be done directly using a dissecting microscope; measurement of colony length requires a software program.

As described, Rosa26 or GFP mice are good donors because cell staining or fluo-rescence can detect colonization. The following protocol is provided for LacZ staining of recipient testes following Rosa26 donor spermatogonial stem cell colonization.

1. Remove the testis from the animal and gently remove the tunica albuginea. Avoid damaging the testis.
2. Removal of the tunica is not absolutely necessary.
3. Fix testes in 4% PFA for 1 h, rocking on ice or at 4°C.
4. Dispose of the 4% PFA and rinse the testis in LacZ rinse buffer for 30 min, rock-ing on ice or at 4°C.
5. Change LacZ rinse buffer and wash testis for 30 min, rocking at room temperature.
6. Remove rinse buffer and add approx. 2–5 mL LacZ staining solution. Incubate at 37°C for a minimum of 6–8 h, rocking.
7. After staining, wash samples with LacZ rinse buffer at room temperature for 15 min followed by washing with PBS for 15 min at room temperature.
8. Remove PBS and begin a series of graded PBS–ethanol washes starting with PBS/30% ethyl alcohol (EtOH), PBS/50% EtOH, and so on.
9. Final solution to store testis samples should be 70% EtOH in water. Testes can be stored in 70% EtOH for extended periods without the blue stain fading.

11.4   Notes

1. Some transgenic mouse strains produce small litter sizes and have difficulty maintaining lit-ters. In our experience with Rosa26 and GFP mice husbandry, we maintain breeding pairs of heterozygous mice and screen for transgenic offspring. GFP mice can be screened as pups using an ultraviolet illuminator. Rosa26 mice can be screened by polymerase chain reaction (PCR) or by staining tail snips using the X-gal staining protocol described in **Subheading 3.4**. Tail snips do not need to be fixed in 4% PFA prior to staining.
2. Usually, W/W^v mice (available from Jax; WBB6F1/J-Kit^W/Kit^{W-v}) can be used as recipients because they do not have any endogenous germ cells. The mutation in W mice does not affect Sertoli cell function, and the seminiferous tubule is a suitable environment for supporting germ-cell differentiation. At the time of analysis, testes from recipient animals have to be fixed and

sectioned to determine the extent of donor-derived spermatogenesis. The presence of differentiating germ cells indicates that donor cells colonized the testis and are capable of undergoing spermatogenesis. No spermatogenesis in the testes indicates the mutation affects the ability of germ cells to either colonize or differentiate and thus requires more research. If donor cell colonization cannot be visualized by microscopy, another approach to assay for the presence of donor cells is PCR. If donor cells have a specific gene sequence that can be detected by PCR, DNA can be purified from the testes and PCR performed to determine if donor cells colonized the testis.

3. Some X-gal staining protocols use 20 mM potassium ferrocyanide and 20 mM potassium ferricyanide. However, we have found that a 5 mM concentration is sufficient for both and reduces background staining, which can create difficulty for analysis.

4. Glass needles can be pulled using a flame from a Bunsen burner or similar if a micropipet puller is not available. The challenge with using a flame is the consistency of the needles.

5. Researchers should discuss survival rodent surgery with the institutional veterinarian to determine special or additional requirements specific for their institution.

6. The concentration of the cells for injection depends on the number of spermatogonial stem cells present in the sample. If the proportion of spermatogonial stem cells in the cell suspension is high, then high rates of colonization may make it difficult to count individual colonies. On the other hand, very low numbers of spermatogonial stem cells in the cell suspension may result in few colonies and create problems for analysis.

7. The procedure generally takes 30–45 min so the anesthesia dose should be calculated so the mouse recovers in approx. 1–1.5 h.

8. Monitoring mice for discomfort should be discussed with the institutional veterinarian. Mice can be administered analgesics for discomfort following surgery. If suturing is used to close the incision, the suture should be inspected several days following surgery because mice will chew sutures out.

9. To induce hypothermia, neonatal mice can be cooled using wet ice or placed on a sponge soaked in ice water. After surgery, the mice should be warmed to 37°C in an incubator and returned to the cage approx. 15 min prior to adding the mother. The mice should be rubbed in the bedding to acquire the mother's scent. Researchers should consult **ref. 38** for details about hypothermia-induced anesthesia and mouse surgery in general.

Acknowledgments I gratefully acknowledge the support from the National Institutes of Health (HD046521). I thank Drs. Daniel Johnston and Jon Oatley for helpful discussion on protocol development.

References

1. Sharpe, R. (1994) Regulation of spermatogenesis, in *The physiology of reproduction* (E. K. Knobil and J. D. Neill, eds.), Raven, New York, pp. 1363–1434.

2. de Rooij, D. G., and Russell, L. D. (2000) All you wanted to know about spermatogonia but were afraid to ask. *J. Androl.* **21,** 776–798.

3. Spradling, A., Drummond-Barbosa, D., and Kai, T. (2001) Stem cells find their niche. *Nature.* **414,** 98–104.

4. Shinohara, T., Orwig, K. E., Avarbock, M. R., and Brinster, R. L. (2001) Remodeling of the postnatal mouse testis is accompanied by dramatic changes in stem cell number and niche accessibility. *Proc. Natl. Acad. Sci. U. S. A.* **98,** 6186–6191.

5. McLean, D. J., Russell, L .D., and Griswold, M. D. (2002) Biological activity and enrichment of spermatogonial stem cells in vitamin A-deficient and hyperthermia-exposed testes from mice based on colonization following germ cell transplantation. *Biol. Reprod.* **66,** 1374–1379.

6. Brinster, R. L. (2002) Germline stem cell transplantation and transgenesis. *Science*. **296,** 2174–2176.

7. McLean, D. J. (2005) Spermatogonial stem cell transplantation and testicular function. *Cell Tissue Res*. **322,** 21–31.

8. Brinster, R. L., and Zimmermann, J. W. (1994) Spermatogenesis following male germ-cell transplantation. *Proc. Natl. Acad. Sci. U. S. A*. **91,** 11298–11302.

9. Kanatsu-Shinohara, M., Toyokuni, S., and Shinohara, T. (2004) CD9 is a surface marker on mouse and rat male germline stem cells. *Biol. Reprod*. **70,** 70–75.

10. Kubota, H., Avarbock, M. R., and Brinster, R. L. (2003) Spermatogonial stem cells share some, but not all, phenotypic and functional characteristics with other stem cells. *Proc. Natl. Acad. Sci. U. S. A*. **100,** 6487–6492.

11. Shinohara, T., Avarbock, M.R., and Brinster, R.L. (1999) beta1- and alpha6-integrin are surface markers on mouse spermatogonial stem cells. *Proc. Natl. Acad. Sci. U. S. A*. **96,** 5504–5509.

12. Shinohara, T., Orwig, K. E., Avarbock, M. R., and Brinster, R. L. (2000) Spermatogonial stem cell enrichment by multiparameter selection of mouse testis cells. *Proc. Natl. Acad. Sci. U. S. A*. **97,** 8346–8351.

13. Hamra, F. K., Gatlin, J., Chapman, K. M., et al. (2002) Production of transgenic rats by lentiviral transduction of male germ-line stem cells. *Proc. Natl. Acad. Sci. U. S. A*. **99,** 14931–14936.

14. Nagano, M., Brinster, C. J., Orwig, K. E., Ryu, B. Y., Avarbock, M. R., and Brinster, R. L. (2001) Transgenic mice produced by retroviral transduction of male germ-line stem cells. *Proc. Natl. Acad. Sci. U. S. A*. **98,** 13090–13095.

15. Kubota, H., Avarbock, M. R., and Brinster, R. L. (2004) Growth factors essential for self-renewal and expansion of mouse spermatogonial stem cells. *Proc. Natl. Acad. Sci. U. S. A*. **101,** 16489–16494.

16. Kubota, H., and Brinster, R. L. (2006) Technology insight: In vitro culture of spermatogonial stem cells and their potential therapeutic uses. *Nat. Clin. Pract. Endocrinol. Metab*. **2,** 99–108.

17. Boettger-Tong, H.L., Johnston, D.S., Russell, L.D., Griswold, M.D., and Bishop, C.E. (2000) Juvenile spermatogonial depletion (jsd) mutant seminiferous tubules are capable of supporting transplanted spermatogenesis. *Biol. Reprod*. **63,** 1185–1191.

18. Buaas, F. W., Kirsh, A. L., Sharma, M., et al. (2004) Plzf is required in adult male germ cells for stem cell self-renewal. *Nat. Genet*. **36,** 647–652.

19. Johnston, D. S., Russell, L. D., Friel, P. J., and Griswold, M. D. (2001) Murine germ cells do not require functional androgen receptors to complete spermatogenesis following spermatogonial stem cell transplantation. *Endocrinology*. **142,** 2405–2408.

20. Herrid, M., Vignarajan, S., Davey, R., Dobrinski, I., and Hill, J. R. (2006) Successful transplantation of bovine testicular cells to heterologous recipients. *Reproduction*. **132,** 617–624.

21. Honaramooz, A., Megee, S. O., and Dobrinski, I. (2002) Germ cell transplantation in pigs. *Biol. Reprod*. **66,** 21–28.

22. Ogawa, T., Dobrinski, I., Avarbock, M. R., and Brinster, R. L. (2000) Transplantation of male germ line stem cells restores fertility in infertile mice. *Nat. Med*. **6,** 29–34.

23. Shinohara, T., Avarbock, M. R., and Brinster, R. L. (2000) Functional analysis of spermatogonial stem cells in Steel and cryptorchid infertile mouse models. *Dev. Biol*. **220,** 401–411.

24. Brinster, C. J., Ryu, B. Y., Avarbock, M. R., Karagenc, L., Brinster, R. L., and Orwig, K. E. (2003) Restoration of fertility by germ cell transplantation requires effective recipient preparation. *Biol. Reprod*. **69,** 412–420.

25. Zhang, Z., Shao, S., and Meistrich, M. L. (2006) Irradiated mouse testes efficiently support spermatogenesis derived from donor germ cells of mice and rats. *J. Androl*. **27,** 365–375.

26. Oatley, J. M., de Avila, D. M., McLean, D. J., Griswold, M. D., and Reeves, J. J. (2002) Transplantation of bovine germinal cells into mouse testes. *J. Anim. Sci*. **80,** 1925–1931.

27. Ogawa, T., Dobrinski, I., Avarbock, M. R., and Brinster, R. L. (1999) Xenogeneic spermatogenesis following transplantation of hamster germ cells to mouse testes. *Biol. Reprod*. **60,** 515–521.

28. Oatley, J. M., Reeves, J. J., and McLean, D. J. (2004) Biological activity of cryopreserved bovine spermatogonial stem cells during in vitro culture. *Biol. Reprod*. **71,** 942–947.

29. Franca, L. R., Ogawa, T., Avarbock, M. R., Brinster, R. L., and Russell, L. D. (1998) Germ cell genotype controls cell cycle during spermatogenesis in the rat. *Biol. Reprod.* **59,** 1371–1377.
30. Hill, J. R., and Dobrinski, I. (2006) Male germ cell transplantation in livestock. *Reprod. Fertil. Dev.* **18,** 13–18.
31. McLean, D. J., Friel, P. J., Johnston, D. S., and Griswold, M. D. (2003) Characterization of spermatogonial stem cell maturation and differentiation in neonatal mice. *Biol. Reprod.* **69,** 2085–2091.
32. Kanatsu-Shinohara, M., Toyokuni, S., and Shinohara, T. (2005) Genetic selection of mouse male germline stem cells in vitro: offspring from single stem cells. *Biol. Reprod.* **72,** 236–240.
33. Honaramooz, A., Behboodi, E., Megee, S. O., et al. (2003) Fertility and germline transmission of donor haplotype following germ cell transplantation in immunocompetent goats. *Biol. Reprod.* **69,** 1260–1264.
34. Oatley, J. M., Avarbock, M. R., Telaranta, A. I., Fearon, D. T., and Brinster, R. L. (2006) Identifying genes important for spermatogonial stem cell self-renewal and survival. *Proc. Natl. Acad. Sci. U. S. A.* **103,** 9524–9529.
35. Geijsen, N., Horoschak, M., Kim, K., Gribnau, J., Eggan, K., and Daley, G. Q. (2004) Derivation of embryonic germ cells and male gametes from embryonic stem cells. *Nature.* **427,** 148–154.
36. Oatley, J. M., and Brinster, R. L. (2006) Spermatogonial stem cells. *Methods Enzymol.* **419,** 259–282.
37. Zhang, Z., Shao, S., and Meistrich, M. L. (2006) The radiation-induced block in spermatogonial differentiation isdue to damage to the somatic environment, not the germ cells. *J. Cell Physiol.* **211,** 149–158.
38. Hendrich, H. (2004) *The laboratory mouse*, Elsevier, Amsterdam.

Chapter 12
Isolating Highly Pure Rat Spermatogonial Stem Cells in Culture

F. Kent Hamra, Karen M. Chapman, Zhuoru Wu, and David L. Garbers

Contents

Summary Methods are detailed for isolating highly pure populations of spermatogonial stem cells from primary cultures of testis cells prepared from 22- to 24-day-old rats. The procedure is based on the principle that testicular somatic cells bind tightly to plastic and collagen matrices when cultured in serum-containing medium, whereas spermatogonia and spermatocytes do not bind to plastic or collagen when cultured in serum-containing medium. The collagen-non-binding testis cells obtained using these procedures are thus approx. 97% pure spermatogenic cells. Stem spermatogonia are then easily isolated from the purified spermatogenic population during a short incubation step in culture on laminin matrix. The spermatogenic cells that bind to laminin are more than 90% undifferentiated, type A spermatogonia and are greatly enriched in genetically modifiable stem cells that can develop into functional spermatozoa. This method does not require flow cytometry and can also be applied to obtain enriched cultures of mouse spermatogonial stem cells. The isolated spermatogonia provide a highly potent and effective source of stem cells that have been used to initiate in vitro and in vivo culture studies on spermatogenesis.

Keywords Germ cell; germline; multipotent; pluripotent; spermatogonia; spermatogonial; spermatogenesis; stem cell; testis.

From: *Methods in Molecular Biology, Vol. 450, Germline Stem Cells*
Edited by S. X. Hou and S. R. Singh © Humana Press, Totowa, NJ

12.1 Introduction

The ability to conditionally stimulate highly pure cultures of mammalian spermatogonial stem cells to undergo spermatogenesis in vitro would provide a long-sought experimental system for studying how spermatozoa are produced in vivo *(1)*. Although such an in vitro culture system has yet to be established, methods are available to isolate and culture highly pure fractions of rodent spermatogonial stem cells through spermatogenesis in vivo by transplanting them into the seminiferous tubules of recipient animals *(2–6)*. Methods to derive and propagate rat and mouse spermatogonial lines for long periods of time in culture have been established *(7–11)*. Spermatogonia from these cultures are also able to reestablish colonies of spermatogenesis and form functional spermatozoa after transplantation into recipient animals. Now, such highly pure populations of spermatogonial stem cells should be used to discover culture conditions that will support their development through spermatogenesis in vitro. This chapter provides a cell culture protocol to isolate rat spermatogonial stem cells in a highly pure state for studies of spermatogenesis *(5,7,12–15)*.

12.1.1 Procedures A–C

The protocol for isolating rat spermatogonial stem cells is divided into three separate procedures: A, B, and C, which are outlined in **Fig. 12.1**. Procedure A

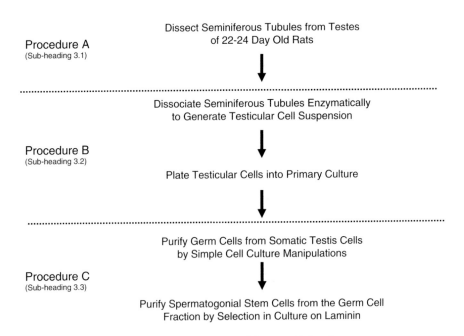

Fig. 12.1 Procedures for purifying rat spermatogonial stem cells in culture

(**Subheading 12.3.1**) describes how to isolate seminiferous tubules from rat testes. Procedure B (**Subheading 12.3.2**) describes how to prepare testis cell cultures from isolated seminiferous tubules. Procedure C (**Subheading 12.3.3**) describes how to isolate spermatogonial stem cells from testis cell cultures.

12.2 Materials

Make sure to receive and prepare all of the following reagents and supplies at least a day ahead of time before proceeding with procedures A–C unless indicated otherwise (i.e., prepare fresh).

12.2.1 Isolating Rat Seminiferous Tubules (Procedure A)

12.2.1.1 Surgical Supplies and Animals

1. Microdissecting scissors (cat. no. RS-5852, Roboz Surgical Inc.).
2. Operating sharp-sharp scissors (cat. no. RS-6802, Roboz Surgical Inc.).
3. 5/45° angle tweezers (cat. no. RS-5005, Roboz Surgical Inc.).
4. Male Sprague–Dawley rats are received at 15–16 d of age (Harlan Co. Inc.).

12.2.1.2 Chemicals and Supplies

1. N-2-Hydroxyethylpiperazine-N'-2-ethanesulfonic acid (HEPES).
2. Glycine.
3. Hemacytometer (cat. no. 02-671-6, Fisher Inc.)

12.2.1.3 Media and Solutions

1. DHF12 medium: Dulbecco's modified Eagle's medium:Ham's F12 medium 1:1 (DHF12) (cat. no. 11330-032, Invitrogen Inc.). The solution is received complete from the manufacturer in 500-mL bottles. Store at 4°C until the expiration date listed by the manufacturer.
2. PBS: Dulbecco's phosphate-buffered saline (PBS; cat. no. 14190-250, Invitrogen Inc.), 200 mg/L KCl (w/v), 200 mg/L KH_2PO_4 (w/v), 8 g/L NaCl (w/v), 1.15 g/L Na_2HPO_4 (w/v). The solution is received complete from the manufacturer in 500-mL bottles. Store at 22°C–24°C.
3. Antibiotic–antimycotic solution: 10,000 U/mL penicillin G sodium (U/v), 10,000 μg/mL streptomycin sulfate (w/v), and 25 μg/mL amphotericin B (w/v) (cat. no. 15240-062, Invitrogen Inc.). The solution is received complete from the manufacturer in 100-mL bottles. Make 12.5-mL aliquots in sterile 15-mL tubes and then store at −20°C until the expiration date listed by the manufacturer.

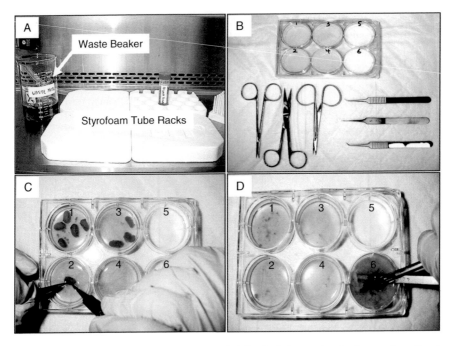

Fig. 12.2 Some materials and methods needed for isolating and culturing testis cells. **A** Preparation of laminar flow biosafety cabinet (culture hood); **B** preparation of dissection instruments and medium for isolating testes from rats; **C** isolation of seminiferous tubules from rat testes; **D** washing and mincing of isolated seminiferous tubules

4. PBS+: PBS, 1% (v/v) antibiotic–antimycotic solution: To prepare, add 5 mL antibiotic–antimycotic solution to a 500-mL bottle of PBS. Store at 4°C for up to 1 mo.
5. Glycine buffer: $1M$ glycine , 20 mM HEPES, pH 7.2. To make 200 mL, add 15 g glycine and 950 mg HEPES to 180 mL deionized water. Adjust the pH of the solution to 7.2 by titrating in 0.1N NaOH. Adjust the volume to 200 mL with deionized water. Sterilize the solution by passing it through a 0.2-µm filter into a sterile container. Store at 4°C for up to 4 mo.
6. Medium A: DHF12 medium, 1% (v/v) antibiotic–antimycotic solution. To make, add 5 mL antibiotic–antimycotic solution to a 500-mL bottle of DHF12 medium. Gently swirl the bottle by hand to mix. Store at 4°C for up to 2 wk.

12.2.2 Preparing Primary Cultures of Rat Testis Cells (Procedure B)

12.2.2.1 Chemicals and Supplies

1. Sterile mesh: Sterile, 30-cm^2 sheets of 41-µm pore, nylon spectra-mesh (cat. no. 08-670-202, Fisher Inc.) are obtained from the manufacturer. To prepare, use

sterile scissors to cut approx. five 6-cm² pieces of mesh from the 30-cm² sheets. Store the 6-cm² pieces in a sterile, 10-cm culture dish until **step 12** of the procedure B (**Subheading 12.3.2**).

2. Modified filtration system (MFS): Use a sterile, disposable no. 10 scalpel (cat. no. 08-927-5A, Fisher Inc.) to cut the cellulose filter out of a disposable, 250-mL, 0.2-μ filtration system (cat. no. 430767, Fisher Inc.) received from the manufacturer (*see* **Fig. 12.3A,C** for illustrations of the MFS). Discard the cellulose filter. Prepare just prior to use in **Subheading 12.3.2, step 12**.

3. Steriflip filters, 0.2-μm (cat. no. SCGP00525, Fisher Inc.).

4. Rattail collagen I-coated cell culture dishes (10-cm dishes, cat. no. 08-772-75, Fisher Inc.).

5. Falcon brand 5-mL pipets (cat. no. 357543, Falcon Inc.) (*see* **Note 1**).

12.2.2.2 Solutions

1. Heat-inactivated FBS: To prepare, thaw a 500-mL bottle of FBS (cat. no. S11550, Atlanta Biologicals Inc.) received from the manufacturer (takes ~ 5 h at 22°C–24°C) and then place it into a clean water bath equilibrated to 56°C.

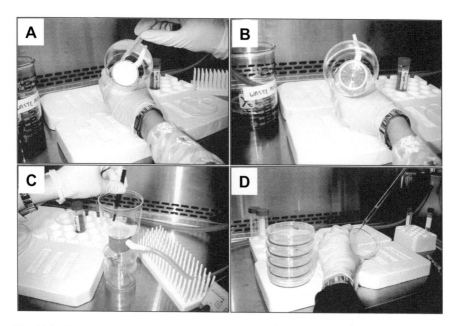

Fig. 12.3 Preparation of a modified filtration system (MFS) and method to isolate germ cells from cultures of testis cells. **A** Dissection of the 0.2-μm cellulose filter out of the manufacturer's filtration system; **B** the exposed filter grid of the manufacturer's filtration system after removal of the 0.2-μm cellulose filter; **C** the MFS attached to a vacuum source with sterile mesh (41-μm opening) placed on top of the exposed filter grid; **D** harvesting the germ cell population from cultures of testis cells by pipeting medium A over the surface area of each culture

Submerge the bottle enough to cover 100% of its contents. Do not get bath water on the lid of the bottle. Heat treat at 56°C for 45 min. After heat treating, clean the bottle using 70% ethanol, wipe dry, and cool the contents of the bottle at 4°C. Gently swirl the bottle by hand to mix its contents and then prepare 45-ml aliquots of the FBS in sterile, 50-mL plastic tubes. Store the aliquots at −20°C to −80°C for up to 1 yr.

2. Heat-inactivated horse serum (HS): The solution is received complete as a heat-inactivated solution from the manufacturer in 500-mL bottles (cat. no. 26050-088, Invitrogen Inc.). Thaw a 500-mL bottle of the HS (takes ~ 5 h at 22°C–24°C). Gently swirl the bottle by hand to mix its contents and then prepare 45-mL aliquots of the HS in sterile, 50-mL plastic tubes. Store the aliquots of HS at −20°C to −80°C for up to 1 yr.

3. Medium B: DHF12 medium, 8.2% (v/v) FBS, 1.1% (v/v) antibiotic–antimycotic solution. To make, add 6 mL antibiotic–antimycotic solution and 45 mL FBS to a 500-mL bottle of DHF12 medium. Gently swirl the bottle by hand to mix. Store at 4°C for up to 2 wk.

4. Medium C: DHF12 medium, 5.5% (v/v) HS, 2.4% (v/v) FBS, 1.1% (v/v) antibiotic–antimycotic solution. To make, add 30 mL HS, 13 mL FBS, and 6 mL antibiotic–antimycotic solution to 500 mL DHF12 medium. Gently swirl the bottle by hand to mix. Store at 4°C for up to 2 wk.

5. Dispase: The solution is received complete from the manufacturer in 100-mL bottles (cat. no. CB-40235, Fisher Inc.) that contain 50 caseinolytic units/mL.

12.2.3 Purifying Spermatogonial Stem Cells from Cultures of Rat Testis Cells (Procedure C)

12.2.3.1 Media and Solutions

1. Medium D: DHF12 medium, 10% (v/v) FBS, 1% (v/v) antibiotic–antimycotic solution. To make 100 mL, add 10 mL FBS and 1 mL antibiotic–antimycotic solution to 89 mL DHF12 medium. Store at 4°C for up to 2 wk.

2. Medium E: Medium D, 0.00024% (v/v) 2-mercaptoethanol. To make 50 mL, add 7.8 µL 2-mercaptoethanol (98+%, cat. no. M3148, Sigma Inc.) to 10 mL medium A, which makes 0.078% (v/v) 2-mercaptoethanol solution. Then, add 150 µL of the freshly prepared 0.078% (v/v) 2-mercaptoethanol solution to 50 mL medium D and sterilize by passing through a 0.2-µm Steriflip filter. Prepare solution fresh before use in **Subheading 12.3.3, step 14**.

3. PBS–BSA: PBS, 0.5% bovine serum albumin (BSA) (w/v). To make 40 mL, add 200 mg BSA (cat. no. A4503, Sigma Inc.) to 39.8 mL PBS in a 50-mL tube. Let BSA go into solution without mixing by letting tube contents stand at 22°C–28°C for approx. 30 min and then filter the solution through a 0.2-µm Steriflip filter. Prepare solution fresh before use in **Subheading 12.3.3, step 14**.

4. Laminin stocks: Vials containing 1 mg/mL laminin solution (cat. no. L2020, Sigma Inc.) are received frozen from the manufacturer. To make frozen aliquots, thaw one vial of mouse laminin on ice (requires ~1 h). Once thawed, make approx. six 150-μL aliquots of the laminin solution in sterile microfuge tubes on ice. Store the laminin stocks at −80°C for up to 1 yr.

12.3   Methods

Aseptic technique and work within a laminar airflow biosafety cabinet (culture hood) should be employed during preparation and manipulation of all the solutions and cultures for procedures A–C unless indicated otherwise, that is, except for **steps 4–9** of procedure A (**Subheading 12.3.1**) and **steps 1–3** of procedure B (**Subheading 12.3.2**).

12.3.1   Isolating Rat Seminiferous Tubules (Procedure A)

Procedures A and B should be performed on the same day and approx. 3 d prior to performing procedure C. For example, procedures A and B are performed on a Monday, then continue with procedure C on Thursday morning of that week. Because of the use of animals, **steps 4–9** of procedure A (**Subheading 12.3.1**) and **steps 1–3** of procedure B (**Subheading 12.3.2**) should be performed in a room that is separate from the cell culture room.

1. Clean and then sterilize one pair of operating scissors, two pairs of microdissecting scissors, and three pairs of 5/45° angle tweezers by soaking them in 70% ethanol or by heating them in a bead sterilizer (*see* **Note 2**).
2. Equilibrate the temperature of the room in which the cell culture hood is located (culture room) to 26°C–28°C if possible. The elevated room temperature will help to reduce heat shock received by the testis cells throughout the steps in procedures A–C. Prewarm bottles of medium B and medium C in a cell culture water bath equilibrated to 33°C–35°C (*see* **Note 3**).
3. Prepare the culture hood ahead of time for work during procedure B, as illustrated in **Fig. 12.2A**. Place a sterile, 1-L beaker in the left or right rear corner of the hood. The beaker will be used for discarding used medium. Place Styrofoam racks (obtained from packages of 50-mL disposable tubes) upside down in the culture hood to use as a working surface (optional; *see* **Fig. 12.2A**) (*see* **Note 4**).
4. Place a sterile diaper pad in a clean area on top of a laboratory bench located *outside* the culture room (24°C–26°C) and then place a sterile six-well cell culture plate on the diaper pad (*see* **Fig. 12.2B**). Add 6 mL/well of medium A to wells 1–4 of the culture plate. Add 6 mL PBS+ to well 5 of the culture plate. Add 6 mL glycine buffer to well 6 of the culture plate (*see* **Fig. 12.2B**).

5. Anesthetize five 22- to 24-d-old Sprague–Dawley rats (*see* **Note 5**). Once the rats have reached a surgical plane of anesthesia, euthanize the rats by cervical dislocation. Swab the lower abdominal area with 70% ethanol. Open the abdominal skin just rostral to the pelvis using a pair of operating scissors (**step 5** should be conducted under the guidelines of an animal protocol approved by the researcher's institution).

6. Use microdissecting scissors and tweezers to dissect out ten testes as cleanly as possible from all five rats. Transfer all isolated testes into well 1 of the six-well culture dish (*see* **Fig. 12.2C**) (*see* **Note 6**).

7. Use tweezers to transfer a single testis from well 1 into well 2 of the culture dish. Use two sets of tweezers (*see* **Fig. 12.2C**) to dissect away the tunica albuginea (thin clear membrane enclosing seminiferous tubules of each testis) and the major blood vessels from the seminiferous tubules of the testis in well 2. Discard the tunica albuginea and the blood vessels. Use tweezers to transfer the isolated seminiferous tubules from well 2 into well 3 of the culture dish (*see* **Fig. 12.2C**).

8. Repeat **step 7** for each testis until the seminiferous tubules from all ten of the isolated testes are pooled into well 3 of the culture dish (*see* **Fig. 12.2C**).

9. Proceed immediately to Procedure B.

12.3.2 Preparing Primary Cultures of Rat Testis Cells (Procedure B)

1. Wash the tubules in fresh medium A by using tweezers to transfer them from well 3 into well 4 of the culture dish. Then, wash the tubules in PBS by using tweezers to transfer them from well 4 into well 5 of the culture dish.

2. Use tweezers to immediately transfer the tubules from well 5 into well 6 of the culture dish and then proceed to mince the tubules in well 6 for 3 min using a pair of microdissecting scissors (*see* **Fig. 12.2D**). After mincing the tubules, continue incubating in well 6 for 5 min (8 min total incubation time in well 6). Proceed immediately to **step 3** (*see* **Note 7**).

3. Using a 5-mL pipet (*see* **Note 1**), immediately transfer tubules into a 50-mL plastic tube containing 25 mL glycine buffer. Bring the tube volume to 35 mL with glycine buffer and then close the lid of the tube. Mix the tubules by gently inverting the tube five times.

4. Pellet the suspension by centrifugation at 200*g* for 5 min in a table-top centrifuge (*see* **Note 3**). Use a 25-mL pipet to remove the supernatant solution from the pellet. Discard the supernatant solution into the waste beaker that is located in the culture hood.

5. Bring the volume of the tube containing the pellet to 35 mL with medium B and then disaggregate the pellet into a suspension by gently pipeting up and down approx. five times using a 5-mL pipet.

6. Repeat **step 4**.

7. Add 1000 caseinolytic units of the dispase solution (100 caseinolytic units/testis = ~ 2 mL/testis) to the tube containing the pelleted tubules (prethaw two 12-mL aliquots of the dispase solution in the 33°C–35°C water bath just prior to **step 1**). Using a 5-mL pipet, gently suspend the tubules in the dispase solution by pipeting up and down approx. five times and then close the lid of the tube. Lay the tube containing the tubules in dispase solution (digest) horizontally on a shelf in a cell culture incubator equilibrated to 32.5°C and then incubate the digest for a total of 30 min. During this 30-min period, remove the digest from the incubator every 5 min and then gently rock the digest by hand ten times. Use gentle back-and-forth motions of approx. 45° angle to rock the digest while keeping the tube in a horizontal position.

8. Bring the digest volume to 35 mL with medium B and position the tube vertically on an environmental shaker set to 220 rpm. Agitate the digest for 10 s at 220 rpm (26°C–28°C) to further disaggregate the testis cells (*see* **Note 8**).

9. Pellet the digest by centrifugation at 200*g* for 5 min in a table-top centrifuge. Use a 25-mL pipet to remove the supernatant solution from the pellet. Discard the supernatant solution into the waste beaker.

10. Bring the volume of the tube containing the pellet to 35 mL with medium B and then disaggregate the pellet into a suspension by gently pipeting up and down approx. five times using a 5-mL pipet.

11. Pellet the cell suspension by centrifugation at 200*g* for 5 min in a table-top centrifuge. Use a 25-mL pipet to remove the supernatant solution from the pellet. Discard the supernatant solution into the waste beaker.

12. At this point, suspend the pellet to 30 mL in medium C and then thoroughly mix the suspension by pipeting up and down five to ten times using a 10-mL pipet. Use the 10-mL pipet to remove any large aggregated clumps of genomic DNA (clearly visible) from the suspension and then discard the DNA into the waste beaker. Using a 10-mL pipet, filter the suspension by pipeting it through sterile mesh (41-μm opening, 5–6 cm²) placed in the 250-mL MFS, which is connected to a vacuum source (*see* **Fig. 12.3A,C** for illustration of the MFS). Leave the top of the MFS partially unscrewed to a loose position (*see* **Note 9**). Use a blunt set of sterile forceps to handle the mesh. If the mesh becomes clogged during filtration because of genomic DNA that is released from broken cells, use the sterile forceps to discard the clogged mesh from the MFS and replace with a fresh piece of mesh. Continue filtering the suspension until all 30 mL have been filtered (*see* **Note 10**). Discard all used mesh.

13. Transfer the cellular filtrate from the bottom chamber of the MFS into a 50-mL tube and then adjust the volume in the tube to 35 mL with medium C. Discard the used MFS.

14. Pellet the cell suspension by centrifugation at 200*g* for 5 min in a table-top centrifuge. Use a 25-mL pipet to remove the supernatant solution from the pellet. Discard the supernatant solution into the waste beaker.

15. Suspend the pellet to 35 mL in medium C and then mix the suspension by gently pipeting up and down five to ten times using a 10-mL pipet. Determine the cellular concentration of the mixed suspension using a hemocytometer.

16. Plate the testis cells into 10-cm plastic culture dishes containing 10 mL medium C at approx. 1.25×10^7 cells/dish ($\sim 2.25 \times 10^5$ cells/cm^2) and then place the cultures in a humidified cell culture incubator equilibrated to 32.5°C, 5.5% CO_2. Incubate the testis cell cultures at 32.5°C, 5.5% CO_2 undisturbed, for approx. 65 h prior to use in procedure C. The final cell suspension should yield enough testis cells to seed 8–12 cultures in 10-cm dishes.

12.3.3 Purifying Spermatogonial Stem Cells From Cultures of Rat Testis Cells (Procedure C)

Procedure C should be started in the morning. Until familiar with procedure C, a full day should be dedicated to purifying the spermatogonia from the testis cell cultures established by procedures A and B. Unless indicated otherwise (i.e., pre-pare fresh), it is most efficient to prepare reagents and materials specifically listed for procedure C a day or two ahead of time.

1. The day prior to performing procedure C, coat the wells of a sterile, 12-well culture dish with laminin ($\sim 5\,\mu g/cm^2$). To prepare, thaw one 150-μL aliquot from the 1 mg/mL laminin stocks on ice (requires ~ 30 min) and then add the entire volume of the aliquot into a sterile 50-mL tube containing 8 mL ice-cold medium A. Gently swirl the tube by hand to mix the contents. Add 1 mL of the diluted laminin/well into 8 wells of a 12-well plastic culture dish (*see* **Note 11**). Wrap the plate with Parafilm and incubate the laminin in the wells of the culture dish overnight at 4°C. The next morning, remove the plate from the 4°C cooler and place it in one corner of the cell culture hood (out of the main working area) until **step 17** (*see* **Note 11**).

2. The next morning, warm medium A, medium C, and PBS+ to 34°C–35°C in the cell culture water bath. Prepare the cell culture hood as described in **Subheading 12.3.1, step 3** (*see* **Fig. 12.2A**).

3. Add 5 mL medium C to prerinse each of two 10-cm, collagen I-coated culture dishes. Rock the medium back and forth by hand in each dish two to three times at approx. 25° to 35° angles. Use a pipet to completely remove the medium from each culture dish. Discard the rinse medium into a waste beaker located in the culture hood (*see* **Fig. 12.2A**). Add 4 mL fresh medium C to each of the two collagen I-coated culture dishes and then place the dishes into a humidified cell culture incubator equilibrated to 32.5°C, 5.5% CO_2 for use in **step 11**.

4. Use a 25-mL pipet to remove all medium from four to six dishes of testis cell cultures that were established in Procedure A. Discard the spent medium into the waste beaker.

5. Wash the cultures by gently adding 4 mL of warm medium A to each culture (pipet the wash medium near to one edge of the dish to prevent washing germ cells loose at this step). Gently rock each culture back and forth once at approx. 25°–35° angles by hand. Then, use a 25-mL pipet to remove the medium from each culture. Discard the wash medium into the waste beaker.

6. Wash the cultures a second time by adding 4 mL warm PBS+ to each culture, as in **step 5**. Then, use a 25-mL pipet to remove the wash medium from each culture. Discard the wash medium into the waste beaker (*see* **Note 12**).

7. Add 4 mL warm medium A to each culture. Then, harvest the germ cell population from the cultures by repeatedly pipeting the added medium A over the surface area of the dish (**Fig. 12.3D**). To ensure near-complete recovery of the germ cell population, try to cover the surface area of the dish approx. twice by pipeting of medium A. View cultures under a phase-contrast microscope to check that the germ cells are in suspension and are no longer attached to the somatic cells. Also, check to see that the somatic cell monolayer is still bound to the dish (see **Fig. 12.4**, TC and SC) (*see* **Note 13**).

8. Pool the germ cell-enriched suspensions obtained from a total of four to six cultures into two sterile 15-mL tubes. The somatic cell-enriched monolayers that are left attached to the culture dishes (*see* **Fig. 12.4**, SC) can be used for further experimentation if desired or discarded.

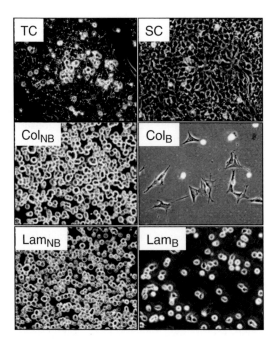

Fig. 12.4 Purifying germ cells and somatic cells from testis cell cultures. TC, culture of testis cells established on plastic using procedures A and B; spermatogenic cells overlie somatic cells. SC, somatic cell-enriched population on plastic after removal of spermatogenic cells from a culture of testis cells. Col_{NB}, spermatogenic cell-enriched population that does not bind to collagen that was isolated from the testis cell cultures. Col_{B}, a somatic cell-enriched population that binds to collagen that was isolated from the testis cell cultures. Lam_{NB}, a spermatogenic cell-enriched population that does not bind to laminin that was isolated from the Col_{NB} population. Lam_{B}, a spermatogenic cell-enriched population that binds to laminin that was isolated from the Col_{NB} population

9. Pellet the germ cell-enriched suspensions at 200*g* in a table-top centrifuge. Carefully pour off the supernatant solutions into the waste beaker and retain the cellular pellets.

10. Pool the germ cell pellets from both 15-mL tubes by suspending them in a total of 4 mL medium C using a 5-mL pipet.

11. Remove one collagen I-coated culture dish (prepared in **step 3**) from the cell culture incubator. Plate the cell suspension from **step 10** directly into the collagen I-coated culture dish, which already contains 4 mL medium C (i.e., 8 mL final volume medium C in the culture dish). Place the culture back into the cell culture incubator (*see* **Note 14**).

12. Repeat **steps 4–11** with the remaining testis cell cultures.

13. After all the testis cell cultures from procedure A have been processed through **step 11**, incubate the germ cell suspensions in the collagen I-coated culture dishes for approx. 2 h at 32.5°C, 5.5% CO_2. Remove the cultures from the incubator and then use a 10-mL pipet to mix the cultures by gently pipeting up and down about five times over the surface area of each dish. Place the cultures back into the incubator and incubate the suspensions in the collagen I-coated culture dishes for approx. 2 h more at 32.5°C, 5.5% CO_2.

14. Start warming medium E and PBS–BSA solutions to 34°C–35°C in a cell culture water bath.

15. Remove both of the cultures on collagen I-coated dishes from the incubator. Use a 5-mL pipet to mix the suspensions in each culture dish by gently pipeting up and down about times over the surface area of each dish. Use a 10-mL pipet to harvest the collagen-nonbinding (Col_{NB}) germ cell suspensions from each dish (*see* **Fig. 12.4**, Col_{NB}) and pool all of the suspensions into a single 50-mL tube (*see* **Note 15**).

16. Pellet the pooled suspensions of Col_{NB} cells at 200*g* for 5 min. Pour off the supernatant solutions into the waste beaker. Suspend the pellet in 8 mL medium E and then filter the germ cell suspension by pipeting it through a 70-µm, tube-top cell strainer into a fresh 50-mL tube. Immediately proceed to **step 17**.

17. Pipet the laminin solution out of all the wells of the 12-well culture dish prepared in **step 1**. Discard the laminin solution into the waste beaker. Add 1 mL medium A per well to prerinse each well of the laminin-coated culture dish. Then, pipet all the rinse medium off each well and discard the solution into the waste beaker. Immediately proceed to **step 18**.

18. Using a 5-mL pipet, mix the filtered suspension of Col_{NB} cells obtained from **step 16** by gently pipeting up and down four to five times. Plate out 1 mL/well of the mixed suspension into eight wells of the laminin-coated culture dish (*see* **Note 16**).

19. Incubate the Col_{NB} cells in the laminin-coated dish for a total of 40–50 min in a humidified cell culture incubator at 32.5°C, 5.5% CO_2. After the first 20–25 min of the incubation period, remove the laminin-coated dish from the incubator and use a 1-mL pipet tip (P1000 tip) to mix the cells in each well by very gently pipeting the suspension up and down once, near the wall on one side of each well (*see* **Note 17**). Immediately place the cells back into the incubator for the remainder of the incubation period.

20. Remove the laminin-coated culture dish from the incubator and then mix the nonbound cells in each well, termed laminin-nonbinding testis cells (*see* **Fig. 12.4**, Lam$_{NB}$), by gently pipeting up and down once using a 1-mL tip as described in **step 19**.

21. Use a 1-mL tip to transfer the mixed suspensions of Lam$_{NB}$ cells from four wells into a sterile 15-mL tube. Then, immediately add back 0.8 mL fresh medium E to each of the four wells from which the Lam$_{NB}$ cells were removed. Process the remaining four wells of the dish the same way until all the Lam$_{NB}$ cells are harvested from the dish and pooled into the same 15-mL tube (*see* **Note 18**). The Lam$_{NB}$ cells can now be discarded if not required or used for further experimentation if needed.

22. Using a 1-mL tip, mix the medium E in each well of the culture dish by gently pipeting up and down once, as in **step 19**. Then, in sets of four wells, remove the medium E from the wells and discard it into the waste beaker. Immediately add back 0.8 mL fresh medium E to each of the four wells by gently dispensing the medium dropwise evenly over the full surface area of each well (*see* **Note 19**). Repeat this step for the remaining four wells containing bound cells, which are termed laminin-binding (Lam$_{B}$) testis cells (*see* **Fig. 12.4**, Lam$_{B}$).

23. In sets of four wells, use a 1-mL tip to remove the medium E from each well and discard it into the waste beaker. Immediately add back 0.8 mL PBS–BSA solution to each of the wells from which medium E was removed. Repeat this step for the remaining four wells containing cells. After the PBS–BSA solution has been added to all eight of the wells containing cells, place the dish back into the incubator (*see* **Note 20**). Incubate the culture dish for 4 min at 32.5°C, 5.5% CO_2. Proceed immediately to **step 24**.

24. Add 4 mL medium E into each of two sterile 15-mL tubes. Detach the Lam$_{B}$ cells from the dish using a 1-mL tip to gently pipet the PBS–BSA solution up and down about five times in each well in a set of four wells. Take care to cover the full surface area of each well while pipeting to detach all the Lam$_{B}$ cells in each well from the laminin matrix. Harvest the detached Lam$_{B}$ cells from the four wells using a 1-mL tip to transfer them into one of the 15-mL tubes containing medium E. Process the remaining four wells until all the wells containing cells on the dish have been pooled into the two 15-mL tubes containing medium E (i.e., Lam$_{B}$ cells from a total of four wells/15 mL tube). View the empty culture wells under a microscope to make sure that the Lam$_{B}$ cells were harvested.

25. Pellet the suspensions of Lam$_{B}$ cells at 200g for 5 min. Carefully pour off the supernatant solution into the waste beaker by inverting the tube and holding it upside down. Keep the tube upside down and gently shake out the residual medium before reverting it. The cell pellet will remain bound in the tube (*see* **Note 21**).

26. Pool the pellets from both 15-mL tubes into a single 15-mL tube by suspending them with a total of 2 mL medium E. Determine the cellular concentration of the pooled suspension using a hemoacytometer. This protocol consistently yields approx. 10^6 Lam$_{B}$ cells (i.e., $9.9 \times 10^5 \pm 2.8 \times 10^5$ cells/mL/well, SD,

n = 34 primary cultures) per ten dishes of primary testis cell cultures prepared by procedures A and B. The concentration of stem cells in the isolated Lam_B cell population is 100 times that of the Lam_{NB} population and is essentially free of somatic testis cells.

12.3.4 Concluding Remarks

The isolated Lam_B spermatogonia can now be used as a highly enriched source of stem cells to produce transgenic rats *(5)*, to study the molecular biology of spermatogenesis *(12–15)*, to derive spermatogonial stem cell lines *(7)*, and to initiate spermatogenesis during culture in vitro or in vivo *(7,13,14)*. **Figure 12.5** illustrates in vitro and in vivo spermatogenesis colony-forming assays that were initiated with freshly isolated Lam_B spermatogonia. Note that even after removal of somatic cells, Lam_B spermatogonia are enriched greater than 100-fold in stem cell activity compared to the Lam_{NB} spermatogenic cells.

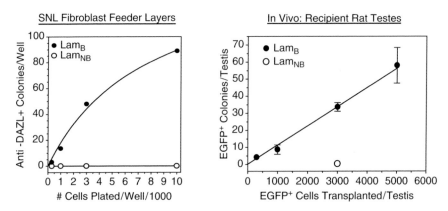

Fig. 12.5 Spermatogenesis colony-forming assays initiated with freshly isolated Lam_B spermatogonia and Lam_{NB} spermatogenic cells. **A** The number of spermatogenic colonies formed/well ($0.32\,cm^2$) by different numbers of freshly isolated wild-type Lam_B and Lam_{NB} germ cells after culturing for 9 d on feeder layers of the SNL 96/7 mouse fibroblast line (SNL 96/7 cells were provided by Allan Bradley). Spermatogenic colonies were scored in each well after labeling with an antibody to the germ cell-specific marker protein, Deleted in Azoospermia-like (DAZL). Data represent the average number of colonies formed per well from triplicate wells. **B** The number of enhanced green fluorescent protein-positive ($EGFP^+$) spermatogenic colonies formed per testis by different numbers of freshly isolated $EGFP^+$, Lam_B, and Lam_{NB} germ cells after transplantation into the seminiferous tubules of busulfan-treated, wild-type rats. Colonies were scored at 32 d following their transplantation. Mean ± SD, n = 4–5 testes transplanted/point. The $EGFP^+$, Lam_B, and Lam_{NB} germ cells were isolated from Germ Cell Specific (GCS)-EGFP transgenic rats by procedures A, B, and C, as reported here. The GCS-EGFP transgenic rat line was provided by Robert E. Hammer and expresses EGFP specifically in germ cells during all stages of male and female gametogenesis *(18)*

12.4 Notes

1. Freshly isolated rat seminiferous tubules stick to plastic easily and are lost within several types of plastic pipets, including 10-mL Falcon brand pipets. However, we find that the tubules stick the least to Falcon brand 5-mL pipets (cat. no. 357543, Falcon Inc.); this reduces the loss of tubules during the steps in procedures A and B. This is not a problem after the tubules have been disaggregated into a cellular suspension with dispase.
2. The tips of the 5/45° angle tweezers are extremely delicate and thus can be easily damaged by hitting/bumping them on the cell culture dish or other surfaces. Care should be taken not to damage the tweezers, which will make dissecting out and manipulating the seminiferous tubules much easier.
3. The water bath and table-top centrifuge should be located in the culture room and as close to the culture hood as possible.
4. The Styrofoam tube racks function to reduce heat shock to the cultures from the work surface.
5. Testes from rats of these ages contain spermatogonia and spermatocytes but have yet to develop spermatids *(16)*.
6. Avoid transferring rodent hair into the culture dish with the isolated testes to reduce risk of contamination of the subsequent cultures with microorganisms.
7. The short incubation step in the hypertonic glycine buffer selectively destroys the interstitial somatic cells prior to culture *(17)*. The germ cells, peritubular myoid cells, and Sertoli cells do not seem to be affected by this short treatment. However, care should be taken to minimize this incubation step to no more than 10 min as longer incubation times may become harmful for the other cell types *(17)*.
8. Optionally, if a shaking device is not near the cell culture room, gently agitate the digest in a vertical position by hand for 10 s to disaggregate the testis cells.
9. This will allow the researcher to use the free hand to manually control the strength of the vacuum to the lowest required for filtering the suspension. High vacuum pressure will kill cells.
10. Use the entire surface area of the mesh in the MFS while filtering the suspension to reduce clogging. Use one piece of mesh for every 10–12 mL cell suspension.
11. As a general estimate, one should coat 2 wells of a 12-well culture dish with laminin for every three 10-cm culture dishes of testis cells that are plated out in procedure B. If required, the laminin-coated culture plate can be prepared the same morning that procedure C is started; however, the laminin solution should incubate in the culture plate for at least 4 h while within the culture hood before use.
12. This wash step helps to further remove debris and dead cells from the cultures and loosens the attachment of live germ cells to the somatic cells so that the germ cells can be easily harvested in **step 7**. However, too long an incubation period in PBS+ during **step 6** can also loosen the attachment of the somatic cells to the culture dish, and **step 6** should be omitted if one observes by using the microscope that somatic cells are clearly being rinsed from the plate during **step 7**.
13. In this step, bound germ cells are pipeted free from off the top of the tightly adherent somatic cells on the culture dish (**Fig. 12.4**, TC and SC). A germ cell-enriched suspension (>80% germ cells) is thus obtained in medium A. If it appears that parts of the somatic cell monolayers are also washing off of the plate, reduce the strength of the washes by pipeting more slowly during **step 7** or by omitting the PBS+ wash in **step 6**. The germ cells should detach from the somatic cells with only gentle pipeting. Reducing the number of somatic cells harvested at this point is important for obtaining a highly pure culture of spermatogonia at the end of procedure C.
14. Processing two sets of four to six testis cell cultures (i.e., 8–12 total 10-cm cultures from procedure B) through **steps 4–11** of procedure C (**Subheading 12.3.3**) will yield germ cell suspensions to be plated into the two individual 10-cm collagen I-coated culture dishes that were prepared in **step 3**.

15. This selection step on collagen I further depletes somatic cells from the germ cell-enriched suspension and yields a highly pure suspension of germ cells (>96%), termed collagen-non-binding testis cells (*see* **Fig. 12.4**, Col$_{NB}$), based on their lack of adherence to the collagen plates *(5,14)*. The cells that bind to the collagen I-coated dishes are predominantly somatic and are termed collagen-binding testis cells (*see* **Fig. 12.4**, Col$_B$) *(5,14)*. The Col$_B$ somatic cells can be used for further experimentation or discarded after this step.

16. On average, the suspension of Col$_{NB}$ cells plated on laminin is $1.8 \times 10^6 \pm 7.2 \times 10^5$ cells/mL/well (±SD, n = 34 primary cultures).

17. At this point, the undifferentiated spermatogonia are bound to laminin. Care should be taken not to dislodge them from the culture dish during this step by too stringent pipeting. The cultures can be briefly checked under the microscope to see if the laminin-binding cells (Lam$_B$ cells) are still present (*see* **Fig. 12.4**, Lam$_B$) and then immediately placed back into the incubator.

18. The suspension of Lam$_{NB}$ cells obtained in this step (*see* **Fig. 12.4**, Lam$_{NB}$) contains approx. 98% germ cells that consist of differentiating spermatogonia and spermatocytes and is depleted of the spermatogonial stem cells *(5,14)*.

19. The washing and mixing in **steps 19–22** are critical for obtaining a high yield of relatively pure undifferentiated spermatogonia (>90%).

20. The testis cells that remain bound to the laminin-coated dish during this step are termed laminin-binding testis cells (*see* **Fig. 12.4**, Lam$_B$) *(5,14)*. The Lam$_B$ cell population contains approx. 96% germ cells, of which approx. 90% are undifferentiated type A spermatogonia that are highly enriched in germline stem cell activity *(5,13,14)*.

21. *Do not revert* the tube until all of the media has been poured out; this could create backwash that could knock the pellet loose. To be safe, the final supernatant can be poured off into a sterile 50-mL tube, which can be saved until the final yield of Lam$_B$ cells has been determined by cell counts.

Acknowledgments This work was supported by the Cecil H. and Ida Green Center for Reproductive Biology Sciences and the Howard Hughes Medical Institute.

References

1. Steinberger, A., and Steinberger, E. (1966) In vitro culture of rat testicular cells. *Exp. Cell. Res.* **44**, 443–452.
2. Brinster, R. L., and Avarbock, M. R. (1994) Germline transmission of donor haplotype following spermatogonial transplantation. *Proc. Natl. Acad. Sci. U. S. A.* **91**, 11303–11307.
3. Brinster, R. L., and Zimmermann, J. W. (1994) Spermatogenesis following male germ-cell transplantation. *Proc. Natl. Acad. Sci. U. S. A.* **91**, 11298–11302.
4. Ryu, B. Y., Orwig, K. E., Kubota, H., Avarbock, M. R., and Brinster, R. L. (2004) Phenotypic and functional characteristics of spermatogonial stem cells in rats. *Dev. Biol.* **274**, 158–170.
5. Hamra, F. K., Gatlin, J., Chapman, K. M., et al. (2002) Production of transgenic rats by lentiviral transduction of male germ-line stem cells. *Proc. Natl. Acad. Sci. U. S. A.* **99**, 14931–14936.
6. Kubota, H., Avarbock, M. R., and Brinster, R. L. (2003) Spermatogonial stem cells share some, but not all, phenotypic and functional characteristics with other stem cells. *Proc. Natl. Acad. Sci. U. S. A.* **100**, 6487–6492.
7. Hamra, F. K., Chapman, K. M., Nguyen, D. M., Williams-Stephens, A. A., Hammer, R. E., and Garbers, D. L. (2005) Self renewal, expansion, and transfection of rat spermatogonial stem cells in culture. *Proc. Natl. Acad. Sci. U. S. A.* **102**, 17430–17435.
8. Kanatsu-Shinohara, M., Miki, H., Inoue, K., et al. (2005) Long-term culture of mouse male germline stem cells under serum-or feeder-free conditions. *Biol. Reprod.* **72**, 985–991.
9. Kanatsu-Shinohara, M., Ogonuki, N., Inoue, K., et al. (2003) Long-term proliferation in culture and germline transmission of mouse male germline stem cells. *Biol. Reprod.* **69**, 612–616.

10. Kubota, H., Avarbock, M. R., and Brinster, R. L. (2004) Growth factors essential for self-renewal and expansion of mouse spermatogonial stem cells. *Proc. Natl. Acad. Sci. U. S. A.* **101,** 16489–16494.

11. Ryu, B. Y., Kubota, H., Avarbock, M. R., and Brinster, R. L. (2005) Conservation of spermatogonial stem cell self-renewal signaling between mouse and rat. *Proc. Natl. Acad. Sci. U. S. A.* **102,** 14302–14307.

12. Diederichs, S., Baumer, N., Schultz, N., et al. (2005) Expression patterns of mitotic and meiotic cell cycle regulators in testicular cancer and development. *Int. J. Cancer* **116,** 207–217.

13. Hamra, F. K., Chapman, K. M., Nguyen, D., and Garbers, D. L. (2006) Identification of Neuregulin as a factor required for formation of aligned spermatogonia. *J. Biol. Chem.* **282,** 721–730.

14. Hamra, F. K., Schultz, N., Chapman, K. M., et al. (2004) Defining the spermatogonial stem cell. *Dev. Biol.* **269,** 393–410.

15. Schultz, N., Hamra, F. K., and Garbers, D. L. (2003) A multitude of genes expressed solely in meiotic or postmeiotic spermatogenic cells offers a myriad of contraceptive targets. *Proc. Natl. Acad. Sci. U. S. A.* **100,** 12201–12206.

16. Malkov, M., Fisher, Y., and Don, J. (1998) Developmental schedule of the postnatal rat testis determined by flow cytometry. *Biol. Reprod.* **59,** 84–92.

17. Mather, J. P., and Phillips, D. M. (1984) *Primary culture of testicular somatic cells*, Vol. 2, Liss, New York.

18. Cronkhite, J. T., Norlander, C., Furth, J. K., Levan, G., Garbers, D. L., and Hammer, R. E. (2005) Male and female germline specific expression of an EGFP reporter gene in a unique strain of transgenic rats. *Dev. Biol.* **284,** 171–183.

Chapter 13
Deriving Mouse Spermatogonial Stem Cell Lines

Ilaria Falciatori, Kate Lillard-Wetherell, Zhuoru Wu, F. Kent Hamra, and David L. Garbers

Contents

Summary   In adult males, spermatogonial stem cells function to replenish developing gametes that are continuously released from the testes as mature spermatozoa. Because of their potential importance to research, medicine, industry, and conservation, numerous attempts have been made in the past to cultivate spermatogonial stem cells in vitro. However, only recently have culture methods been established that effectively promote the proliferation of mammalian spermatogonial stem cells in vitro. We describe a simple and reproducible protocol for the derivation and maintenance of mouse spermatogonial stem cell lines that proliferate for long periods of time in culture.

Keywords   Germ cells; germline stem cells; mouse; spermatogonia; spermatogonial stem cells; spermatogonial stem cell culture; SSC; testis.

13.1   Introduction

Spermatogonial stem cells (SSCs) self-renew in the testes and provide differentiated daughter cells that are committed to developing into mature spermatozoa by the process of spermatogenesis. Methods for culturing mouse and rat SSCs in vitro have now been established in different laboratories (*1–3*). After proliferating for long periods of time in culture, SSCs can develop into functional spermatozoa when transplanted back into the testes of an immunocompatible recipient (*1–3*). Moreover, SSCs can be genetically modified with DNA constructs during culture in vitro and can then develop into transgenic spermatozoa in recipient testes. Therefore, such recipient males can be used to transmit experimentally induced germline modifications to progeny by natural mating. For these reasons, SSCs are a powerful tool for the creation of transgenic animals, including animals with targeted genomic mutations *(1,4–6)*. Several labs have identified pluripotent, embryonic-like stem cells in cultures of SSCs that were established from neonatal and adult mice *(7,8)*. This finding has spurred interest in SSCs by the greater stem cell community because, to date, pluripotent cells have not been identified in any other nonembryonic tissue.

The protocol outlined describes all the steps necessary to derive and culture mouse SSC lines in vitro. The derivation and culture procedure has been divided into four protocols, which are outlined in **Fig. 13.1**. In **Subheading 13.3.1**, we describe the isolation of mouse testicular cells and their behavior in culture during the derivation of SSC lines. In **Subheading 13.3.2**, we describe how to maintain and passage established SSCs lines in culture on feeder layers of mouse embryonic fibroblasts (MEFs). In addition, methods for propagating MEFs in culture and for blocking their proliferation with mitomycin C (MMC) prior to use as feeder layers

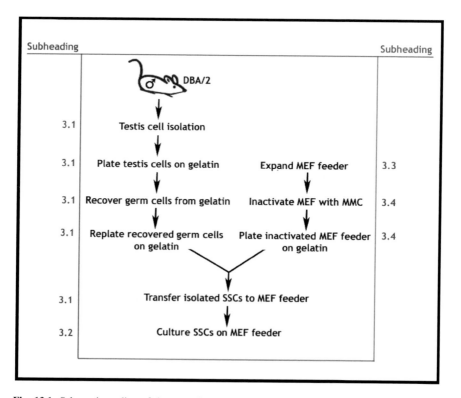

Fig. 13.1 Schematic outline of the procedures described in the chapter. Major steps required for derivation and culturing of spermatogonial stem cells (SSCs) as described in this chapter are outlined. The subheadings under which each step is detailed are indicated to the left and right of the outline

are presented in **Subheadings 13.3.3** and **13.3.4**, respectively. This protocol is based on that published by Kanatsu-Shinohara et al. *(1)* but includes several modifications established in our laboratory.

13.2   Materials

13.2.1   *Derivation and Maintenance of a Spermatogonial Stem Cell Culture*

1. DBA/2 mice (Harlan) (*see* **Note 1**).
2. Sterile surgical instruments: Two fine tweezers (Roboz, cat. no. RS-4918), scissors (Roboz, cat. no. RS-5912).
3. 35-mm tissue culture dishes.
4. 0.2% gelatin solution (*see* **Note 2**). Store at 4°C for up to 1 mo.
5. Tissue culture plates coated with 0.2% gelatin (*see* **Note 3**).
6. 40-μm nylon mesh cell strainers (Becton Dickinson, cat. no. 352340).
7. Trypsin 0.05%-EDTA (ethylenediaminetetraacetic acid; Invitrogen, cat. no. 25300-054).
8. Dulbecco's phosphate-buffered saline (D-PBS; Invitrogen, cat. no. 14190-136).
9. Dispase solution (Becton Dickinson, cat. no. 354235). The solution is received complete from the manufacturer; 12-mL aliquots are prepared and stored at −20°C. Sterile aliquots can be refrozen when necessary.
10. Shinohara Medium with FBS (SF): *See* **Table 13.1** for medium composition and preparation (*see* **Notes 4–18**).
11. DMEM/F12-PS: Dulbecco's modified Eagle's medium nutrient mixture F12 (Ham), 1X (Invitrogen cat. no. 11330-032) supplemented with 1% antibiotic–antimycotic solution (Invitrogen cat. no. 15240–062). Store at 4°C for up to 1 mo.
12. DMEM/F12-FBS: DMEM/F12 (Invitrogen cat. no. 11330-032) supplemented with 10% heat-inactivated fetal bovine serum (FBS) (*see* **Note 15**) and 1% antibiotic–antimycotic solution (Invitrogen cat. no. 15240-062). Store at 4°C for up to 1 mo.
13. MMC-treated MEFs (*see* **Subheadings 13.2.3** and **13.2.4**).

13.2.2   *Culturing and Cryopreservation of Established SSC Lines*

1. Trypsin 0.05%–EDTA (Invitrogen cat. no. 25300-054).
2. MMC-treated MEFs (*see* **Subheadings 13.2.3** and **13.2.4**).
3. D-PBS (Invitrogen cat. no. 14190–136).
4. SF: *See* **Table 13.1** for medium composition.
5. Recovery cell culture freezing medium (Invitrogen 12648-010).
6. Nalgene Mr. Frosty 5100 Cryo 1°C freezing container (Nalgene cat. no. 5100-0001).

Table 13.1 SF media components

Step	Component	Supplier, catalog number	Final concentration
1 Base media			
1.1	StemPro34 SFM base	Invitrogen, 10639-011	1X to volume
1.2	40X StemPro34 nutrient supplement	Invitrogen, 10639-011	1X
2 Add non-growth factor components			
2.1	Bovine serum albumin	Calbiochem, 126609	5 mg/mL
2.2	d-(+) Glucose	Sigma, G7021	6 mg/mL
2.3	L-Glutamine	Invitrogen, 25030-149	2 mM
2.4	50X Antibiotic–antimycotic	Invitrogen, 15240-062	1X
2.5	50X MEM vitamins	Invitrogen, 111420-052	1X
2.6	50X Nonessential amino acids	Invitrogen, 11140-050	1X
2.7	d-Biotin (*see* **Note 4**)	Sigma, B4639	10 µg/mL
2.8	Insulin (*see* **Note 5**)	Sigma, I1882	25 µg/mL
2.9	Pyruvic acid, sodium salt (*see* **Note 6**)	Sigma, P4562	30 µg/mL
2.10	dl-Lactic acid (60% solution) (*see* **Note 7**)	Sigma, L7900	0.06%
2.11	Ascorbic acid (*see* **Note 8**)	Sigma, A4034	100 µM
2.12	Sodium selenite (*see* **Note 9**)	Sigma, S5261	30 nM
2.13	Putrescine (*see* **Note 10**)	Sigma, P5780	60 µM
2.14	Bovine Apo-transferrin (*see* **Note 11**)	Sigma, T1428	100 µg/mL
2.15	Progesterone (*see* **Note 12**)	Sigma, P8783	60 ng/mL
2.16	β-Estradiol 17-cypionate (*see* **Note 13**)	Sigma, E8004	30 ng/mL
2.17	2-Mercaptoethanol (*see* **Note 14**)	Sigma, M3148	10 µM
2.18	Fetal bovine serum (*see* **Note 15**)	Hyclone, SH30071	1%
3 Add growth factor components			
3.1	ESGRO (mouse LIF)	Chemicon, ESG1107	10^3 U/mL
3.2	Recombinant mouse EGF (*see* **Note 16**)	R&D Systems, 2028-EG-200	20 ng/mL
3.3	Recombinant human basic FGF (*see* **Note 17**)	Sigma, F0291	10 ng/mL
3.4	Recombinant rat GDNF (*see* **Note 18**)	R&D Systems, 512-GF-050	10 ng/mL

EGF, epidermal growth factor; FGF, fibroblast growth factor; GDNF, glial-cell-line-derived neurotrophic factor; MEM, modified Eagle's medium; LIF, Leukemia Inhibitory Factor

13.2.3 Culturing and Cryopreservation of Mouse Embryonic Fibroblasts

1. T225 culture flasks.
2. DR4-MEFs (American Type Culture Collection [ATCC] cat. no. SCRC-1045).
3. D-PBS (Invitrogen cat. no. 14190–136).
4. MEF-DMEM: 85% DMEM (ATCC cat. no. 30-2002), 15% heat-inactivated FBS (Hyclone cat. no. SH30071) (*see* **Note 15**). Store at 4°C for up to 1 mo.
5. Trypsin 0.25%–EDTA (Invitrogen cat. no. 25200-056).

6. MEF freezing medium: 80% characterized heat-inactivated FBS (Hyclone, cat. no. SH30071), 20% dimethyl sulfoxide (DMSO; ATCC cat. no. 4-X) (*see* **Note 15**). Prepare fresh immediately before use.
7. Nalgene Mr. Frosty 5100 Cryo 1°C freezing container (Nalgene cat. no. 5100-0001).

13.2.4 Mitomycin C Treatment of MEFs

1. Gelatin-coated plates: As described in **Subheading 13.2.1**.
2. DR4-MEFs (ATCC cat. no. SCRC-1045).
3. D-PBS (Invitrogen cat. no. 14190-136).
4. MMC from *Streptomyces* (Sigma cat. no. M0503) diluted to a stock concentration of 1 mg/mL in D-PBS.
5. DMEM (ATCC cat. no. 30-2002).
6. MEF-DMEM: 85% DMEM (ATCC cat. no. 30-2002), 15% heat-inactivated FBS (Hyclone cat. no. SH30071) (*see* **Note 15**). Store at 4°C for up to 1 mo.
7. Trypsin 0.25%–EDTA (Invitrogen cat. no. 25200-056).

13.3 Methods

13.3.1 Derivation and Maintenance of a Spermatogonial Stem Cell Culture

In this subheading, we describe all the steps necessary to establish an SSC line. We first describe the isolation of a heterogeneous cell suspension from the mouse testes. These cells are initially plated on gelatin to remove the majority of somatic cells, resulting in a cell suspension enriched in germ cells. The remaining somatic cells support growth of the germ cells for 2–3 wk before transferring onto monolayers of MMC-treated MEFs.

1. Using sterile surgical instruments, excise testes from three DBA/2 mice (7 d of age; *see* **Note 1**), and place them in a 35-mm dish containing 2 mL DMEM/F12-PS. **Steps 1–3** are accomplished outside a biological safety cabinet.
2. Wash the testes once by transferring to a new 35-mm dish containing 2 mL DMEM/F12-PS.
3. Use sterile tweezers to remove the tunica albuginea (clear membrane surrounding the seminiferous tubules of the testis) and transfer the decapsulated testes to a new dish containing 2 mL DMEM/F12-PS. From this point, all the steps are performed in a biological safety cabinet using aseptic technique.
4. Loosen the testicular tissue by pulling with sterile tweezers such that most of the seminiferous tubules are exposed to medium.

5. Collect the loosened tissue in a 50-mL polypropylene tube containing 20 mL DMEM/F12-FBS and pellet by centrifugation at 270g for 5 min.

6. Discard the supernatant and resuspend the pellet in dispase solution prewarmed to 32°C. The volume of dispase depends on the amount of tissue to be digested (*see* **Note 19**). Incubate the tube horizontally at 32°C for up to 30 min. Gently agitate the digest by inverting the tube four to six times every 5 min. Also during this step, note the progression of the digest at each 5-min interval (*see* **Note 20**).

7. Stop the digestion by adding 30 mL DMEM/F12-FBS and shaking the tube vigorously by hand for 10 s (*see* **Note 21**).

8. Pellet by centrifugation at 270g for 5 min and discard the supernatant.

9. Wash once with 30 mL DMEM/F12-FBS, centrifuge at 270g for 5 min, and discard the supernatant.

10. Resuspend the pellet in 2 mL DMEM/F12-FBS. To remove debris, filter through a nylon mesh cell strainer (40-μm pore) into a new 50-mL polypropylene tube (*see* **Fig. 13.2A,B**).

11. Count the cells in a hemocytometer (*see* **Note 22**).

12. Adjust the volume of SF medium to 2.0×10^5 cells/mL and plate 1 mL into each well of a gelatin-coated 12-well plate.

13. Incubate for 16–24 h at 37°C in a humidified incubator with 5% CO_2.

14. To disrupt binding of germ cells to somatic cells, gently pipet SF medium up and down in a figure-eight motion across the surface of the well approx. ten times using a p1000 pipet (*see* **Fig. 13.2C,D**). Recover floating cells in a polypropylene tube (*see* **Note 23**).

15. Pellet cells by centrifugation at 270g for 5 min, discard the supernatant, and resuspend the cells in 5 mL SF medium. Thoroughly disaggregate cell clumps by pipeting the solution up and down ten times with a 2-mL pipet. Count cells in a hemocytometer.

16. Adjust the volume of SF medium to 2.0×10^5 cells/mL and plate 1 mL into each well of a gelatin-coated 12-well plate. Incubate in a humidified incubator at 37°C with 5% CO_2.

17. Cells should be replenished with fresh SF medium every 3 d (*see* **Note 24** and **Fig. 13.2E**). After 1 wk to 10 d, the cells are ready to be passaged onto a new gelatin-coated plate.

18. Aspirate the medium and wash once with 1 mL D-PBS (*see* **Note 25**).

19. Add 0.5 mL trypsin 0.05%–EDTA prewarmed to 37°C and incubate at 24°C or 37°C until the cells begin to detach.

20. Block trypsinization with 1 mL SF medium and pipet up and down 15–20 times with a p1000 pipet to disaggregate cell clumps. Recover the cells in a polypropylene tube

21. Pellet cells by centrifugation at 270g for 5 min, discard the supernatant, and resuspend cells in 5 mL SF medium. Count cells using a hemocytometer.

22. Adjust the volume of SF medium to 2.0×10^5 cells/mL and plate 1 mL into each well of a gelatin-coated 12-well plate. Incubate in a humidified incubator at 37°C with 5% CO_2.

23. Cells should be replenished with fresh SF medium every 3 d (*see* **Note 24**). After 1 wk to 10 d, the cells are ready to be passaged onto MEFs.
24. Repeat **steps 18–21** to passage for the second time.
25. Adjust the volume of SF medium to 2.0×10^5 cells per mL and plate 1 mL into each well of a 12-well plate containing MMC-treated MEFs (prepared as

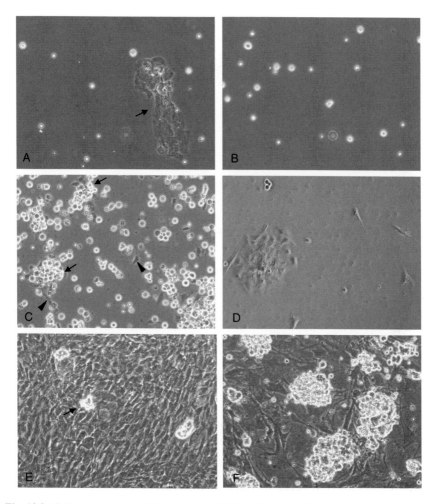

Fig. 13.2 Cell morphology at different stages of SSC derivation and culturing. **A** Testicular cells after enzymatic isolation. Note that single cells and undigested debris (*arrow*) are present in the suspension. **B** Testicular cells after filtration. Only single cells are present in the suspension. **C** Testicular cells after 18-h incubation on gelatin. Somatic cells are attached to the substrate (*arrowheads*), and germ cell clumps are clearly distinguishable (*arrows*). Most somatic cells cannot be clearly seen because they lie under the germ cells. **D** Somatic cells left in the gelatin-coated plate in **C** after germ cell recovery. **E** Small germ cell clusters (*arrow*) attached to testicular-derived somatic cells 5 d after isolation. **F** Germ cell clusters on mouse embryonic fibroblasts (MEFs) treated with mitomycin C (MMC) in an established culture. Magnification ×10 for all images

described in **Subheadings 13.3.3** and **13.3.4**). Incubate in a humidified incubator at 37°C with 5% CO_2.

26. Cells should be replenished with fresh SF medium every 3 d (*see* **Note 24**). After approx. 5–7 d in culture, cells are ready to be passaged as an established SSC line (*see* **Subheading 13.3.2, steps 2–8**).

13.3.2 Culturing and Cryopreservation of Established SSC Lines

In this subheading, we describe the routine passaging and cryopreservation of SSCs following their isolation and derivation, as described in **Subheading 13.3.1**.

1. Once SSCs are established on MEFs, they are passaged every 5–7 d; after this amount of time, they have typically reached a density of approx. 5–10×10^4 cells/cm^2 (*see* **Fig. 13.2F**). Volumes given in this section are for passaging in 12-well culture dishes and should be scaled according to the size of the culture vessel used.
2. Aspirate the medium and wash once with 1 mL D-PBS (*see* **Note 25**).
3. Add trypsin 0.05%–EDTA prewarmed to 37°C and incubate at 24°C or 37°C until the cells begin to detach.
4. Block trypsinization with 1 mL SF medium and pipet up and down 15–20 times with a p1000 pipet to disaggregate cell clumps. Recover cells into a polypropylene tube.
5. Pellet cells by centrifugation at 270g for 5 min and discard the supernatant.
6. Resuspend cells in 5 mL D-PBS and repeat **step 5**.
7. Resuspend cells in 5 mL SF medium and count cells in a hemocytometer (*see* **Note 26**).
8. Plate cells onto MMC-treated MEFs (*see* **Subheadings 13.3.3** and **13.3.4**) at a density of 1.3×10^4 cells/cm^2 in SF media (*see* **Note 27**).
9. Cells should be replenished with fresh SF medium every 2–3 d between passages (*see* **Note 24**).
10. To cryopreserve cells, trypsinize and collect cells according to **steps 2–7**.
11. Pellet cells by centrifugation at 270g for 5 min and discard SF medium.
12. Resuspend cells in recovery cell culture freezing medium at a concentration of 5.0×10^5 cells/mL and aliquot 1 mL into individual 1.0- to 2.0-mL cryogenic vials.
13. Place cryogenic vials into a Mr. Frosty freezing container or other styrofoam-insulated container and freeze for 16–24 h at −80°C.
14. Transfer cryogenic vials to a cryogenic liquid nitrogen vapor phase freezer for long-term storage.
15. To establish a culture from cryopreserved cells, thaw cryovials in a 37°C water bath for approx. 90 s and immediately transfer to a polypropylene tube with 9 mL SF medium.
16. Pellet cells by centrifugation at 270g for 5 min and discard supernatant.
17. Resuspend cells in 1–2 mL SF medium and plate onto MMC-treated MEFs in 1 well of a 12-well culture dish.

13.3.3 Culturing and Cryopreservation of Mouse Embryonic Fibroblasts

MMC-treated MEFs are necessary for routine culturing of SSCs, as described in **Subheading 13.3.2**. In this subheading, methods are presented for maintenance and cryopreservation of MEFs based on guidelines established by the supplier (ATCC).

1. Thaw a fresh vial of DR4-MEFs from ATCC in a 37°C water bath for approx. 90 s and immediately transfer to a polypropylene tube with 9 mL MEF-DMEM.
2. Pellet cells by centrifugation at 270g for 5 min and discard the supernatant.
3. To completely remove trace DMSO present in freezing medium, wash cells by resuspending the cell pellet with 10 mL MEF-DMEM and then repeating **step 2**.
4. Resuspend the cell pellet in 50 mL MEF-DMEM and transfer to a T225 flask. Incubate in a humidified incubator at 37°C with 5% CO_2.
5. Medium should be changed every 3 d. Cells are split once they reach confluency (~5–7 d).
6. Aspirate media from cells and wash once with D-PBS.
7. Add 5 mL trypsin 0.25%–EDTA prewarmed to 37°C to the flask and incubate at 24°C until cells begin to detach. Gently tap the flask to detach cells and monitor using an inverted microscope.
8. Block trypsinization with 10 mL MEF-DMEM and recover cells in a polypropylene tube.
9. Pellet the cells by centrifugation at 270g for 5 min and discard the supernatant.
10. Resuspend the cell pellet in 10 mL D-PBS and repeat **step 9**.
11. Resuspend the cell pellet in 5 mL MEF-DMEM and count cells using a hemocytometer.
12. Plate cells at a density of 5.0 × 10^6 cells per T225 flask. This generally corresponds to a 1:5 to 1:7 split ratio. Incubate in a humidified incubator at 37°C with 5% CO_2.
13. Medium should be changed every 3 d. Cells are split once they reach confluency (~5–7 d).
14. At passage 2 (P2), MEFs can be cryopreserved for future use as described in **steps 15–19**. One may also repeat **steps 6–12** up to two times and cryopreserve the resulting P3 or P4 MEFs (*see* **Note 28**).
15. For cryopreservation, cells are trypsinized and collected as described in **steps 6–11**.
16. Resuspend cells in MEF freezing medium at a density of 5.0 × 10^6 cells/mL and aliquot 1 mL into individual 1.0- to 2.0-mL cryogenic vials.
17. Place cryogenic vials into a Mr. Frosty freezing container or other styrofoam-insulated container and freeze for 16–24 h at −80°C.
18. Transfer cryogenic vials to a cryogenic liquid nitrogen vapor phase freezer for long-term storage.
19. Cryopreserved P2 to P4 MEFs are thawed and established according to **steps 1–4**.

13.3.4 Mitomycin C Treatment of MEFs

In this subheading, methods are presented for preparing feeder layers of MEFs to support the growth of SSC lines, as described in **Subheading 13.3.2**. In particular, this includes treating the MEFs with MMC to block their proliferation in culture prior to their use as feeder layers.

1. Freshly confluent vessels of MEFs are required for the following procedure (*see* **Subheading 13.3.3**). Volumes given in this section are for MMC treatment of MEFs cultured in T225 flasks and should be scaled according to the size of the culture vessel used.
2. Aliquot 25 mL MEF-DMEM for each T225 flask to be treated. Dilute the MMC stock 1:100 in DMEM for a final concentration of 10 μg/mL MMC (*see* **Note 29**).
3. Aspirate MEF-DMEM from the required number of T225 flasks containing freshly confluent monolayers of MEFs and then add 25 mL MEF-DMEM containing 10 μg/mL MMC to each flask. Incubate in a humidified incubator at 37°C with 5% CO_2 for 2 h and 20 min. Typically, 1.8–2.0 × 10^7 MEFs are obtained from one T225 flask after the cells have reached confluence.
4. Pipet off MMC-containing DMEM and dispose appropriately into biohazard waste.
5. Wash cells with fresh DMEM three times and aspirate media to completely remove MMC.
6. Add 5 mL trypsin 0.25%–EDTA prewarmed to 37°C to the flask and incubate at 24°C until cells begin to detach. Gently tap the flask to detach cells and monitor using an inverted microscope.
7. Block trypsinization with 10 mL MEF-DMEM and recover cells into a polypropylene tube.
8. Pellet the cells by centrifugation at 270 g for 5 min and discard the supernatant.
9. Resuspend the cell pellet in 10 mL D-PBS and repeat **step 8**.
10. Resuspend the cell pellet in MEF-DMEM and plate at a density of 4.25 × 10^4 cells/cm^2 on gelatin-coated plates or flasks. Incubate in a humidified incubator at 37°C with 5% CO_2.
11. MMC-treated MEFs may be used for SSC culturing once they have attached to the gelatin-coated plates (generally 4–6 h after plating) and up to 2 wk after treatment.
12. Excess MMC-treated MEFs may be cryopreserved (according to the protocol given in **Subheading 13.3.3**) for future use (*see* **Note 30**).

13.4 Notes

1. DBA/2 mice 7–15 d of age are recommended. If other strains or ages are used, this protocol may need to be modified by the user. We recommend starting with the strain and age described to establish the procedure before attempting modifications.

2. Weigh out 0.2 g type A gelatin from porcine skin (Sigma, cat. no. G-1890) and add 100 mL water. Autoclave to dissolve gelatin and sterilize.

3. For 12-well plates, dispense 0.5 mL gelatin solution per well; incubate at 24°C for at least 1 h or at 37°C for at least 30 min. Wash twice with 1 mL PBS before use. Plates can also be coated in advance, washed twice with sterile water, air dried in the culture hood, and stored at 4°C for up to 1 mo. Volumes should be adjusted according to the size of culture vessels used.

4. d-Biotin stock: Dilute to 10 mg/mL in 1N NaOH in tissue culture grade water (Sigma, cat. no. W3500). Store at −20°C.

5. Insulin stock: Dilute to 10 mg/mL with acidified water (100 μL glacial acetic acid in 10 mL tissue culture grade water). Store at 4°C.

6. Pyruvic acid, sodium salt stock: Prepare 30 mg/mL in tissue culture grade water immediately prior to use.

7. dl-Lactic acid: Store 60% stock solution at 4°C and warm to 24°C prior to use.

8. Ascorbic acid stock: Prepare 19.8 mg/mL stock in tissue culture grade water immediately prior to use.

9. Sodium selenite stock: Dilute to 1 mM in StemPro34 SFM base medium (Invitrogen cat. no. 10639–011). Store at −20°C.

10. Putrescine stock: Dilute to 80 mM in StemPro34 SFM base medium. Store at −20°C.

11. Bovine apo-transferrin stock: Dilute to 20 mg/mL in StemPro34 SFM base medium. Store at −20°C.

12. Progesterone stock: Dilute to 60 mg/mL in ethanol and store at −20°C (concentrated stock). Dilute concentrated stock to 60 μg/mL in StemPro34 SFM base medium immediately prior to use.

13. β-Estradiol 17-cypionate stock: Dilute to 30 mg/mL in ethanol and store at −20°C (concentrated stock). Dilute concentrated stock to 30 μg/mL in StemPro34 SFM base medium immediately prior to use.

14. 2-Mercaptoethanol stock: Dilute to 10 mM in StemPro34 SFM base medium immediately prior to use.

15. FBS: FBS is heat inactivated prior to use by incubating thoroughly thawed and mixed FBS at 56°C for 30 min and transferring to ice. Heat-inactivated FBS can be aliquoted and frozen for future use.

16. Recombinant mouse epidermal growth factor (EGF) stock: Reconstitute in D-PBS (Invitrogen cat. no. 14190-144) with 0.1% BSA (Calbiochem cat. no. 126609). Store at −20°C.

17. Recombinant human basic fibroblast growth factor (FGF) stock: Reconstitute in D-PBS with 0.1% BSA. Store at −20°C.

18. Recombinant rat glial-cell-line-derived neurotrophic factor (GDNF) stock: Reconstitute in D-PBS with 0.1% BSA. Store at −20°C.

19. Determine the wet weight of the testicular tissue. Use 1 mL dispase for every 100 mg of tissue. It is essential to use the appropriate volume of dispase to achieve optimal digestion.

20. As digestion progresses, one will observe the seminiferous tubules dissociating from each other, and then the walls of the tubules losing their integrity. It is important to gently agitate the digestion mixture every 5 min to facilitate tissue dissociation. It is normal to see some incompletely digested seminiferous tubules remaining at the end of the digestion.

21. Addition of medium at this step is important to reduce exposure of the dissociated cells to concentrated dispase. After vigorous shaking, the seminiferous tubules will almost completely disappear, and the suspension will contain isolated cells and undigested debris.

22. The cell yield varies depending on age of the mice. We recover an average of $5.5–7.5 \times 10^5$ cells per testis using 7-d-old mice.

23. Somatic cells bind to gelatin; germ cells do not. Cells detached by gentle pipeting are enriched for germ cells. Most somatic cells remain attached to the gelatin-coated plate and can be discarded.

24. SSCs frequently become detached and form floating cell clumps. If you observe floating cell clumps, do not discard them. Instead collect used media with cell clumps, centrifuge at 270g

for 5 min to pellet cells, discard supernatant, and resuspend cells in fresh SF. Use this suspension to feed the cells.

25. SSCs frequently become detached and form floating cell clumps. Do not discard floating clusters if present. Collect floaters and add to cells recovered after trypsinization. Floating single cells are generally not healthy; therefore, it is advisable to discard used media if floating cells are primarily single cells.

26. MEFs are considerably larger than SSCs; therefore, one can easily distinguish these cells by size when counting with a hemocytometer. It may be useful to compare cultures of MEFs alone to SSCs mixed with MEFs until accustomed to differentiating these cell types by size using the hemocytometer.

27. SSCs maintained on MEFs are not particularly sensitive to low density; therefore, the user may plate them according to their needs. A split ratio between 1:3 and 1:10 is recommended.

28. It is not advisable to cryopreserve P5 or P6 MEFs. MEFs should be discarded after P6.

29. MMC stock (1 mg/mL) can be stored at 4°C for up to 1 mo. Discard stock if precipitation occurs.

30. Frozen MMC-treated MEFs may need to be plated at a slightly higher density than fresh MMC-treated MEFs because of cell loss during freezing.

References

1. Kanatsu-Shinohara, M., Ogonuki, N., Inoue, K., et al. (2003) Long-term proliferation in culture and germline transmission of mouse male germline stem cells. *Biol. Reprod.* **69,** 612–616.

2. Ryu, B. Y., Kubota, H., Avarbock, M. R., and Brinster, R. L. (2005) Conservation of spermatogonial stem cell self-renewal signaling between mouse and rat. *Proc. Natl. Acad. Sci. U. S. A.* **102,** 14302–14307.

3. Hamra, F. K., Chapman, K. M., Nguyen, D. M., Williams-Stephens, A. A., Hammer, R. E., and Garbers, D. L. (2005) Self renewal, expansion, and transfection of rat spermatogonial stem cells in culture. *Proc. Natl. Acad. Sci. U. S. A.* **102,** 17430–14735.

4. Hamra, F. K., Gatlin, J., Chapman, K. M., et al. (2002) Production of transgenic rats by lentiviral transduction of male germ-line stem cells. *Proc. Natl. Acad. Sci. U. S. A.* **99,** 14931–14936.

5. Dann, C. T., Alvarado, A. L., Hammer, R. E., and Garbers, D. L. (2006) Heritable and stable gene knockdown in rats. *Proc. Natl. Acad. Sci. U. S. A.* **103,** 11246–11251.

6. Kanatsu-Shinohara, M., Ikawa, M., Takehashi, M., et al. (2006) Production of knockout mice by random or targeted mutagenesis in spermatogonial stem cells. *Proc. Natl. Acad. Sci. U. S. A.* **103,** 8018–8023.

7. Kanatsu-Shinohara, M., Inoue, K., Lee, J., et al. (2004) Generation of pluripotent stem cells from neonatal mouse testis. *Cell.* **119,** 1001–1012.

8. Guan, K., Nayernia, K., Maier, L. S., et al. (2006) Pluripotency of spermatogonial stem cells from adult mouse testis. *Nature.* **440,** 1199–1203.

Chapter 14
Production of Knockdown Rats by Lentiviral Transduction of Embryos with Short Hairpin RNA Transgenes

Christina Tenenhaus Dann and David L. Garbers

Contents

Summary The primary method for determining the function of a gene in rodents has been to make a knockout mouse through homologous recombination in embryonic stem cells. However, with the advent of RNA interference (RNAi) technology, new methods for studying gene function are now possible in a wide array of animals. We describe a protocol for knocking down a gene of interest in vivo in rats by stably expressing a short hairpin RNA (shRNA). Transgenic rats are produced using a simple and efficient procedure for transducing single-cell embryos with a lentiviral vector. The vector described is designed to result in ubiquitous expression of shRNA. Thus, it is well suited to study genes expressed specifically in male germ cells in which the predicted phenotype would be male sterility. This system has been used to generate a transgenic line with stable and heritable knockdown of the gene *Deleted in Azoospermia-like* (*Dazl*), resulting in male sterility and germline transmission of the transgene through females.

Keywords Germ cell; lentivirus; one-cell embryo; rat; RNAi; shRNA; transgene.

14.1 Introduction

We have entered an era when the availability of the complete genome sequence of several organisms has resulted in powerful new ways to explore questions in biology. For example, a large number of genes expressed during the process of spermatogenesis have been identified using microarray or other global gene expression profiling approaches *(1–4)*. However, the prospect of determining the functional relevance of each of these genes remains daunting. One potential high-throughput means of screening for the function of such genes would be through the use of RNA interference (RNAi).

RNAi refers to the process by which small (~21 nt) double-stranded RNAs (dsR-NAs) direct the degradation of complimentary cellular messenger RNAs (mRNAs). RNAi can be achieved using small interfering RNA (siRNA) or short hairpin RNA (shRNA) *(5)*. siRNAs are commonly used in vitro; they can be readily transfected into cells, resulting in transient interference. For sustained interference, cells can be engineered to carry a transgene, from which shRNA is continuously or condition-ally produced and then processed by the cellular machinery into siRNA. This has been shown to be an effective method to study gene function in vitro and in vivo. RNAi provides particular promise for study of gene function in the rat. Although rats are widely used as model organisms, knockout technology has not been devel-oped in this species because of the lack of pluripotent rat ES cell lines *(6,7)*.

Several approaches have been used to produce genetically modified rodents expressing shRNA. These include lentiviral transduction of embryos *(8–11)* or the injection of DNA into the pronucleus of single-cell embryos *(12–18)*; both tech-niques result in random transgene insertion and variable transgene copy number. Alternatively, embryos may be derived from embryonic stem (ES) cells that have been genetically modified prior to their injection into blastocysts *(19–23)*. Delivering a transgene to an embryo via lentiviral transduction has several advantages over other methods. First, it is less technically demanding than injecting DNA into the pronucleus, given that the lentivirus only needs to be injected under the zona pelluc-ida. In fact, transgenic mice have even been produced by simply soaking denuded (zona pellucida removed) embryos in media containing lentivirus prior to embryo transfer *(24,25)*. Second, lentiviral transgenes appear to be less susceptible to silenc-ing compared to other retroviral gene transfer agents *(26)*. Third, Vesicular Stomatitis Virus-G glycoprotein (VSVG) pseudotyped lentivirus has a wide host cell range and could be used to transduce embryos from many species, including rats *(27)*.

Here, we describe a method for generating transgenic rats using a lentiviral vector that is designed to express both shRNAs and green fluorescent protein (GFP) from separate cassettes. Expression of GFP is driven by the human ubiquitin C (Ubc) promoter, and expression of the desired shRNA is driven by the U6 promoter; both promoters are active in a wide variety of cell types *(28)*. Because of the dominant nature of RNAi, the method described is best suited for genes that function in a specific tissue or in the adult rat. It is therefore quite suitable to study germ cell-specific genes with putative functions in spermatogenesis. This is exemplified by the phenotype of transgenic rats that stably express shRNA designed to stimulate degradation of tran-scripts that encode for the Deleted in Azoospermia-like (Dazl) protein *(8,29)*.

This protocol is divided into five separate procedures. In **Subheading 14.3.1**, we describe a method for designing shRNA vectors that target specific genes. In **Subheading 14.3.2**, a method for preparing and concentrating lentivirus is presented. A protocol for testing lentiviral shRNA vectors using cell lines is discussed in **Subheading 14.3.3**. Finally, in **Subheadings 14.3.4** and **14.3.5**, we describe a procedure for lentiviral transduction of rat embryos and methods to identify and analyze transgenic animals, respectively.

14.2   Materials

14.2.1   *Preparation of Vector (see* **Fig. 14.1***)*

14.2.1.1   shRNA Design and Oligonucleotide Preparation

1. Polyacrylamide gel electrophoresis (PAGE) purified, 5′ phosphorylated sense and antisense oligonucleotides (IDT or Sigma).
2. 5X annealing buffer: 350 mM Tris-HCl, pH 7.5, 50 mM MgCl$_2$, 500 mM KCl.

14.2.1.2   Oligonucleotide Cloning into pLLU2G Vector

1. The restriction endonuclease, Hpal. (NEB).
2. Alkaline phosphatase.
3. Rapid ligation kit (Roche).
4. Competent cells with mutations in the *endA* and *recA* genes (such as Fusion Blue, BD Biosciences),
5. LB plates and medium with ampicillin (50 μg/mL).

14.2.1.3   Screening of Clones

1. Primer TD143 (CAGTGCAGGGGAAAGAATAGTAGAC).
2. Primer TD144 (GCGGCCGCTTAAGCTTGGAACCC).
3. 10X polymerase chain reaction (PCR) buffer: 100 mM Tris-HCl, pH 8.8, 250 mM KCl, 50 mM (NH$_4$)$_2$SO$_4$, 20 mM MgSO$_4$.
4. PCR master mix (per 25 μL reaction): 0.5 μL 10 μM TD143, 0.5 μL 10 μM TD144, 2.5 μL 10X PCR buffer, 2.5 μL 50% dimethyl sulfoxide (DMSO), 0.25 μL 25 mM dNTPs (deoxynucleotide 5′-triphosphates), 13.5 μL water, 0.25 μL 5U/μL Taq.
5. Thin-wall PCR tubes.

14.2.2   *Preparation of Lentivirus*

14.2.2.1   Production of Lentivirus

1. Plasmid DNAs (pMD2G, pMDLg/pRRE, pRSV-REV, pLLU2G or derivative) *(8,30)*.
2. Ethyl alcohol (EtOH) precipitated and resuspended in sterile TE (pH 7.5) at 500 μg /mL (*see* **Note 1**).

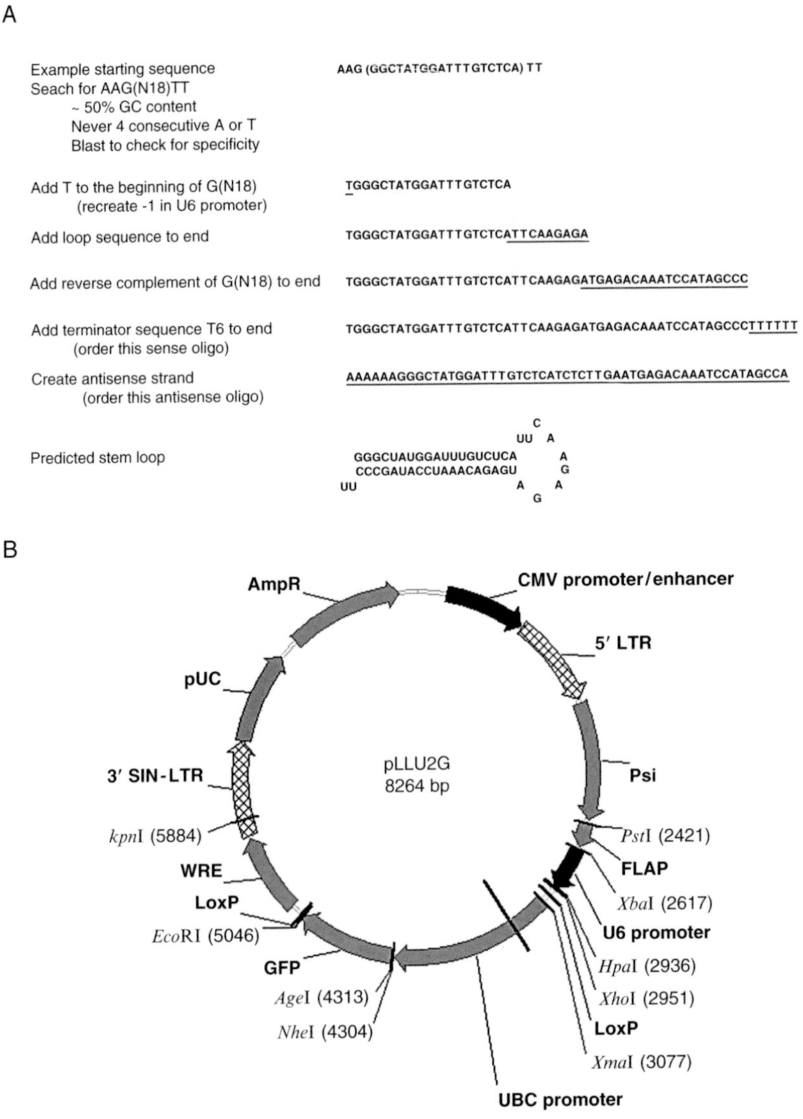
A
Example starting sequence
Seach for AAG(N18)TT
~ 50% GC content
Never 4 consecutive A or T
Blast to check for specificity
AAG (GGCTATGGATTTGTCTCA) TT
Add T to the beginning of G(N18)
(recreate -1 in U6 promoter)
TGGGCTATGGATTTGTCTCA
Add loop sequence to end
TGGGCTATGGATTTGTCTCATTCAAGAGA
Add reverse complement of G(N18) to end
TGGGCTATGGATTTGTCTCATTCAAGAGATGAGACAAATCCATAGCCC
Add terminator sequence T6 to end
(order this sense oligo)
TGGGCTATGGATTTGTCTCATTCAAGAGATGAGACAAATCCATAGCCCTTTTTT
Create antisense strand
(order this antisense oligo)
AAAAAAGGGCTATGGATTTGTCTCATCTCTTGAATGAGACAAATCCATAGCCA
Predicted stem loop
C
UU A
GGGCUAUGGAUUUGUCUCA A
CCCGAUACCUAAACAGAGU G
UU A A
G
B
AmpR
CMV promoter/enhancer
5′ LTR
pUC
Psi
3′ SIN-LTR
pLLU2G
8264 bp
kpnI (5884)
PstI (2421)
FLAP
WRE
XbaI (2617)
LoxP
U6 promoter
EcoRI (5046)
HpaI (2936)
GFP
XhoI (2951)
AgeI (4313)
LoxP
NheI (4304)
XmaI (3077)
UBC promoter

Fig. 14.1 **A** Guidelines for designing short hairpin (shRNA)-encoding oligonucleotides. **B** Map of pLLU2G vector. Useful unique restriction sites are shown. Hpa I is the restriction site for oligonucleotide insertion. pLLU2G is derived from pLL3.7 described in **ref. *10***

14.2.2.2 Lipofectamine 2000 Transfection

1. 0.1% gelatin (cell culture tested) dissolved in phosphate-buffered saline (PBS), autoclaved, and filtered.
2. Low passage number (<20) 293FT cells (Invitrogen) maintained in T150 flasks.

3. Dulbecco's modified Eagle's medium (DMEM) (with 10% FBS, 2 mM glutamine) with antibiotics (penicillin/streptomycin and 500 µg/mL G418) for cell maintenance and without antibiotics for transfection.
4. Lipofectamine 2000 and Opti-Mem (Invitrogen).

14.2.2.3 Calcium Phosphate Transfection

1. 0.5M CaCl$_2$ in ultrapure (Millipore) water; filter and store in 50-mL aliquots at 70°C; thawed aliquots are good for 2 mo at 4°C.
2. 2X HeBS, pH 7.0: 0.28M NaCl, 0.05M HEPES, 1.5 mM Na$_2$HPO$_4$ anhydrous in ultrapure (Millipore) water; filter and store in 50-mL aliquots at −70°C; thawed aliquots are good for 2 mo at 4°C.
3. BW: 2.5 mM HEPES, pH 7.3, in ultrapure (Millipore) water; prepare fresh from 1M HEPES stock and filter before use.

14.2.2.4 Concentration of Lentivirus

1. 50 mL Steriflip-HV filter unit with 0.45-µ PVDF (Millipore).
2. 30-mL Polyallomer Konical tube (Beckman) and adapters (Beckman 358156).
3. SW-28 rotor and ultracentrifuge (Beckman).

14.2.2.5 Titering of Lentivirus

1. 293 FT cells or cells to be used for in vitro knockdown tests.
2. 6 mg/mL Polybrene (hexadimethrine bromide) in water; filter and store in 100-µL aliquots at −20°C. Polybrene is added to media just prior to use at 6 µg/mL.

14.2.3 *Testing for Knockdown In Vitro*

1. Coding sequence of target gene cloned in frame with myc in pCDNA4-myc-HIS (Invitrogen).
2. Transfectable cell line such as Cos7 (CRL-1651, ATCC) or cell line expressing target gene.
3. Lipofectamine 2000 and OptiMem (Invitrogen).
4. Media without antibiotics and media with polybrene (*see* **Subheading 14.2.2.5**).
5. 9E10 monoclonal antibody (anti-myc).

14.2.4   Lentiviral Infection of Rat Embryos

14.2.4.1   Superovulation and Embryo Preparation

1. Ten stud males (proven fertile) and ten females (65–100 g).
2. 40 U/mL PMSG (Sigma G4877) in sterile filtered PBS with 1 mg/mL bovine serum albumin (BSA); store in 3-mL aliquots at −20°C (good for 1 mo).
3. 40 U/mL hCG (Sigma, CG5) in sterile filtered PBS with 1 mg/mL BSA; store in 3-mL aliquots at −20°C (good for 1 mo).
4. M2 medium (embryo tested; Sigma).
5. 3 mg/mL hyaluronidase (Sigma H4272) in M2; store in 100-µL aliquots at −20° C.
6. 1 mg/mL cytochalasin B in DMSO (store in lightproof vial); add 5 µL to 2 mL M2 just before use.
7. Mineral oil (embryo tested).
8. KSOM (Specialty Media MR-020-D).
9. 35-mm Petri dishes.

14.2.4.2   Embryo Injection (*see* **Fig. 14.2**)

1. Inverted microscope equipped with TransferMan NK micromanipulators (Eppendorf).
2. Microloaders (Eppendorf) for pipeting virus into Sterile Transfer Tip needle.
3. CellTram Vario and needle holder equipped with a VacuTip for holding embryo in place (Eppendorf) and loaded with silicone oil.
4. Transjector 5246 and needle holder equipped with a Sterile Transfer Tip-R (ICSI) needle (Eppendorf) and loaded with virus.
5. Aluminum block injection "chamber." This can be custom made in a machine shop (dimensions: 75 × 24 × 3 mm, with a 30 × 20 mm oval hole cut out of the middle). Prior to injection, a 24 × 60 mm coverslip is sealed to the bottom of the block by placing pieces of paraffin wax between the glass and the aluminum and heating in an oven at 56°C for 10 min.

14.2.5   Production of Transgenic Rats

1. 15 vasectomized males (may be purchased from Harlan).
2. 25–30 females (150–175 g).
3. Foster colony (optional).

A

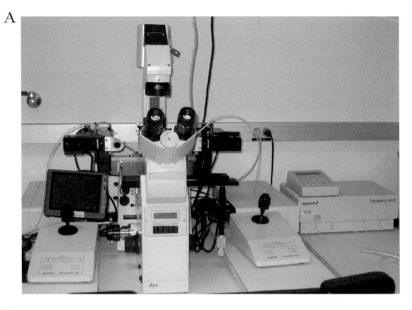

B

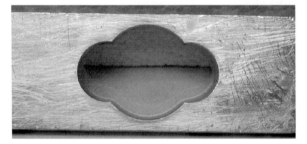

Fig. 14.2 **A** Microscope setup with manipulators for injecting embryos. **B** Aluminum block with glass coverslip attached as described in **Subheading 14.2.4.2**)

14.2.6 Founder Analysis

14.2.6.1 Genotyping

1. SNET: 10 mM Tris-HCl, pH 8.0, 5 mM ethylenediaminetetraacetic acid (EDTA), 1% sodium dodecyl sulfate (SDS), 400 mM NaCl.
2. PK: 20 mg/mL Proteinase K.
3. Primer GFP5′: CTGACCCTGAAGTTCATCTGCACCAC.
4. Primer GFP3′: TCCAGCAGGACCATGTGATC.
5. 10X PCR buffer: 100 mM Tris-HCl, pH 8.8, 250 mM KCl, 50 mM (NH$_4$)$_2$SO$_4$, 20 mM MgSO$_4$.

6. PCR master mix (per 25 μL reaction): 0.5 μL 10 μ*M* GFP5′, 0.5 μL 10 μ*M* GFP3′, 2.5 μL 10X PCR buffer, 2.5 μL 50% DMSO, 0.6 μL 15 m*M* dNTPs, 15.15 μL water, 0.25 μL 5 U/μL Taq.
7. Thin-wall PCR tubes.
8. Iso-freeze PCR chiller block (ISC BioExpress).

14.2.6.2   GFP Expression

For GFP expression, use a stereoscope equipped with a fluorescence attachment, GFP filter set, and 1.6X Plan Apo lens.

14.3   Methods

14.3.1   Preparation of Vector

14.3.1.1   shRNA Design and Oligonucleotide Preparation

The precise requirements for an effective shRNA target sequence remain ill-defined, but the following are recommended (*see* **Fig. 14.1**) *(10)*.

1. Search for the following sequence within the target mRNA: AAG(N18)TT (*see* **Note 2**).
2. Check G(N18) for the following:

   a. No terminators (four or more As or Ts in a row).
   b. 30–50% GC content.
   c. Lower GC content in the 3′ portion of the target mRNA sequence.
   d. Unique (intended) target on BLAST search.

3. Add T to the beginning of G(N18) to re-create −1 position of U6 promoter.
4. Add TTCAAGAGA (loop sequence) to end of G(N18).
5. Add reverse complement of G(N18) after loop.
6. Add TTTTTT (terminator sequence).
7. The resulting sequence [TG(N18)TTCAAGAGA(N18Rev)TTTTTT] and the reverse complement of the entire sequence are required (each should be 54 nucleotides long).
8. Order PAGE-purified oligonucleotides with 5′ phosphate modification.
9. Resuspend the oligonucleotides in water and determine the concentration using a spectrophotometer.
10. Make a 20 μL mixture containing 1 μ*M* of each oligonucleotide and 1X annealing buffer.

11. Anneal in a PCR machine using the following program: 95°C for 30 s, 60°C for 10 min, ramp from 60°C to 20°C at −1°C/15 s.

14.3.1.2 Oligonucleotide Cloning into pLLU2G Vector

1. Linearize approx. 5 μg pLLU2G vector with Hpa I, treat with alkaline phosphatase, and gel purify.
2. Ligate approx. 2 μL vector and 1 μL diluted (1:5 in water) annealed oligonucleotides.
3. Transform 1 μL of a 20 μL ligation reaction in 50 μL Fusion Blue cells.

14.3.1.3 Screening of Clones

Screen approx. 30 bacterial colonies using PCR. This reduces the number of plasmid preparations required by allowing elimination of colonies with no insert or multiple (tandem) inserts. By increasing the number of colonies screened, it allows one to find rare clones with insert if the ligation is inefficient.

1. Draw a grid on a new LB/ampicillin plate that has been warmed to room temperature.
2. Aliquot 5 μL water into each PCR tube.
3. Touch a pipet tip to a colony, mix in the 5 μL of water, and dot the residual liquid left in the tip onto a spot on the grid of the LB/ampicillin plate ("master plate").
4. Aliquot 20 μL PCR master mix (*see* **Subheading 14.2.1.3**) into each PCR tube.
5. Run the following program in a PCR machine: 1X (94°C for 5 min), 35X (94°C for 30 s, 52°C for 30 s, 72°C for 45 s), 1X (72°C for 7 min).
6. During the PCR reaction and gel analysis (**steps 5** and **7**), place the master plate at 37°C for approx. 5 h.
7. Add loading dye to each PCR reaction and run 15 μL of each reaction on a 1% agarose gel.
8. Reactions from colonies with insert should have a strong band that migrates more slowly (50 nt) compared to the positive control (*see* **Note 3**).
9. Perform plasmid preparations on clones with insert (identified in **step 8**) and check the overall integrity of each plasmid clone by restriction digests. Each of the following three different digests gives informative patterns, allowing pLLU2G parent vector to be readily discerned from clones carrying the oligonucleotide insert: Ava I, XbaI/XhoI, Hind III. Check sequences encoding each shRNA for the presence of these restriction sites before restriction analysis.
10. Sequence each clone with TD143 or TD144. This is important for determining the accuracy of the oligonucleotide synthesis and orientation of the insert.

14.3.2   Preparation of Lentivirus

14.3.2.1   Production of Lentivirus

Two transfection methods are presented. The Lipofectamine 2000 transfection is simpler but more costly. Each method should give similar yields of lentiviral particles. Both protocols are written for a single 10-cm plate but may be scaled up (three or four plates is reasonable).

14.3.2.1.1   Lipofectamine 2000 Method

1. Apply 5 mL 0.1% gelatin to a 10-cm plate and incubate at room temperature for at least 1 h. Use of a gelatin-coated plate improves adhesion of the cells.
2. Remove gelatin and apply 3×10^6 293 FT cells in 10 mL media (no antibiotics) to the plate.
3. Gently rock the plate to ensure an even distribution of cells and allow cells to adhere approx. 16 h (overnight) at 37°C.
4. Prepare transfection complexes according to the manufacturer's directions for a 10-cm plate except using 36 µL Lipofectamine 2000 and the following amounts of DNA: 3 µg pMD2G, 5 µg pMDLG/pRRE, 2.5 µg pRS-REV, and 10 µg pLLU2G or derivative.
5. Very gently replace medium (on cultured cells plated in **step 3**) with 5 mL pre-warmed, fresh medium (no antibiotics).
6. Slowly apply transfection complexes one drop at a time over the entire plate by placing the tip of the pipet just above the surface of the medium in the plate to prevent disruption of cell adhesion.
7. Incubate approx. 16 h (overnight) at 37°C.
8. Replace the transfection complexes with 10 mL fresh medium (no antibiotics) (*see* **Note 4**). Use biosafety level 2 guidelines (*see* **Note 5**) for all subsequent steps involving lentivirus.
9. 24 h later (48 h posttransduction), collect the supernatant medium (A collection) and replace with 10 mL fresh medium (no antibiotics). Store supernatant medium A at 4°C overnight.
10. 24 h later (72 h posttransduction) collect the supernatant medium (B collection).
11. Observe cells to confirm efficient transfection. Bright GFP expression should be present in the cells throughout the plate.
12. Cover the surface of the plate with bleach to kill the cells, transfer the bleach to a waste beaker, and discard the plate (*see* **Note 5**).

14.3.2.1.2   Calcium Phosphate Method

1. Follow **Subheading 14.3.2.1.1, steps 1–3**.
2. Mix DNAs (3 µg pMD2G, 5 µg pMDLG/pRRE, 2.5 µg pRS-REV, and 10 µg pLLU2G or derivative) with BW to give a final volume of 250 µL (*see* **Note 1**) *(8,30)*.

3. Add 250 μL 0.5M CaCl$_2$ to DNA/BW and mix well.
4. Add the DNA/BW/CaCl$_2$ mixture slowly dropwise (1 drop every 2 s) to 500 μL 2X HeBS in a 15-mL tube while vortexing at a medium speed.
5. Allow transfection complexes to form undisturbed at room temperature for exactly 30 min.
6. Very gently replace medium (on cultured cells plated in **Subheading 14.3.2.1.1, step 3**) with 10 mL prewarmed, fresh medium (no antibiotics).
7. Follow **Subheading 14.3.2.1.1, steps 6–12**.

14.3.2.2 Concentration of Lentivirus

1. Centrifuge collected media at 2500g for 15 min at 4°C to remove cellular debris.
2. Filter supernatant media using a SteriFlip.
3. Place conical bottom tubes into the rotor tubes fitted with adapters (SW28 rotor with Beckman ultracentrifuge) and pipet supernatant into each tube, taking care to place equal volumes into matched rotor tube sets (e.g., 1 and 4) for balancing.
4. Centrifuge at 25,000 rpm (82,705g) for 90 min at 4°C to pellet lentivirus particles.
5. Pour off supernatant medium into the bleach waste (*see* **Note 5**) and use a pipet or Kimwipes to remove as much residual liquid as possible from the side of the tube.
6. Apply 30–50 μL PBS to the pellet (which may be visibly yellow from precipitated GFP) and incubate for 2–16 h (overnight) at 4°C.
7. Resuspend pellet by pipeting up and down 20 times and aliquot (5–20 μL each) prior to storing at −80°C.

14.3.2.3 Titering of Lentivirus

1. Plate 4 × 10^5 293 FT cells in each well of a six-well plate and allow cells to adhere approx. 16 h (overnight) at 37°C (*see* **Note 6**).
2. Aliquot 1 mL medium/polybrene into 1.5-mL tubes.
3. Add virus (typically 1 μL lentivirus diluted 1:10) to each tube. Perform duplicates for each amount of virus tested and include one well with no virus.
4. Replace medium in each well of the six-well plate with 1 mL medium/polybrene/virus.
5. Incubate approx. 16 h (overnight) at 37°C.
6. Replace medium containing virus with fresh medium.
7. Determine the percentage of GFP-positive cells by FACS analysis 48–72 h posttransduction.
8. Calculate the titer as follows: % positive × cells plated/amount virus (e.g., 10% × 4 × 10^5/0.1 μL virus = 4 × 10^5 transducing units (TU)/μL).

14.3.3   Testing for Knockdown In Vitro

14.3.3.1   Knockdown of Exogenously Expressed myc-Tagged Target

1. Plate 5×10^4 cells in each well of a 24-well plate and allow cells to adhere approx. 16h (overnight) at 37°C.
2. Transduce cells with lentivirus from multiplicity of infection (MOI) 5–50 (e.g., for MOI 50: 5×10^4 cells × MOI 50/4 × 10^5 TU/μL = 6.25 μL virus) in 200 μL medium/polybrene.
3. Incubate approx. 16h (overnight) at 37°C.
4. Replace medium containing virus with fresh medium (no antibiotics).
5. Approximately 48h posttransduction, transfect the cells with 200 ng plasmid containing myc-tagged target using Lipofectamine 2000. Follow Invitrogen's guidelines for 24-well format Lipofectamine 2000 transfection.
6. At 24–48h posttransfection, harvest cells by trypsinizing and resuspending the cell pellet in sample loading buffer.
7. Assess knockdown of myc-tagged target by Western analysis using 9E10 antibody *(8)*.

14.3.3.2   Knockdown of Endogenous Expression of Target Gene

1. Plate 5×10^4 cells in each well of a 24-well plate and allow cells to adhere approx. 16h (overnight) at 37°C.
2. Transduce cells with lentivirus from MOI 5–50 (e.g., for MOI 50: 5×10^4 cells × MOI 50/4 × 10^5 TU/μL = 6.25 μL virus) in 200 μL medium/polybrene.
3. Approximately 72h posttransduction, the degree of knockdown at the protein level can be analyzed by immunostaining if an antibody is available. Alternatively, the degree of knockdown at the mRNA level can be analyzed by quantitative reverse transcriptase polymerase chain reaction (RT-PCR). It is preferable to sort GFP positive cells (FACS) prior to RNA isolation to ensure that the knockdown is only evaluated in transduced cells.

14.3.4   Lentiviral Transduction of Rat Embryos

14.3.4.1   Superovulation and Embryo Preparation

1. Inject five females intraperitoneally (65–100 g) with 20 U PMSG (*see* **Note 7**).
2. At 46–48h later, inject females intraperitoneally with 20 U hCG and place one female in a cage with one stud male.

3. The following morning, check for the presence of a vaginal plug (expect at least four of five females to be plugged).
4. At 20–22 h following the hCG injection, sacrifice the females that were found to be plugged.
5. Isolate one-cell embryos using 25-gauge needles attached to two syringes as tools to rip open the swollen ampulla of each oviduct. Embryos will be enveloped in masses of cumulus cells. Useful guidelines for embryo manipulation techniques may be found in **ref. 31**.
6. Treat embryos with hyaluronidase (100-µL aliquot added to 3 mL M2 in a 35-mm Petri dish) to remove cumulus cells and wash embryos by transferring through three dishes of KSOM using a mouth pipet (*see* **Note 8**).

14.3.4.2 Embryo Injection

1. Transfer embryos (in batches of ~30) to a 50-µL drop of M2 with cytochalasin B under mineral oil in an aluminum injection chamber block (*see* **Fig. 14.2B**). The addition of cytochalasin B improves the durability of the embryo while puncturing with the needle.
2. Centrifuge a newly thawed aliquot of virus in a microfuge at 14,000 g for 10 min at 4°C.
3. In a biosafety cabinet, load approx. 3 µL concentrated lentivirus into a needle with a microloader.
4. Set the pressure such that there is a slow, steady stream of liquid expelled from the needle.
5. While holding a fertilized embryo (pronuclei visible) in place with a VacuTip, puncture the zona pellucida with the needle. The embryo should be slightly displaced within the zona pellucida by the pressure of the liquid as it enters, eventually reaching a state of equilibrium. When no further liquid appears to enter the perivitelline space, remove the needle. The embryo should bounce back into place, possibly expelling some liquid through the hole left by the needle. However, if the pressure is set too high, the embryo will be lysed by the sudden increase in volume under the zona pellucida.
6. Inject all fertilized embryos, separating successfully injected embryos from those that were lysed or unfertilized (*see* **Note 9**).
7. Place the injected embryos into a 100-µL drop of KSOM under mineral oil in a 35-mm dish and place in a 37°C incubator until ready for transfer into pseudopregnant females.
8. During initial experiments, it is helpful to set aside 5–10 injected embryos and culture them in vitro. At least 80% of the injected, fertilized embryos should undergo the first cell division. GFP expression should be detected in embryos that are cultured past the two-cell stage, when zygotic transcription begins (*see* **Note 10**).

14.3.5 Production of Transgenic Pups

1. The day before injecting embryos, place two females (150–175 g) into each of 15 cages containing a vasectomized male.
2. The following morning, identify pseudopregnant females by the presence of a vaginal plug.
3. Transfer injected embryos (~40 per female) into the oviduct of pseudopregnant females using approved surgical procedures *(32)*.
4. Pups should be born 22 d after transfer. Litter sizes should be between 5 and 15. If the pups are not born, the embryos should be recovered by cesarian section and fostered to another mother (*see* **Note 11**).

14.3.6 Analysis of Founders

14.3.6.1 Genotyping

1. Isolate genomic DNA from d 23 pups (~0.5-cm tail snip lysed in 4 mL SNET + 75 μL PK overnight at 55°C; 600 μL lysate subjected to two rounds of phenol/chloroform extraction, ethanol precipitated, and resuspended in 50 μL water).
2. Set up PCR by adding 3 μL of DNA (or water for negative control) to 23-μL aliquots of PCR master mix (**Subheading 14.2.6.1**) in 0.2-mL thin-wall tubes on a cold block.
3. Place tubes in preheated PCR machine and run the following program: 1X (94° C for 5 min), 30X (94° C for 30 s, 52° C for 30 s, 72° C for 30 s), 1X (72° C for 7 min).
4. Add loading dye and run half of the PCR reaction out on a 1% gel.
5. Founders may have varied levels of the PCR product because of different degrees of somatic mosaicism (*see* **Note 12**). It is essential to compare PCR products amplified from a founder animal's DNA to a negative control reaction using DNA from a wild-type animal's tail snip as a template.

14.3.6.2 GFP Expression in Ear Punch

Overall transgene expression levels can be estimated in each founder by observing GFP expression present in a 2-mm ear punch *(8)*. A known wild type should be used for comparison to determine background autofluorescence.

14.3.6.3 Breeding

Each founder may be bred to wild-type rats to establish lines. Mosaicism has been observed in founders generated by lentiviral infection of embryos *(8,25,33)*. Therefore, the transgene may only segregate to a small percentage of the progeny and at least 50 F1 pups should be genotyped before concluding that the transgene has failed to transmit. Alternatively, the transgene may integrate in such a way that every diploid germ cell carries the transgene, in which case the expected Mendelian segregation pattern would be 50% transgenic progeny. Finally, multiple copies of the transgene may integrate into the genome of an injected embryo, typically because of higher lentiviral titers, in which case even greater than 50% of the progeny may be transgenic. Cage numbers maybe reduced by only breeding those founders with detectable GFP expression (unless it is desirable to obtain a nonexpressing control line).

14.4   Notes

1. Descriptions of these viral packaging plasmids and their use can be found at http://tronolab. epfl.ch/page58114.html. Also, a helpful searchable discussion database can be found at lenti-web.com. Virapower (Invitrogen) is a cocktail of ready-to-use viral packaging plasmids that can be substituted for those listed above.
2. If AAG(N18)TT is not present in the target mRNA, then relax the criteria and search for AAG(N18)T.
3. It is important to run a positive control using pLLU2G vector as this provides a size comparison for vector without insert.
4. The volume per plate may be varied from 7.5 to 10 mL such that the final supernatant volume to be concentrated is 30 mL (the volume constraint of the centrifuge tubes used in ultracentrifugation).
5. The use of lentiviral vectors described here is considered biosafety level 2. The biosafety office at your institution should be notified prior to using these reagents. The production of virus and handling of concentrated viral stocks should be performed in a biosafety cabinet (tissue culture hood) while wearing two pairs of gloves and a lab coat. All pipets and plates exposed to infectious virus should be decontaminated with bleach prior to disposal as standard biohazardous material. This may be accomplished by placing a 500-mL beaker with about 300 mL 50% bleach (in tap water) in the tissue culture hood and using this as a waste beaker as well as a source of bleach for decontaminating plates or pipets.
6. Use 293 FT cells to determine titer if the virus is to be injected into embryos. Based on published literature, the range of viral titers (as determined in 293 cells) used for embryo injections is $4–100 \times 10^5$ TU/µL *(8,10,25,33)*. If the virus is to be used for transducing another cell line, it is appropriate to titer in that cell line instead of in 293 FT cells.
7. Rats housed in a 6 AM/6 PM light/dark cycle may be successfully superovulated by injecting PMSG between 3 and 6 PM and hCG 46–48 h later.
8. Typical recovery after hyaluronidase treatment is at least 100 embryos from five females.
9. A variable number of the isolated embryos will be unfertilized. These can be distinguished based on the following properties: The cumulus cells tend to resist hyaluronidase treatment on unfertilized embryos and therefore remain tightly bound even after the described treatment with hyaluronidase. The surface of fertilized embryos transiently becomes highly irregular while in M2; unfertilized embryos remain perfectly spherical. The pronuclei of fertilized embryos are visible using Nomarski optics.
10. Rat embryo development arrests at the two-cell stage when cultured in vitro in KSOM. Conditions for promoting in vitro development beyond the two-cell stage are published in **ref.** *34*.
11. It is helpful to split the injected embryos (preferably a total of ~ 80 embryos) between two pseudopregnant females. This way, if a cesarean section is required for one of the mothers, the other mother may be used as a foster.

12. Genetic mosaicism refers to the presence of the transgene in the DNA of only a subset of cells within the founder animal. This may be caused by the integration of the transgene after the one-cell stage. Mosaicism following lentiviral transduction of embryos has been reported by several groups *(8,25,33,35)*.

Acknowledgments We thank Robert Hammer for assisting in the testing and development of techniques described in this chapter and Kent Hamra for critically reading the manuscript.

References

1. Hamra, F.K., Schultz, N., Chapman, K.M., et al. (2004) Defining the spermatogonial stem cell. *Dev. Biol.* **269**, 393–410.
2. Schultz, N., Hamra, F.K., and Garbers, D.L. (2003) A multitude of genes expressed solely in meiotic or postmeiotic spermatogenic cells offers a myriad of contraceptive targets. *Proc. Natl. Acad. Sci. U. S. A.* **100**, 12201–12206.
3. Small, C.L., Shima, J.E., Uzumcu, M., Skinner, M.K., and Griswold, M.D. (2005) Profiling gene expression during the differentiation and development of the murine embryonic gonad. *Biol. Reprod.* **72**, 492–501.
4. Wang, P.J., McCarrey, J.R., Yang, F., and Page, D.C. (2001) An abundance of X-linked genes expressed in spermatogonia. *Nat. Genet.* **27**, 422–426.
5. Dillon, C.P., Sandy, P., Nencioni, A., Kissler, S., Rubinson, D.A., and Van Parijs, L. (2005) RNAi as an experimental and therapeutic tool to study and regulate physiological and disease processes. *Annu. Rev. Physiol.* **67**, 147–173.
6. Hedrich, H. (2000) *History, strains and models,* Academic Press, London.
7. Lazar, J., Moreno, C., Jacob, H.J., and Kwitek, A.E. (2005) Impact of genomics on research in the rat. *Genome Res.* **15**, 1717–1728.
8. Dann, C.T., Alvarado, A.L., Hammer, R.E., and Garbers, D.L. (2006) Heritable and stable gene knockdown in rats. *Proc. Natl. Acad. Sci. U. S. A.* **103**, 11246–11251.
9. Lu, W., Yamamoto, V., Ortega, B., and Baltimore, D. (2004) Mammalian Ryk is a Wnt coreceptor required for stimulation of neurite outgrowth. *Cell.* **119**, 97–108.
10. Rubinson, D.A., Dillon, C.P., Kwiatkowski, A.V., et al. (2003) A lentivirus-based system to functionally silence genes in primary mammalian cells, stem cells and transgenic mice by RNA interference. *Nat. Genet.* **33**, 401–406.
11. Tiscornia, G., Singer, O., Ikawa, M., and Verma, I.M. (2003) A general method for gene knockdown in mice by using lentiviral vectors expressing small interfering RNA. *Proc. Natl. Acad. Sci. U. S. A.* **100**, 1844–1848.
12. Coumoul, X., Shukla, V., Li, C., Wang, R.H., and Deng, C.X. (2005) Conditional knockdown of Fgfr2 in mice using Cre-LoxP induced RNA interference. *Nucleic Acids Res.* **33**, e102.
13. Fedoriw, A.M., Stein, P., Svoboda, P., Schultz, R.M., and Bartolomei, M.S. (2004). Transgenic RNAi reveals essential function for CTCF in H19 gene imprinting. *Science.* **303**, 238–240.
14. Hasuwa, H., Kaseda, K., Einarsdottir, T., and Okabe, M. (2002) Small interfering RNA and gene silencing in transgenic mice and rats. *FEBS Lett.* **532**, 227–230.
15. Hasuwa, H., and Okabe, M. (2004) RNAi in living mice. *Methods Mol. Biol.* **252**, 501–508.
16. Rao, M.K., Pham, J., Imam, J.S., et al. (2006) Tissue-specific RNAi reveals that WT1 expression in nurse cells controls germ cell survival and spermatogenesis. *Genes Dev.* **20**, 147–152.
17. Shinagawa, T., and Ishii, S. (2003) Generation of Ski-knockdown mice by expressing a long double-strand RNA from an RNA polymerase II promoter. *Genes Dev.* **17**, 1340–1345.
18. Xia, X.G., Zhou, H., Samper, E., Melov, S., and Xu, Z. (2006) Pol II-expressed shRNA knocks down Sod2 gene expression and causes phenotypes of the gene knockout in mice. *PLoS Genet.* **2**, e10.

19. Carmell, M.A., Zhang, L., Conklin, D.S., Hannon, G.J., and Rosenquist, T.A. (2003) Germline transmission of RNAi in mice. *Nat. Struct. Biol.* **10,** 91–92.
20. Kunath, T., Gish, G., Lickert, H., Jones, N., Pawson, T., and Rossant, J. (2003) Transgenic RNA interference in ES cell-derived embryos recapitulates a genetic null phenotype. *Nat. Biotechnol.* **21,** 559–561.
21. Seibler, J., Kuter-Luks, B., Kern, H., et al. (2005) Single copy shRNA configuration for ubiquitous gene knockdown in mice. *Nucleic Acids Res.* **33,** e67.
22. Ventura, A., Meissner, A., Dillon, C.P., et al. (2004) Cre-lox-regulated conditional RNA interference from transgenes. *Proc. Natl. Acad. Sci. U. S. A.* **101,** 10380–10385.
23. Yu, J., and McMahon, A.P. (2006) Reproducible and inducible knockdown of gene expression in mice. *Genesis.* **44,** 252–261.
24. Ikawa, M., Tanaka, N., Kao, W.W., and Verma, I.M. (2003) Generation of transgenic mice using lentiviral vectors: a novel preclinical assessment of lentiviral vectors for gene therapy. *Mol. Ther.* **8,** 666–673.
25. Lois, C., Hong, E.J., Pease, S., Brown, E.J., and Baltimore, D. (2002) Germline transmission and tissue-specific expression of transgenes delivered by lentiviral vectors. *Science.* **295,** 868–872.
26. Pfeifer, A., Ikawa, M., Dayn, Y., and Verma, I.M. (2002) Transgenesis by lentiviral vectors: lack of gene silencing in mammalian embryonic stem cells and preimplantation embryos. *Proc. Natl. Acad. Sci. U. S. A.* **99,** 2140–2145.
27. Burns, J.C., Friedmann, T., Driever, W., Burrascano, M., and Yee, J.K. (1993) Vesicular stomatitis virus G glycoprotein pseudotyped retroviral vectors: concentration to very high titer and efficient gene transfer into mammalian and nonmammalian cells. *Proc. Natl. Acad. Sci. U. S. A.* **90,** 8033–8037.
28. Nenoi, M., Mita, K., Ichimura, S., et al. (1996) Heterogeneous structure of the polyubiquitin gene UbC of HeLa S3 cells. *Gene.* **175,** 179–185.
29. Ruggiu, M., Speed, R., Taggart, M., et al. (1997) The mouse Dazla gene encodes a cytoplasmic protein essential for gametogenesis. *Nature.* **389,** 73–77.
30. Dull, T., Zufferey, R., Kelly, M., et al. (1998) A third-generation lentivirus vector with a conditional packaging system. *J. Virol.* **72,** 8463–8471.
31. Nagy, A., Gertsenstein, M., Vintersten, K., and Behringer, R. (eds.) (2003) *Manipulating the mouse embryo,* 3rd ed., Cold Spring Harbor Laboratory Press, Cold Spring Harbor, NY.
32. Hammer, R.E., Maika, S.D., Richardson, J.A., Tang, J.P., and Taurog, J.D. (1990) Spontaneous inflammatory disease in transgenic rats expressing HLA-B27 and human beta 2m: an animal model of HLA-B27-associated human disorders. *Cell.* **63,** 1099–1112.
33. van den Brandt, J., Wang, D., Kwon, S.H., Heinkelein, M., and Reichardt, H.M. (2004) Lentivirally generated eGFP-transgenic rats allow efficient cell tracking in vivo. *Genesis.* **39,** 94–99.
34. Zhou, Y., Galat, V., Garton, R., Taborn, G., Niwa, K., and Iannaccone, P. (2003) Two-phase chemically defined culture system for preimplantation rat embryos. *Genesis.* **36,** 129–133.
35. Dann, C. T. (2007) New technology for an old favorite: lentiviral transgenesis and RNAi in rats. *Transgenic Res.* **16,** 571–580.

Chapter 15
Testicular Germ Cell Tumors in Mice

New Ways to Study a Genetically Complex Trait

Jason D. Heaney and Joseph H. Nadeau

Contents

Summary Testicular germ cell tumors (TGCTs) are the most common cancer affecting young men. Although TGCTs are common and the genetic component of susceptibility is unusually strong, discovery of TGCT susceptibility genes in humans has been challenging. The 129/Sv inbred mouse strain is an important experimental model for studying the genetic control of TGCT susceptibility. It is the only inbred mouse strain with an appreciable frequency of spontaneous TGCTs. TGCTs in 129/Sv males share various developmental and histological characteristics with human pediatric TGCTs. As in humans, susceptibility in 129/Sv is a genetically complex trait that is too complex for conventional genetic approaches. However, several genetic variants, when congenic or isogenic on the 129/Sv background, act as genetic modifiers of TGCT susceptibility. Alternative experimental approaches based on these modifier genes can be used to unravel the complex genetic control of TGCT susceptibility. We discuss the application of modifier genes in genetic interaction tests and sensitized polygenic trait analyses toward the understanding of the complex genetics and biology of TGCT susceptibility in mice.

Keywords Consomic; interaction; Mouse; Polygene; QTL; sensitized; susceptibility; testicular tumor, modifier.

From: *Methods in Molecular Biology, Vol. 450, Germline Stem Cells*
Edited by S. X. Hou and S. R. Singh © Humana Press, Totowa, NJ

15.1 Introduction

15.1.1 Susceptibility to Testicular Germ Cell Tumors Is a Genetically Complex Trait in Humans

Testicular germ cell tumors (TGCTs) are the most common cancer affecting 20- to 35-y-old males, and the incidence has increased dramatically since 1957 (1). When detected early, TGCTs are treatable with orchiectomy followed by radiation or chemotherapy (2). However, treatment often results in sterility, and TGCTs readily metastasize if not detected early (2). Because the social, emotional, and medical costs remain high, progress in early diagnosis prior to metastasis remains important.

Two groups of human TGCTs result from abnormalities in primordial germ cell (PGC) development (see **Subheading 15.1.2**) (3,4). The most common group of TGCTs occurs in adolescent boys and young men. These tumors, which are classified as seminomas and nonseminomas, arise from carcinoma in situ (CIS) of the seminiferous tubules. The other PGC-derived TGCTs are infantile (or pediatric) TGCTs, which present before puberty. These tumors are histologically classified as teratomas, teratocarcinomas (teratomas with embryonal carcinoma stem cell [EC cell] elements), and yolk sac tumors (3,4). Infantile TGCTs are often classified as a unique entity with etiology and pathology distinct from adult tumors.

The genetic component of TGCT susceptibility is unusually strong in humans, as indicated by the high familial index and significantly elevated relative risk among sons and brothers of affected individuals (5,6). A quarter of all early-onset TGCTs could be caused by genetic factors, making testicular cancer the third most heritable form of cancer (7). However, the genetic control of susceptibility is poorly understood. Linkage studies have been difficult because of the lack of multigenerational pedigrees with affected individuals and the complexity of genetic control. Despite considerable work, no TGCT susceptibility genes have been identified in humans (1,8–11). An appropriate animal model of human TGCTs may help to unravel the complex genetics of the human disease.

15.1.2 The Origin of TGCTs

PGCs are totipotent stem cells that normally differentiate into mature gametes and ultimately the cells and tissues of an adult organism. PGCs are one of the first embryonic cell lineages to be specified and in mice are first detected in the primitive streak around embryonic d 7 (E7) (12). The nascent PGCs then migrate through the dorsal mesentery of the gut and arrive at the genital ridges by E11.5 (12). While migrating, the PGCs proliferate, expanding from 100 cells at E7 to approx. 25,000 cells at E13.5 (13). Concomitant with sex determination at E13.5, male PGCs enter G1 mitotic arrest and remain mitotically inactive until a few days after birth (14).

TGCTs occur spontaneously only in the 129/Sv family of inbred mice (3–10% of males affected by 3–4 wk of age) and result from abnormalities in the development of PGCs *(8,15)*. In a series of experiments involving single-cell transplants and PGC-deficient mice (*Steel* and *White* mutants), Stevens demonstrated the PGC origin of TGCTs *(16,17)*. Experimental induction of TGCTs by grafting 129/Sv genital ridges into adult 129/Sv testes revealed a critical period from E11.5 to E12.5 during which PGCs are susceptible to experimental teratocarcinogenesis *(18)*. At least three fundamental events occur during this critical time of development: (1) sex determination and activation of testis and ovary-specific developmental pathways (E12.5–E13.5); (2) female germ cells enter meiosis, and male germ cells enter mitotic arrest (E13.5–E14.5); and (3) genomic imprints are erased (E10.5–E11.5) and are later reestablished in a sex-specific manner *(19–23)*. A defect in any of these processes may contribute to tumorigenesis. Specifically, it has been hypothesized that TGCTs arise from PGCs that fail to enter mitotic arrest and undergo neoplastic transformation *(15,16,18,24,25)*.

15.1.3 TGCTs of Mice are Similar to Human Pediatric Tumors

TGCTs in mice are first evident microscopically at E15 as foci of pluripotent tumor stem cell (EC cells) and macroscopically at 3–4 wk after birth as disorganized cell masses consisting of embryonic and extraembryonic tissue types at various stages of differentiation *(15,16,26–28)*. These teratomas and teratocarcinomas share various developmental and histological characteristics with human pediatric TGCTs *(29)*. As in humans, TGCTs in 129/Sv mice are predominantly unilateral, with 70% of tumors occurring in the left testis and 30% of TGCTs in the right testis *(8)*. Bilateral tumors are rare, affecting approx. 0.025% of males *(30)*.

The similarities in developmental origin and progression make the 129/Sv inbred mouse strain an excellent animal model for unraveling the complex genetics of human TGCTs. Furthermore, little is known about the cellular process involved in early TGCT development in humans because tumorigenesis often initiates during fetal development, and early tumor progression is difficult to detect clinically. Studies of TGCTs in mice will provide insights into the fundamental developmental processes (i.e., sex determination, the mitotic:meiotic switch, and imprinting) that may contribute to tumor susceptibility.

15.1.4 Susceptibility to TGCTs Is a Genetically Complex Trait in Mice

As in humans, the genetic component of TGCT susceptibility in mice is strong but complex. TGCTs occur in the 129 family of mice but not in other strains, demonstrating that susceptibility is a heritable trait. However, in segregating crosses between 129/Sv and other strains, only 1 affected male was found among more than 11,000 progeny tested, which is consistent with as many as six to eight 129/Sv-derived genes contributing to TGCT susceptibility *(15,31)*. The low frequency of affected males (0.01%) in the segregating population precludes analysis of TGCT susceptibility with standard genetic approaches.

Alternative approaches based on modifier genes are needed to understand the genetic control of TGCT susceptibility. A modifier gene, unlike a susceptibility gene, is not required or sufficient to induce a phenotype. Instead, modifiers interact with susceptibility genes to alter the phenotypic penetrance (the proportion of mice with a particular mutation, which manifest that genotype at a phenotypic level). As an example, several genetic variants, when congenic or isogenic on the 129/Sv background, act as modifiers of TGCT susceptibility (*see* **Table 15.1**). Although many of these variants have distinct phenotypic effects regardless of genetic background, most

Table 15.1 Published genetic variants that affect testicular germ cell tumor (TGCT) susceptibility in 129/Sv mice

Gene/locus[a,b]	Variant	% Affected males	Function (in wild type)	Reference
129/Sv	Control	5–10%	Genes and functions unknown	*(29)*
Kitl[sl]	Deletion	15%, lethal[c]	Ligand for KIT receptor	*(15)*
Kitl[SlJ]	Deletion	9%, lethal[c]	Ligand for KIT receptor	*(15)*
Oct4	Transgene	40–70%	Transcription factor	*(32)*
Chr 19[MOLF]	Consomic	24%, 80%[c]	Unknown	*(30)*
Dnd1[Ter]	Point	17%, 94%[c]	RNA-binding protein	*(33, 34)*
Trp53[null]	Knockout	15%, 35%[c]	Tumor suppressor	*(35)*
A[y]	Deletion	1%, lethal[c]	RNA binding, translation, coat color	*(15)*
Pten [null & floxed]	Knockout	ND, 100%[c]	Lipid phosphatase	*(36, 37)*

ND, not determined.

[a] With the exception of the *Pten*[null] and *Pten*[floxed] alleles, all genes and loci are modifiers of TGCT susceptibility. *Pten* is a TGCT susceptibility gene because inactivation induces TGCTs on several inbred backgrounds.

[b] Gene and variant definitions:

Kit[Sl] and *Kit*[SlJ] = Steel mutations. Approx 810 kb and 650 kb deletions, respectively, of the Kit cell surface receptor locus.

Oct4 = A transgene expressing the stem cell transcription factor *Pou2f1*.

Chr 19[MOLF] = A chromosome substitution (consomic) strain in which both copies of 129-derived chromsome 19 are replaced with MOLF-derived chromosome 19.

Dnd1[Ter] = A nonsense point mutation in *dead end*.

Trp53[null] = Engineered deletion of *transcription related protein 53*.

A[y] = The agouti-yellow mutation. A deletion resulting in the ectopic expression of *agouti* and deletion of *hnRNP-associated with lethal yellow* and *eukaryotic translation initiation factor 2 subunit 2*.

Pten[null & floxed] = Engineered deletions of *Phosphatase and tensin homolog*. The floxed allele contains *loxP* sites for tissue-specific deletion with Cre-recombinase.

[c] Frequency of affected heterozygote and homozygote males, respectively.

affect the frequency of TGCT-bearing males exclusively on the 129/Sv genetic background *(36–38)*. Therefore, these variants act as modifiers of 129-derived TGCT susceptibility genes. In contrast, *Pten (phosphatase and tensin homolog)* is a susceptibility gene because inactivation of *Pten* is sufficient to induce TGCTs on several genetic backgrounds *(36,37)*.

15.2   General Principles

We describe the application of modifier genes in genetic interaction tests and sensitized polygenic trait analyses toward the understanding of the complex genetics and biology of TGCT susceptibility in mice. The subsequent sections describe in detail the theory, logistics, and analysis of these tests.

An interaction test determines the manner in which two genetic variants act together to influence TGCT susceptibility. The first step in an interaction test is to produce 129/Sv-congenic lines for each mutation (*see* **Subheading 15.3.2.1**). The second step is to design single-mutant test crosses to determine the frequency of affected males in single-mutant mice (*see* **Subheading 15.3.2.2**). The third step is to design and implement the interaction crosses (*see* **Subheading 15.3.2.3**). The fourth step is to survey males for TGCTs and genotype (*see* **Subheading 15.3.3.1**). The fifth step is to analyze the interaction test results for additive or epistatic interactions that influence TGCT incidence, interactions that affect tumor laterality (unilateral vs bilateral tumor incidence), and parent-of-origin effects of TGCT modifiers on tumor susceptibility (*see* **Subheadings 15.3.3.2–15.3.3.4**).

In a sensitized polygenic trait analysis, a genetic modifier is used to increase the number of affected individuals in a segregating population, thereby increasing the statistical power of linkage analysis. Whereas genes with subtle effects may be undetectable in conventional crosses, susceptibility loci that interact with the sensitizing mutation are more readily detectable. The first step in a sensitized polygenic trait analysis is to select the parental strains for the cross (*see* **Subheading 15.4.2.1**). The second step is to select the sensitizing mutation and transfer it with crosses onto one or both of the inbred parental backgrounds (*see* **Subheading 15.4.2.2**). The third step is to select (i.e., a backcross or intercross) and implement the breeding scheme (*see* **Subheading 15.4.2.3**). The fourth step is to survey for affected males and genotype for the sensitizing mutation and genetic markers (*see* **Subheadings 15.4.3.1–15.4.3.3**). The fifth step is to analyze the genotype data for chromosomal regions that associate with TGCT susceptibility (*see* **Subheading 15.4.3.4**). The final step is to use chromosome substitution strains (CSSs or consomic strains) and congenic strains to verify the linage analysis and narrow the region of linkage toward the identification of TGCT susceptibility genes (*see* **Subheading 15.4.5**).

15.3   Interaction Tests with Single-Gene Mutations

15.3.1   Principles of TGCT Susceptibility Gene Interaction Tests

In various species, interaction tests with double-mutant individuals have successfully characterized and identified interactions between different proteins and cellular pathways *(39–41)*. Suppressed, enhanced, or novel phenotypes in double mutants compared to single-mutant and wild-type control mice provide clues to the nature of these interactions and the underlying developmental processes and pathogenesis of disease. Our laboratory has been studying the manner in which pairs of single-gene mutations interact to influence TGCT susceptibility in mice *(42)*. Various mutations increase or decrease the frequency of TGCT-bearing males on the 129/Sv inbred background, perhaps through the activation or repression of novel pathways involved in tumorigenesis. In double-mutant mice, the types of interactions between TGCT modifiers provide insight into the network of developmental pathways that control TGCT susceptibility.

In a standard interaction test for TGCT susceptibility, the observed TGCT incidence in double-mutant males is compared to the expected combined (additive) incidence for the individual mutations. If the genetic variants function independently to modify susceptibility, the net effect of the mutations on TGCT incidence is the sum of the frequency of affected males in the single-mutant mice. However, if modifiers act within the same pathway or pathways with overlapping function, the effect on TGCT penetrance will be epistatic (phenotypic expression of a genotype at one locus is influenced by the genotype at another locus), and TGCT incidence will deviate from the expected additive values. In our interaction tests of TGCT modifier genes, both additive and epistatic (enhancer and suppressor) effects on TGCT susceptibility were observed *(42)*. In addition, in some interaction tests effects on tumor laterality and parental inheritance effects were observed. Such observations provided additional clues to the genetic and molecular pathways controlling TGCT susceptibility.

15.3.2   Performing Interaction Crosses with TGCT Modifiers

15.3.2.1   129/Sv Mutant Congenic Strains

Typically, mutations affect TGCT susceptibility only in 129/Sv inbred mice. Unfortunately, many mutations and transgenes that may be of interest to TGCT biology are not maintained on the 129/Sv inbred background *(43)*. Therefore, it is often necessary to transfer candidate TGCT modifiers onto the 129/Sv background. In our laboratory, we backcross mutations of interest onto the 129S1/SvImJ inbred substrain (Jackson Laboratory stock number JR002448). This substrain of 129/Sv does not carry the Steel-Jackson ($Kitl^{Sl-J}$) mutation, which increases TGCT incidence *(44)*. 129/SvImJ mice are agouti and good breeders compared to other 129/Sv substrains.

To make a congenic strain, genetic variants should be backcrossed to 129/SvImJ for at least ten generations. In mathematical terms, for each generation of backcrossing, the contribution of the donor genome decreases by $[(1/2)^{N-1}]$, where N is the backcross generation number, and the F1 generation is counted as N1. By the tenth generation, 99.8% of the genome will be identical to the 129/Sv inbred background, thereby establishing a congenic strain. Alternatively, a polymorphic marker-assisted breeding strategy can be employed to reduce the number of backcrosses *(45)*.

15.3.2.2   Single-Mutant Control Crosses

The average frequency of TGCT-bearing males in the 129/Sv family has slightly increased since 1967, and tumor incidence can fluctuate among individual colonies and the various 129/Sv substrains over time. Therefore, the frequency of affected males should be evaluated in both the 129/Sv colony and single-mutant mice prior to the interaction tests. These frequencies are necessary for calculating the expected additive TGCT frequency in the interaction tests (*see* **Subheading 15.3.3.2**). In addition, because parental genotype can sometimes influence the frequency of affected male progeny, reciprocal crosses (using maternal and paternal inheritance of the mutation in separate crosses) should be used to test for parent-of-origin effects (*see* **Subheading 15.3.3.4**).

15.3.2.3   Interaction Crosses

The mutations in interaction crosses can be inherited from different parents or from a single parent (*see* **Fig. 15.1A,B**). The genotype ratios in these crosses are similar, with a fourth of the progeny falling within each genotype class (one wild type, two single mutants, and one double mutant), when the mutations are not located on the same chromosome. Therefore, the number of double-mutant progeny available for TGCT surveys in the interaction test will be similar in the two breeding schemes. Deviations from the expected 1:2:1 ratio may suggest reduced viability in double-mutant mice. In addition, both breeding schemes generate wild-type and single-mutant

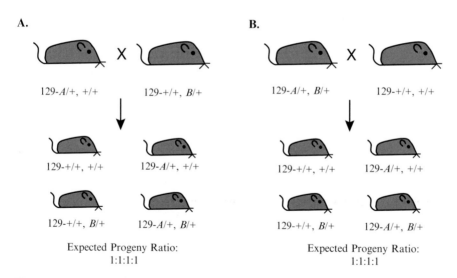

Fig. 15.1 Breeding strategies for interaction tests. **A** Two mutations, congenic on the 129/Sv background, are inherited from different parents. Both 129/Sv parents are heterozygous for a single mutation (*A*/+, +/+ or +/+, *B*/+). **B** Two mutations are inherited from one parent. One 129/Sv parent is wild type at both alleles (+/+, +/+), and one is heterozygous at both alleles (*A*/+, *B*/+). Both strategies produce offspring of similar genotypes, in a 1:1:1:1 progeny ratio, assuming that mutations *A* and *B* are not genetically linked. Reciprocal crosses can be used to test for parent-of-origin effects on tumor frequency

siblings that serve as internal controls for statistical tests of TGCT incidence, tumor laterality, and parental effect (*see* **Subheadings 15.3.3.2, 15.3.3.4**). Progeny genotype can be verified using methods specific for the mutations.

15.3.3   Data Collection and Analysis of Interaction Crosses

15.3.3.1   Surveying for TGCTs

TGCTs are often difficult to observe macroscopically prior to weaning but rarely escape visual detection by 3–4 wk of age *(30,46)*. Small TGCTs appear as discolored areas within an otherwise-normal testis (*see* **Fig. 15.2A**). Histological evaluation of TGCTs should reveal areas of differentiation consisting of tissue types from embryonic and extraembryonic lineages at various stages of differentiation. In testes with small TGCTs, tumors will be found adjacent to seminiferous tubules supporting normal spermatogenesis (*see* **Fig. 15.2B**). With time, the tumor can progress, eventually encompassing most of the testis and expanding in considerable volume or regressing and leaving a brown spot (*see* **Fig. 15.2A**).

A subset of normal (as a control) and tumor-burdened testes from all segregating genotypes should be histologically examined to confirm macroscopic observations. Typically, only one or two hematoxylin and eosin (H&E) stained sections are needed to verify a TGCT. Some testes may appear abnormal (e.g., relatively small, ischemic, or with red blood cell infiltrates). These anomalies are not TGCTs and can be classified as abnormal testes. H&E stained sections can also be used to validate the abnormal classification (*see* **Fig. 15.2C**).

15.3.3.2   Statistical Testing for Effects on TGCT Susceptibility

The χ^2 goodness-of-fit tests are used to determine whether the observed number of affected double-mutant males significantly differs from expected additive values. The following equations can be used to calculate the expected additive TGCT frequencies for interaction crosses.

- The single mutation frequency [$f(m)$] is calculated as the frequency of TGCT-bearing males in single-mutant males on the 129/Sv background [$f(m + b)$] minus the background incidence [$f(b)$] or

$$f(m) = f(m + b) - f(b) \qquad (15.1)$$

- The expected additive TGCT incidence [$f(a)$] is the sum of the TGCT incidence of the single mutations [$f(m1) + f(m2)$] plus the 129/Sv background incidence [$f(b)$] or

$$f(a) = f(m1) + f(m2) + f(b) \qquad (15.2)$$

A.   **B.**

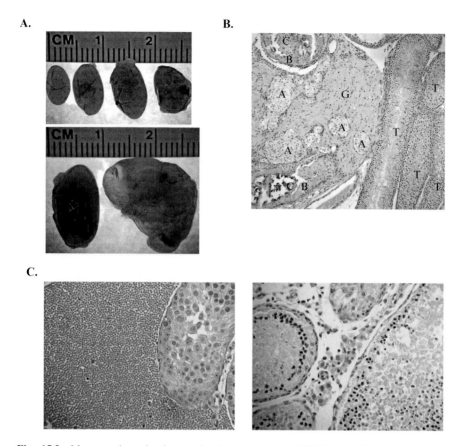

C.

Fig. 15.2 Macroscopic and microscopic characteristics of TGCTs and abnormal testes. **A** Macroscopic stages of mouse TGCT progression. *Top left* is a normal testis. TGCTs start as small areas of discoloration at 3–4 wk of age (*top*). With time, the tumor may encompass most of the testis and expand in volume (*bottom*). **B** hematoxylin and eosin (H&E)-stained section of a 129/Sv teratoma (×40 magnification). The tumor (*left*) consists of various embryonic cell lineages, including bone (*B*), cartilage (*C*), adrenal (*A*), and ganglia (*G*). To the right of the tumor, normal-appearing seminiferous tubules support spermatogenesis (*T*). **C** H&E-stained sections of abnormal 129/Sv mouse testes (×200 magnification). *Left*, blood cell infiltrates in the testis. *Right*, an ischemic testis. Few spermatocytes and spermatids are present. The remaining spermatogonia have pycnotic nuclei and are undergoing cell death

For example, control crosses with single-mutant mice showed that the frequency of affected males is 15% in *Trp53^{null}* heterozygous males and 11% in *Kitl^{Sl-J}* heterozygotes *(42)*. Using Eq. 15.2 and a background TGCT incidence of 5%, the additive TGCT incidence in double-mutant males is expected to be 21% [= (15% − 5%) + (11% − 5%) + 5%]. If 75 double-mutant males are surveyed, 16 (21%) would be expected to have at least one TGCT.

15.3.3.3   Statistical Testing for Effects on Laterality

In some instances, mutations may interact to preferentially alter the incidence of unilateral or bilateral tumors. To test for interactions that affect tumor laterality, the formulas of Youngren et al. can be used to calculate the expected frequencies of unaffected, unilateral, or bilateral cases in the double-mutant population *(42)*. These formulas are based on a model that bilateral TGCTs in 129/Sv mice result from random cooccurrence of unilateral tumors rather than from the action of a distinct set of TGCT susceptibility genes *(42)*.

- The frequency of mice with a tumor in the left testis, regardless of the presence of a tumor in the right testis or [(number of observed unilateral left + bilateral)/n], is $f(L)$. The frequency of mice without a tumor in the left testis is $1 - f(L)$.
- The frequency of mice with a tumor in the right testis, regardless of the presence of a tumor in the left testis or [(number of observed unilateral left + bilateral)/n], is $f(R)$. The frequency of mice without a tumor in the left testis is $1 - f(R)$.
- The expected frequency of mice without a TGCT $f(NT)$ is

$$f(NT) = [1 - f(L)] \times [1 - f(R)] \tag{15.3}$$

- The expected frequency of mice with a TGCT in the left but not the right testis $f(UL)$ is

$$f(UL) = f(L) \times [1 - f(R)] \tag{15.4}$$

- The expected frequency of mice with a TGCT in the left but not the right testis $f(UR)$ is

$$f(UR) = f(R) \times [1 - f(L)] \tag{15.5}$$

- The expected frequency of mice with a unilateral TGCT $f(U)$ is

$$f(U) = f(UL) + f(UR) \tag{15.6}$$

- The expected frequency of mice with bilateral TGCT $f(B)$ is

$$f(B) = f(L) \times f(R) \tag{15.7}$$

These formulas can be used to determine the expected number of males without a TGCT (Eq. 15.3), with a unilateral TGCT (Eq. 15.6), or with bilateral TGCTs (Eq. 15.7). The χ^2 goodness-of-fit tests can determine whether two mutations interact to preferentially influence unilateral or bilateral TGCT frequency or bilateral tumors result from the independent occurrence of TGCTs in the left and right testes.

15.3.3.4   Statistical Testing for Parental Effects

Nutritional or hormonal defects in the mother during pregnancy or nursing, maternal inheritance of mitochondria, or inheritance of parent-specific epigenetic modifications from either parent have been shown to influence a variety of progeny phenotypes *(47–54)*. As with other pathophysiological states, the parent of origin of some mutations influences TGCT incidence *(42)*. Therefore, prior to the interaction tests, reciprocal crosses with single mutants should be used to test whether the maternal or paternal inheritance of a mutation preferentially influences the frequency of affected males.

Even in the absence of parental effects in the single-mutant test crosses, reciprocal crosses should be used to test for parental effects in the interaction crosses. Mutations may interact to produce parental effects that are absent in the single-mutant crosses. When mutations are inherited from different parents, reciprocal crosses can reveal a parental effect of a mutation that interacts with an effect of a second mutation in the offspring. This interaction will only affect TGCT incidence in male progeny harboring the second mutation. Alternatively, reciprocal crosses of single-mutant parents may reveal parent-of-origin effects that are only observed when both parents harbor a mutation. This interaction can affect TGCT incidence in all genotype classes. When both mutations are inherited from a single parent, reciprocal crosses can reveal a parent-of-origin effect resulting from the interaction of the two mutations in the double-mutant parent. Interactions in a double-mutant parent can affect TGCT incidence in all progeny, regardless of their genotype.

Parent-of-origin effects on the frequency of affected males can be determined using χ^2 contigency tables. The two categories are the number of TGCT-bearing male offspring with maternal or paternal origin of a mutation. The frequency of affected males of any genotype is expected to be independent of the parent-of-origin of the mutation and equally distributed between the two categories. Therefore, each genotype class should be independently tested for parental effects. Parental effects skew the TGCT incidence so that tumor susceptibility is higher in progeny with either maternal or paternal inheritance of a mutation.

15.3.4   An Interaction Test with TGCT Modifiers

The interaction cross between *Trp53^{null}* heterozygous and *Kitl^{Sl-J}* heterozygous mutant mice is an excellent example of an interaction test that produced insightful results *(42)*. In both single-mutant test crosses and interaction cross, parental effects of the mutations on TGCT incidence were not observed for any progeny genotype classes. In addition, unilateral and bilateral tumor frequencies did not significantly deviate from the calculated expected values in the interaction crosses. Statistical analysis for interactions that affect TGCT susceptibility, however, produced surprising results. The frequency of affected double-mutant males (7%) was significantly lower than the calculated additive value of 21% ($\chi^2 = 9.6$; $p < 0.005$) *(see* **Subheading 15.3.3.2**). These results suggest that the two mutations complement each other or interact to activate tumor suppressor pathways and restore TGCT incidence to 129/Sv background levels.

15.4 Sensitizing Polygenic Trait Analysis with Mendelian Mutations

15.4.1 Principles of Sensitized Polygenic Trait Analysis

An important goal in genetics is to understand how genetic variation contributes to normal and pathophysiological states. Some traits result from simple Mendelian inheritance of a single genetic variant. However, most traits result from the combined action of multiple genes, often referred to as polygenes or quantitative trait loci (QTLs) *(55,56)*. Polygenes each have a small quantitative effect on a phenotype and together generate a wide range of phenotypic variation.

Genetic linkage analysis is often used as a first step in the ultimate discovery of gene identity. Such analyses are dependent on genetically defined, usually inbred, organisms with differences in trait penetrance (i.e., differences in the frequency of affected individuals or, for a quantitative trait, the extent of the phenotype). By characterizing genotype–phenotype associations in segregating populations derived from crosses between these strains, chromosomal locations for QTLs can be determined. Application of specially constructed strains, such as congenic mice, can be used to refine the genetic intervals and identify candidate genes *(43)*.

A challenge in polygenic trait analysis is the multiplicity of polygenes and the subtlety of their individual effect on the phenotype. Linkage analyses often uncover evidence for a few loci with strong effects that contribute to only a portion of the total genetic variation. It is suspected that genetic variants with small effects escape detection with conventional approaches *(55)*. For example, inheritance of six to eight 129/Sv-derived genes is required for TGCT penetrance as only 1 TGCT-bearing male was observed in the 11,000 mice surveyed in segregating crosses between 129/Sv and other strains *(15,31)*. With the involvement of so many genes, analysis of inheritance is difficult because too few affected mice are found to adequately detect linkage.

Single-gene mutations can be used to simplify the genetic analysis of complex traits. In sensitized polygenic trait analysis, one of the two strains included in the linkage cross has a mutation in a gene known to affect the phenotype. The mutation may (1) increase penetrance in individuals that inherit all required susceptibility genes, (2) reduce the number of genes required for trait penetrance, or (3) activate unique pathways that affect the same trait *(55)*. Although different genetic backgrounds are often used to reveal phenotypes associated with a genetic variant, a sensitized cross targets background susceptibility genes using the mutation as a sensitizer. If sensitization works as predicted, the frequency of affected individuals should be higher than in crosses that are not sensitized, thereby increasing the statistical power of the analysis. Where genes with small effect sizes are undetected in conventional crosses, chromosomal locations of polygenes that interact with the sensitizing mutation are more readily detectable.

15.4.2   Designing Sensitized Polygenic Trait Crosses for TGCT Susceptibility

15.4.2.1   Selection of Parental Strains

Several considerations influence selection of parental strains for the sensitized polygenic linkage analysis. First, a TGCT-resistant parental strain needs to be selected for crosses to the 129/Sv parental strain. Because the 129 family of inbred mice has the only strains with appreciable TGCT frequencies, any other inbred strain could be used for the parental cross. Second, the genetic similarity of the parental strains must be considered. In the past, it was necessary to use genetically distant mouse strains because a limited number of simple sequence length polymorphisms (SSLPs) were informative for alleles of common inbred laboratory mouse strains. Our laboratory has taken advantage of the genetic differences between common inbred strains such as 129/Sv and the wild-derived inbred strain MOLF/Ei (*Mus musculus molossinus*) for sensitized polygenic trait crosses (*57*). However, with the current single-nucleotide polymorphism (SNP) databases (*58,59*), many more informative polymorphic markers are available, allowing linkage analyses using less-divergent strains of mice.

15.4.2.2   Selection of the Sensitizing Mutation

Mutations with strong effects on TGCT susceptibility on the 129/Sv background may also be strong modifiers of tumor frequency in segregating crosses. Progeny with tumors should be more readily obtained in sensitized crosses, thereby increasing the statistical power of the linkage analysis. A mutation such as $Dnd1^{Ter}$, which increases TGCT incidence to 94% on the 129/Sv background, is likely to be a more powerful sensitizing mutation than a variant such as $Kitl^{Sl-J}$, which increases TGCT susceptibility to approx. 10%. However, the interaction of the mutation with the 129/Sv genome may be complex, perhaps involving several 129/Sv genes, and may not be a strong sensitizing mutation in a segregating genetic background. In addition, there may be differences in the way in which sensitizing mutations interact with different TGCT-resistant genome. Therefore, it is difficult to predict if a mutation will have a strong sensitizing effect on a segregating genetic background.

15.4.2.3   Selection of a Breeding Scheme

A backcross or intercross breeding strategy can be used for a sensitized cross. With both strategies, 129/Sv inbred mice, congenic for the sensitizing mutation, are bred to a TGCT-resistant parental strain to generate F1 offspring with the mutant. Alternatively, if the mutation is already congenic on the desired TGCT-resistant strain, such as C57BL/6J, the mutation could also be inherited from the resistant strain.

For the backcross strategy, the resulting F1 offspring with the sensitizing mutation are bred to 129/Sv mice to produce the N2 generation (*see* **Fig. 15.3A**). N2 males are surveyed for TGCTs, and males with the sensitizing mutation and a tumor are genotyped for the linkage analysis. The main advantage of this strategy is the relative simplicity of inheritance and the linkage analysis. Only one parent (the F1) contributes a meiotic recombination event in the N2 generation. Unfortunately, using this breeding scheme N2 offspring will only be homozygous or heterozygous for 129/Sv-derived alleles, and association of TGCT susceptibility with a recessive allele from the TGCT-resistant parental strain will not be detected (*see* **Subheading 15.4.4**) *(30,57)*. In addition, the backcross strategy generates fewer recombination events in the test progeny than an intercross strategy.

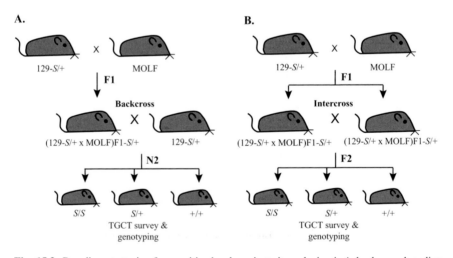

Fig. 15.3 Breeding strategies for sensitized polygenic trait analysis. **A** A backcross breeding strategy. A TGCT sensitizing mutation (*S*) is congenic on the 129/Sv background. The 129/Sv heterozygous sensitized mouse (129-*S*/+) is bred to a genetically divergent wild-derived inbred mouse (MOLF/Ei). The resulting F1 progeny are backcrossed to 129/Sv mice with the sensitizing mutation. The N2 generation will be 129/Sv homozygous or heterozygous at each allele. **B** An intercross breeding strategy. F1 progeny are intercrossed to generate the F2 generation. F2 progeny will be homozygous for 129/Sv-derived alleles or MOLF/Ei-derived alleles or heterozygous at each allele. The N2 and F2 generations are surveyed for TGCTs and genotyped for the sensitizing mutation. N2 and F2 TGCT-bearing males, heterozygous or homozygous for the mutation, are included in the linkage analysis, and unaffected males are genotyped for segregation controls

In an intercross strategy, F1 offspring with the sensitizing mutation are crossed to each other to produce the F2 generation. F2 males are surveyed for TGCTs, and males with the sensitizing mutation are included in the linkage analysis (**Fig. 15.3B**). An advantage of an intercross over the backcross strategy is that mice in the F2 generation can be homozygous for either parental allele. Therefore, association of recessive alleles from either parental strain with TGCT susceptibility can be detected. Another advantage of the intercross strategy is the informative meiotic recombination events that occur in both F1 parents. Therefore, twice as much recombination information is available in the intercross F2 generation compared to the backcross N2 generation. The advantages and disadvantages of parental backgrounds and breeding strategies in linkage analyses have been previously discussed (*60*).

15.4.3 Data Collection and Analysis of Sensitized Crosses

15.4.3.1 Surveying for TGCTs

Male progeny should be surveyed for TGCTs as discussed in **Subheading 15.3.3.1**.

15.4.3.2 Genotyping for the Sensitizing Mutation

N2 and F2 progeny should be genotyped for the sensitizing mutation. Tail DNA can be isolated from affected males using standard DNA purification protocols and kits (e.g., Qiagen's DNAeasy Tissue Kit) or protocols recommended for a specific high-throughput genotyping system.

15.4.3.3 Marker Selection and Genotyping

Both SSLPs and SNPs (single nucleotide polymorphisms) can be used as markers for the linkage analysis. Databases for informative SSLP and SNP markers for various strains are available on the Internet (e.g., www.informatics.jax.org and www.ncbi.nlm.nih.gov/SNP) (*59*). Several large-scale genotyping approaches are available to facilitate genotype determination (*61–66*).

15.4.3.4 Linkage Analysis

Several software packages have been designed to map QTLs. Mapmaker/QTL, Map Manager QTX, or R/QTL are suggested programs for detecting polygenes in a sensitized linkage analysis in mice (*59,67–70*). The guidelines of Lander et al. or permutation tests should used to set statistical thresholds (*71,72*).

Equal numbers of unaffected and TGCT-bearing individuals homozygous or heterozygous for the sensitizing mutation are included in the linkage analysis for TGCT susceptibility loci. Genotyping of unaffected, sensitized males will test for Mendelian segregation of the markers and verify that associations between markers and TGCT susceptibility in the affected population are not genotyping or segregation anomalies.

15.4.4   Sensitized Polygenic Trait Analyses of TGCT Susceptibility

Two mutations have been used to map TGCT susceptibility genes in mice. An engineered mutation of $Trp53$, $Trp53^{null}$, was used to sensitize linkage crosses between 129/Sv and C57BL/6J (73). Two approaches were employed. First, an intercross strategy using $Trp53^{null}$ mutant mice from a 129/Sv and C57BL/6J mixed genetic background identified four 129-derived recessive alleles on chromosomes 6, 9, 12, and 13 that associate with TGCT susceptibility. Following the identification of these critical regions, $Trp53^{null}$ heterozygous F1 progeny from a 129/Sv and C57BL/6J sensitized cross were backcrossed to $Trp53^{null}$ heterozygous 129/Sv. This backcross strategy was used to further evaluate the genetic intervals identified in the intercross. In these linkage crosses, the $Trp53^{null}$ mutation increased TGCT susceptibility and revealed the first linkage for a 129/Sv-derived susceptibility locus ($Pgct1$).

A sensitized linkage analysis using a spontaneous mutation in $Dnd1$, $Dnd1^{Ter}$, best illustrates the power of modifier genes as sensitizers. This mutation was used to sensitize linkage analysis of TGCT susceptibility in crosses between 129/Sv mice heterozygous for $Dnd1^{Ter}$ and MOLF/Ei (34,57). For this analysis, a backcross strategy between $Dnd1^{Ter}$ heterozygous F1 progeny and $Dnd1^{Ter}$ heterozygous 129/Sv was employed to identify critical regions associated with TGCT susceptibility. Sensitization of the linkage cross with $Dnd1^{Ter}$ had a dramatic affect on TGCT incidence. Whereas only 0.01% (1 in 11,000) of males in conventional segregating crosses had a TGCT, 16.5% (39 in 236) of males in the $Dnd1^{Ter}$-sensitized cross had a tumor (30,31,57). This represents a 1700-fold increase in TGCT incidence in a segregating population. With the added statistical power provided by the increased number of affected males, a genome scan revealed evidence for at least one susceptibility loci on MOLF/Ei chromosome 19. Although the MOLF/Ei alleles on chromosome 19 do not cause TGCTs on this tumor-resistant background, the influence on TGCT susceptibility was revealed in the context of the TGCT-susceptible 129/Sv genome.

The mechanism by which the $Trp53^{null}$ and $Dnd1^{Ter}$ mutations increased TGCT susceptibility is unknown. However, the increased frequency of affected mice in the sensitized crosses allowed for the detection of novel TGCT susceptibility loci. It is remarkable that different combinations of mutations and parental strains revealed distinct susceptibility loci. Such a finding suggests that the systematic application of TGCT modifiers and inbred strains in sensitized polygenic trait crosses may be a powerful approach to identify numerous TGCT susceptibility genes.

15.4.5   Consomic Mice

Unfortunately, association of chromosome 19 with TGCT susceptibility in the $Dnd1^{Ter}$-sensitized cross did not reach the suggestive level of significance *(57)*. Instead of collecting more backcross progeny to obtain sufficient power to detect linkage, our laboratory transferred chromosome 19 from MOLF/Ei, in its entirety, onto the 129/Sv background to construct the first mammalian, autosomal consomic strain (129-Chr 19^{MOLF}) *(30)*.

CSSs are made using methods similar to congenic strains. The first step is to select the donor and recipient strains (MOLF/Ei and 129/Sv, respectively). The second step is to transfer the desired chromosome (MOLF/Ei chromosome 19) from the donor strain to the recipient strain with repeated backcrosses to the recipient strain. At each generation, mice inheriting a nonrecombinant donor chromosome are identified by genotyping and are selected for breeding. Based on 17,000 mice genotyped for the B6-Chr$^{A/J}$ consomic panel, the frequency of mice with a nonrecombinant donor chromosome is approx. 20% at each generation *(74)*. Therefore, in a litter of five mice, on average one will have the desired nonrecombinant chromosome. This process is repeated until at least the N10 generation.

The third step in generating CSSs is a genome scan to verify that the donor chromosome was transferred intact and no unselected MOLF/Ei (donor strain) chromosomal segments are present in the genome. Unselected chromosome segments may persist by chance or because of functional interactions with the selected chromosome. Further breeding may eliminate these undesired segments; however, they may be required for survivability. The final step is to homozygous the nonrecombinant donor chromosome with intercrosses of heterozygous progeny. Once homozygous on the recipient background, the donated nonrecombinant chromosome is stable, and a CSS has been established. Detailed methods for constructing consomic strains and the benefits of these mice in linkage analyses have been previously described *(74–76)*.

Heterosomic and homosomic 129-Chr 19^{MOLF} mice have a TGCT incidence of 24% and 82%, respectively. Where linkage analysis in the $Dnd1^{Ter}$-sensitized cross previously discussed did not reach the suggestive level of significance, intercrosses of heterosomic 129-Chr 19^{MOLF} mice revealed a highly significant association between MOLF/Ei chromosome 19 and TGCT susceptibility using fewer than 30 males in the linkage analysis *(30,57)*. Interestingly, the likelihood curve for chromosome 19 was broad and suggested that several susceptibility alleles are present on the substituted MOLF/Ei chromosome. Analysis of a panel of single and double congenic strains derived from 129-Chr 19^{MOLF} identified at least five enhancers and suppressors of TGCT susceptibility on the substituted chromosome *(46)*.

15.5   Conclusion

The genetic control of TGCT susceptibility is complex in both mice and humans. To date, efforts to identify TGCT susceptibility genes in humans have been unsuccessful, and little is known about the mechanisms of TGCT progression. The developmental and genetic similarities between TGCTs in 129/Sv inbred mice and pediatric TGCTs in humans make the laboratory mouse an excellent model for studying the complex genetics of tumor susceptibility. Several genetic variants act as modifiers of TGCT susceptibility in mice and are the basis for new and powerful methods for unraveling the genetic component of this disease. Determining the manner in which TGCT modifier genes interact provides clues into the network of cellular pathways that controls TGCT susceptibility. In addition, sensitizing polygenic trait analyses with TGCT modifiers simplifies linkage analysis of this genetically complex trait and reveals otherwise undetectable susceptibility QTLs. Importantly, application of modifier genes in interaction tests and sensitized polygenic trait analysis is not limited to the study of testicular cancer. Such approaches are applicable to the study of any genetically complex trait for which modifier genes have been identified.

References

1. Bishop, T. D., and Wing, A. (1998) Candidate regions for testicular cancer susceptibility genes. *APMIS Suppl.* **106**, 64–72.
2. Bosl, G. J., and Motzer, R. J. (1997) Testicular germ-cell cancer. *N. Engl. J. Med.* **337**, 242–253.
3. Oosterhuis, J. W., and Looijenga, L. H. (2005) Testicular germ-cell tumours in a broader perspective. *Nat. Rev. Cancer.* **5**, 210–222.
4. Oosterhuis, J. W., Looijenga, L. H., van, E. J., and de, J. B. (1997) Chromosomal constitution and developmental potential of human germ cell tumors and teratomas. *Cancer Genet. Cytogenet.* **95**, 96–102.
5. Forman, D., Oliver, R. T., Brett, A. R., et al. (1992) Familial testicular cancer: a report of the UK family register, estimation of risk and an HLA class 1 sib-pair analysis. *Br. J. Cancer.* **65**, 255–262.
6. Lindelof, B., and Eklund, G. (2001) Analysis of hereditary component of cancer by use of a familial index by site. *Lancet.* **358**, 1696–1698.
7. Heimdal, K., Olsson, H., Tretli, S., Fossa, S. D., Borresen, A. L., and Bishop, D. T. (1997) A segregation analysis of testicular cancer based on Norwegian and Swedish families. *Br. J. Cancer.* **75**, 1084–1087.
8. Stevens, L., and Little, C. C. (1954) Spontaneous testicular teratomas in an inbred strain of mice. *Proc. Natl. Acad. Sci. U. S. A.* **40**, 1080–1087.
9. Rapley, E. A., Crockford, G. P., Teare, D., et al. (2000) Localization to Xq27 of a susceptibility gene for testicular germ-cell tumours. *Nat. Genet.* **24**, 197–200.
10. Nathanson, K. L., Kanetsky, P. A., Hawes, R., et al. (2005) The Y deletion gr/gr and susceptibility to testicular germ cell tumor. *Am. J. Hum. Genet.* **77**, 1034–1043.
11. Crockford, G. P., Linger, R., Hockley, S., et al. (2006) Genome-wide linkage screen for testicular germ cell tumour susceptibility loci. *Hum. Mol. Genet.* **15**, 443–451.
12. Kunwar, P. S., Siekhaus, D. E., and Lehmann, R. (2006) In vivo migration: a germ cell perspective. *Annu. Rev. Cell Dev. Biol.* **22**, 237–265.

13. Tam, P. P., and Snow, M. H. (1981) Proliferation and migration of primordial germ cells during compensatory growth in mouse embryos. *J. Embryol. Exp. Morphol.* **64**, 133–147.
14. McLaren, A. (1984) Meiosis and differentiation of mouse germ cells. *Symp. Soc. Exp. Biol.* **38**, 7–23.
15. Stevens, L. C. (1967) The biology of teratomas. *Adv. Morphog.* **6**, 1–31.
16. Stevens, L. C. (1967) Origin of testicular teratomas from primordial germ cells in mice. *J. Natl. Cancer Inst.* **38**, 549–552.
17. Stevens, L. C. (1962) Testicular teratomas in fetal mice. *J. Natl. Cancer Inst.* **28**, 247–267.
18. Stevens, L. C. (1966) Development of resistance to teratocarcinogenesis by primordial germ cells in mice. *J. Natl. Cancer Inst.* **37**, 859–867.
19. Park, S. Y., and Jameson, J. L. (2005) Minireview: transcriptional regulation of gonadal development and differentiation. *Endocrinology.* **146**, 1035–1042.
20. Chaillet, J. R., Vogt, T. F., Beier, D. R., and Leder, P. (1991) Parental-specific methylation of an imprinted transgene is established during gametogenesis and progressively changes during embryogenesis. *Cell.* **66**, 77–83.
21. De, F. M. (2000) Regulation of primordial germ cell development in the mouse. *Int. J. Dev. Biol.* **44**, 575–580.
22. Lee, J., Inoue, K., Ono, R., et al. (2002) Erasing genomic imprinting memory in mouse clone embryos produced from d 11.5 primordial germ cells. *Development.* **129**, 1807–1817.
23. Sato, S., Yoshimizu, T., Sato, E., and Matsui, Y. (2003) Erasure of methylation imprinting of Igf2r during mouse primordial germ-cell development. *Mol. Reprod. Dev.* **65**, 41–50.
24. Stevens, L. C. (1964) Experimental production of testicular teratomas in mice. *Proc. Natl. Acad. Sci. U. S. A.* **52**, 654–661.
25. Matin, A., Collin, G. B., Varnum, D. S., and Nadeau, J. H. (1998) Testicular teratocarcinogenesis in mice—a review. *APMIS.* **106**, 174–182.
26. Rodriguez, E., Mathew, S., Reuter, V., Ilson, D. H., Bosl, G. J., and Chaganti, R. S. (1992) Cytogenetic analysis of 124 prospectively ascertained male germ cell tumors. *Cancer Res.* **52**, 2285–2291.
27. Vos, A., Oosterhuis, J. W., de Jong, B., Buist, J., and Schraffordt, K. H. (1990) Cytogenetics of carcinoma in situ of the testis. *Cancer Genet. Cytogenet.* **46**, 75–81.
28. Looijenga, L. H., Verkerk, A. J., Dekker, M. C., van Gurp, R. J., Gillis, A. J., and Oosterhuis, J. W. (1998) Genomic imprinting in testicular germ cell tumours. *APMIS.* **106**, 187–195.
29. Stevens, L. C., and Hummel, K. P. (1957) A description of spontaneous congenital testicular teratomas in strain 129 mice. *J. Natl. Cancer Inst.* **18**, 719–747.
30. Matin, A., Collin, G. B., Asada, Y., Varnum, D., and Nadeau, J. H. (1999) Susceptibility to testicular germ-cell tumours in a 129.MOLF-Chr 19 chromosome substitution strain. *Nat. Genet.* **23**, 237–240.
31. Stevens, L. C., and Mackensen, J. A. (1961) Genetic and environmental influences on teratocarcinogenesis in mice. *J. Natl. Cancer Inst.* **27**, 443–453.
32. Gidekel, S., Pizov, G., Bergman, Y., and Pikarsky, E. (2003) Oct-3/4 is a dose-dependent oncogenic fate determinant. *Cancer Cell.* **4**, 361–370.
33. Noguchi, T., and Stevens, L. C. (1982) Primordial germ cell proliferation in fetal testes in mouse strains with high and low incidences of congenital testicular teratomas. *J. Natl. Cancer Inst.* **69**, 907–913.
34. Youngren, K. K., Coveney, D., Peng, X., et al. (2005) The Ter mutation in the dead end gene causes germ cell loss and testicular germ cell tumours. *Nature.* **435**, 360–364.
35. Harvey, M., McArthur, M. J., Montgomery, C. A., Jr., Bradley, A., and Donehower, L. A. (1993) Genetic background alters the spectrum of tumors that develop in p53deficient mice. *FASEB J.* **7**, 938–943.
36. Kimura, T., Suzuki, A., Fujita, Y., et al. (2003) Conditional loss of PTEN leads to testicular teratoma and enhances embryonic germ cell production. *Development.* **130**, 1691–1700.
37. Di, V. D., Cito, L., Boccia, A., et al. (2005) Loss of the tumor suppressor gene PTEN marks the transition from intratubular germ cell neoplasias (ITGCN) to invasive germ cell tumors. *Oncogene.* **24**, 1882–1894.

38. Lam, M. Y., and Nadeau, J. H. (2003) Genetic control of susceptibility to spontaneous testicular germ cell tumors in mice. *APMIS*. **111,** 184–190.

39. Lamoreux, M. L., Wakamatsu, K., and Ito, S. (2001) Interaction of major coat color gene functions in mice as studied by chemical analysis of eumelanin and pheomelanin. *Pigment Cell Res*. **14,** 23–31.

40. Parant, J. M., and Lozano, G. (2003) Disrupting TP53 in mouse models of human cancers. *Hum. Mutat*. **21,** 321–326.

41. Moore, K. J., Swing, D. A., Copeland, N. G., and Jenkins, N. A. (1990) Interaction of the murine dilute suppressor gene (dsu) with 14 coat color mutations. *Genetics*. **125,** 421–430.

42. Lam, M. Y., Youngren, K. K., and Nadeau, J. H. (2004) Enhancers and suppressors of testicular cancer susceptibility in single- and double-mutant mice. *Genetics*. **166,** 925–933.

43. Bolivar, V. J., Cook, M. N., and Flaherty, L. (2001) Mapping of quantitative trait loci with knockout/congenic strains. *Genome Res*. **11,** 1549–1552.

44. Simpson, E. M., Linder, C. C., Sargent, E. E., Davisson, M. T., Mobraaten, L. E., and Sharp, J. J. (1997) Genetic variation among 129 substrains and its importance for targeted mutagenesis in mice. *Nat. Genet*. **16,** 19–27.

45. Armstrong, N. J., Brodnicki, T. C., and Speed, T. P. (2006) Mind the gap: analysis of marker-assisted breeding strategies for inbred mouse strains. *Mamm. Genome*. **17,** 273–287.

46. Youngren, K. K., Nadeau, J. H., and Matin, A. (2003) Testicular cancer susceptibility in the 129.MOLF-Chr19 mouse strain: additive effects, gene interactions and epigenetic modifications. *Hum. Mol. Genet*. **12,** 389–398.

47. Reusens, B., and Remacle, C. (2006) Programming of the endocrine pancreas by the early nutritional environment. *Int. J. Biochem. Cell Biol*. **38,** 913–922.

48. Wu, G., Bazer, F. W., Cudd, T. A., Meininger, C. J., and Spencer, T. E. (2004) Maternal nutrition and fetal development. *J. Nutr*. **134,** 2169–2172.

49. Levin, B. E. (2006) Metabolic imprinting: critical impact of the perinatal environment on the regulation of energy homeostasis. *Philos. Trans. R. Soc. Lond. B Biol. Sci*. **361,** 1107–1121.

50. Fleming, T. P., Kwong, W. Y., Porter, R., et al. (2004) The embryo and its future. *Biol. Reprod*. **71,** 1046–1054.

51. Suomalainen, A. (1997) Mitochondrial DNA and disease. *Ann. Med*. **29,** 235–246.

52. Luft, R. (1994) The development of mitochondrial medicine. *Proc. Natl. Acad. Sci. U. S. A*. **91,** 8731–8738.

53. Morgan, H. D., Sutherland, H. G., Martin, D. I., and Whitelaw, E. (1999) Epigenetic inheritance at the agouti locus in the mouse. *Nat. Genet*. **23,** 314–318.

54. Wolff, G. L., Kodell, R. L., Moore, S. R., and Cooney, C. A. (1998) Maternal epigenetics and methyl supplements affect agouti gene expression in Avy/a mice. *FASEB J*. **12,** 949–957.

55. Matin, A., and Nadeau, J. H. (2001) Sensitized polygenic trait analysis. *Trends Genet*. **17,** 727–731.

56. Tanksley, S. D. (1993) Mapping polygenes. *Annu. Rev. Genet*. **27,** 205–233.

57. Collin, G. B., Asada, Y., Varnum, D. S., and Nadeau, J. H. (1996) DNA pooling as a quick method for finding candidate linkages in multigenic trait analysis: an example involving susceptibility to germ cell tumors. *Mamm. Genome*. **7,** 68–70.

58. Smigielski, E. M., Sirotkin, K., Ward, M., and Sherry, S. T. (2000) dbSNP: a database of single nucleotide polymorphisms. *Nucleic Acids Res*. **28,** 352–355.

59. Peters, L. L., Robledo, R. F., Bult, C. J., Churchill, G. A., Paigen, B. J., and Svenson, K. L. (2007) The mouse as a model for human biology: a resource guide for complex trait analysis. *Nat. Rev. Genet*. **8,** 58–69.

60. Silver, L. M. (1995) *Mouse genetics: concepts and applications,* Oxford University Press, New York.

61. Khripin, Y. (2006) High-throughput genotyping with energy transfer-labeled primers. *Methods Mol. Biol*. **335,** 215–240.

62. Stephens, M., Sloan, J. S., Robertson, P. D., Scheet, P., and Nickerson, D. A. (2006) Automating sequence-based detection and genotyping of SNPs from diploid samples. *Nat. Genet*. **38,** 375–381.

63. Romkes, M., and Buch, S. C. (2005) Genotyping technologies: application to biotransformation enzyme genetic polymorphism screening. *Methods Mol. Biol.* **291,** 399–414.
64. Jordan, B., Charest, A., Dowd, J. F., et al. (2002) Genome complexity reduction for SNP genotyping analysis. *Proc. Natl. Acad. Sci. U. S. A.* **99,** 2942–2947.
65. Lindblad-Toh, K., Winchester, E., Daly, M. J., et al. (2000) Large-scale discovery and genotyping of single-nucleotide polymorphisms in the mouse. *Nat. Genet.* **24,** 381–386.
66. Zhao, Z. Z., Nyholt, D. R., Le, L., et al. (2006) KRAS variation and risk of endometriosis. *Mol. Hum. Reprod.* **12,** 671–676.
67. Manly, K. F., and Olson, J. M. (1999) Overview of QTL mapping software and introduction to map manager QT. *Mamm. Genome.* **10,** 327–334.
68. Lander, E. S., Green, P., Abrahamson, J., et al. (1987) MAPMAKER: an interactive computer package for constructing primary genetic linkage maps of experimental and natural populations. *Genomics.* **1,** 174–181.
69. Manly, K. F., Cudmore, R. H., Jr., and Meer, J. M. (2001) Map Manager QTX, cross-platform software for genetic mapping. *Mamm. Genome.* **12,** 930–932.
70. Broman, K. W., Wu, H., Sen, S., and Churchill, G. A. (2003) R/qtl: QTL mapping in experimental crosses. *Bioinformatics.* **19,** 889–890.
71. Lander, E., and Kruglyak, L. (1995) Genetic dissection of complex traits: guidelines for interpreting and reporting linkage results. *Nat. Genet.* **11,** 241–247.
72. Churchill, G. A., and Doerge, R. W. (1994) Empirical threshold values for quantitative trait mapping. *Genetics.* **138,** 963–971.
73. Muller, A. J., Teresky, A. K., and Levine, A. J. (2000) A male germ cell tumor-susceptibility-determining locus, pgct1, identified on murine chromosome 13. *Proc. Natl. Acad. Sci. U. S. A.* **97,** 8421–8426.
74. Singer, J. B., Hill, A. E., Burrage, L. C., et al. (2004) Genetic dissection of complex traits with chromosome substitution strains of mice. *Science.* **304,** 445–448.
75. Nadeau, J. H., Singer, J. B., Matin, A., and Lander, E. S. (2000) Analysing complex genetic traits with chromosome substitution strains. *Nat. Genet.* **24,** 221–225.
76. Hill, A. E., Lander, E. S., and Nadeau, J. H. (2006) Chromosome substitution strains: a new way to study genetically complex traits. *Methods Mol. Med.* **128,** 153–172.

Chapter 16
Study Origin of Germ Cells and Formation of New Primary Follicles in Adult Human and Rat Ovaries

Antonin Bukovsky, Satish K. Gupta, Irma Virant-Klun, Nirmala B. Upadhyaya, Pleas Copas, Stuart E. Van Meter, Marta Svetlikova, Maria E. Ayala, and Roberto Dominguez

Contents

Summary The central thesis regarding the human ovaries is that, although primordial germ cells in embryonal ovaries are of extraovarian origin, those generated during the fetal period and in postnatal life are derived from the ovarian surface epithelium (OSE) bipotent cells. With the assistance of immune system-related cells, secondary germ cells and primitive granulosa cells originate from OSE stem cells in the fetal and adult human gonads. Fetal primary follicles are formed during the second trimester of intrauterine life, prior to the end of immune adaptation, possibly to be recognized as self-structures and renewed later. With the onset of menarche, a periodical oocyte and follicular renewal emerges to replace aging primary follicles and ensure that fresh eggs for healthy babies are always available during the prime reproductive period. The periodical follicular renewal ceases between 35 and 40 yr of age, and the remaining primary follicles are utilized during the premenopausal period until exhausted. However, the persisting oocytes accumulate genetic alterations and may become unsuitable for ovulation and fertilization. The human OSE stem cells preserve the character of embryonic stem cells, and they may produce distinct cell types, including new eggs in vitro, particularly when derived from patients with premature ovarian failure or aging and postmenopausal ovaries. Our observations also indicate that there are substantial differences in follicular renewal between adult human and rat ovaries. As part of this chapter, we present in detail protocols utilized to analyze oogenesis in humans and to study interspecies differences when compared to the ovaries of rat females.

Keywords Adulthood; follicular renewal; humans; mammals; oogenesis.

From: *Methods in Molecular Biology, Vol. 450, Germline Stem Cells*
Edited by S. X. Hou and S. R. Singh © Humana Press, Totowa, NJ

16.1   Introduction

The origin of germ cells in adult females of higher vertebrates (birds and mammals) has been a matter of dispute for over 100 yrs. There are, in principle, two views: the oocyte "storage" and "continued formation" theories *(1)*.

The storage doctrine is based on the opinion that there is never any increase in the number of oocytes beyond those differentiating during fetal or perinatal ovarian development from embryonic (primordial) germ cells *(2)*. This currently prevailing dogma was complemented by a declaration indicating that the process of oogenesis in the animal kingdom follows a uniform pattern, of which there are two main variants. One variant is that the oogenesis appears to continue either uninterruptedly or cyclically throughout the reproductive life (e.g., most teleosts, all amphibians, most reptiles, and conceivably few mammals). The other variant is that oogenesis occurs only in fetal gonads, and oogonia neither persist nor divide mitotically during sexual maturity (e.g., cyclostomes, elasmobranchs, a few teleosts, perhaps some reptiles, all birds, monotremes, and with a few possible exceptions, all eutherian mammals) *(3–5)*.

The advantage of the storage theory is that it is simple and easy to understand: The extragonadal primordial germ cells migrate into developing ovaries, achieve sex-specific properties, multiply, and complete meiotic prophase of oocytes, which form primordial follicles that serve for reproductive function up to the menopause in human females.

The essential disadvantage, however, is a requirement for the storage of female gametes in higher vertebrates for up to several decades prior to their utilization for the development of new progeny. Under such conditions, there is a high probability of accumulation of genetic alterations in stored oocytes because of the long-lasting influence of environmental hazards. On the other hand, the storage doctrine supports evidence that, in invertebrates and lower vertebrates, the oogenesis continues throughout reproductive life. The most problematic issue here is what the advantage of the oocyte storage in higher vertebrates might be from the Darwinian evolutionary theory point of view on the development of animal species toward higher and more adaptive forms of life and reproduction (e.g., frogs vs mammals).

The continued formation theory indicates that primordial germ cells degenerate and new oocytes originate during adulthood from cyclical proliferation of the ovarian surface epithelium (OSE) stem cells *(1,6)*, and that new oocytes are formed throughout life, and in phase with the reproductive cycle, from germinal epithelium (OSE) of the adult mammal at the same time as vast numbers of already-formed oocytes become eliminated through atresia *(7)*.

The advantage of the continued formation theory is that it determines that there is a uniform ability of the oocyte and follicular renewal in all adult females throughout the animal kingdom species, making this doctrine acceptable from the evolutionary point of view. The disadvantage is that it is not easy to follow a distinct pattern of this process between the species, such as apparent formation of new oogonia in adult prosimian primates *(8–10)* vs a more cryptic process in adult human females *(11,12)*.

We attempted to establish a harmony between these two possibilities by proposing the prime reproductive period (PRP) theory *(13)*. According to the PRP theory, the storage doctrine fits to two periods of human life, that between the termination of fetal oogenesis and the late puberty or premenarcheal period (about 10–12 yr), and that following the end of PRP until menopause. On the other hand, the continued formation doctrine accounts for follicular renewal during PRP, which ensures an availability of fresh oocytes for healthy progeny. During PRP, the number of primary follicles does not show a significant change in human *(14)* and mouse females *(15)* because of the replacement of aging primary follicles by follicular renewal *(11,12,16)*. Atresia of primary follicles declines during the premenopausal period *(17)*, allowing a significantly reduced number of persisting primary follicles to remain functional in humans for another 10–12 yr after termination of follicular renewal during PRP. However, although there are no consequences of oocyte aging for the progeny during childhood, because ovulation is absent, the oocytes persisting after termination of PRP accumulate genetic alterations, resulting in the exponentially growing incidence of fetal trisomies and other genetic abnormalities with advanced maternal age after PRP (*see* **Fig. 16.1**; reviewed in **ref. 12**).

16.1.1 Human Ovaries

Human OSE stem cells are bipotent progenitors of oocytes and granulosa cells. The oocytes and primary follicles are formed as new structures during the fetal period and represent follicular renewal during the PRP. There are, however, some differences in structure and in the pathways of primary follicle formation between fetal and adult ovaries. In midpregnancy fetal ovaries, the OSE stem cells

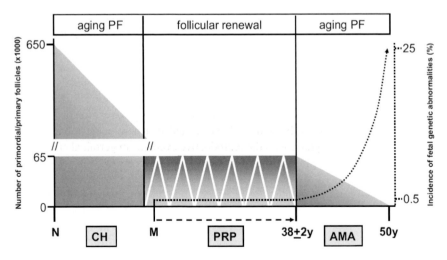

Fig. 16.1 The incidence of trisomic fetuses exponentially increases with age after termination of follicular renewal through the prime reproductive period (PRP). AMA, advanced maternal age; CH, childhood; M, menarche; N, neonate; PF, primary follicles. (Adapted from **ref. 37** with permission of Informa Healthcare, Informa UK Ltd.)

are always present. In normal adult ovaries, however, the OSE cells represent a temporary structure, and new OSE cells originate by mesenchymal–epithelial transition of ovarian tunica albuginea mesenchymal cells.

16.1.1.1 Fetal Ovaries (*see* Fig. 16.2)

The surface of the human fetal ovary is covered by a serous membrane made of OSE that is continuous with the peritoneal mesothelium. During the early stages of ovarian development, there is a rapid proliferation of OSE cells, resulting in cellular stratification, nuclear pleomorphism, and nuclear irregularities *(18)*.

The fetal OSE has been implicated in the formation of oocytes *(19,20)*, and it has been suggested that the OSE is a source of granulosa cells in adult ovaries *(21,22)*. Some immunohistochemical observations, reported and interpreted previously (for details, *see* **refs.** *23* and *24*), are provided here.

Primitive granulosa cells (*see* **Fig. 16.2A**) show a decrease of cytokeratin expression when compared to the OSE cells. They originate from sprouts of OSE cells extending into the ovary between mesenchymal cell cords of rete ovarii. The mesenchymal cell cords are rich in expression of Thy-1 differentiation protein (Thy-1dp), an ancestral member of the immunoglobulin gene superfamily *(25)*, which plays an important role in the stimulation of early cellular differentiation *(26)*. The primitive granulosa cells associate with oocytes in the deeper ovarian cortex to form primary follicles (**Fig. 16.2E**).

Germ cells originate by asymmetric division of OSE cells (**Fig. 16.2B**) and show a diminution of major histocompatibility complex class I (MHC-I) antigen (*see* **Fig. 16.2B**). Subsequently, the germ cells undergo symmetric division, required for crossing over of chromosomes. Next, their size substantially increases under the OSE layer, where they exhibit a tadpole-like shape and enter the ovarian cortex. Note a lack of tunica albuginea.

16.1.1.1.1 Association of Immune System-Related Cells with Fetal Oogenesis

An essential question regards why only some OSE cells are transformed into germ cells. It has been suggested that to become a germ cell, the OSE stem cell should receive an impulse from the ovary-committed monocyte-derived cells (MDCs) and T cells (OCMTs), assuming a milieu of favorable systemic (hormonal) conditions (*see* **Ref.** *23* and **Table 16.1**). During ovarian development, the immune system-related cells migrate through the rete ovarii and interact with resident MDCs (*see* **Fig. 16.2C,G**), and this may result in their ovarian commitment. Apparently, the MDC accompany origination of germ cells from the OSE (*see* **Fig. 16.2D**), as well as T cells (*see* **Fig. 16.2H**) and Thy-1dp (*see* **Fig. 16.2I**). The OSE also shows high binding of immunoglobulins (*see* **Fig. 16.2J**), which may prevent them from a spontaneous transformation into germ cells, and activated MDCs accompany growing primary follicles (*see* **Fig. 16.2F**). Altogether, the origination of the germ cells from

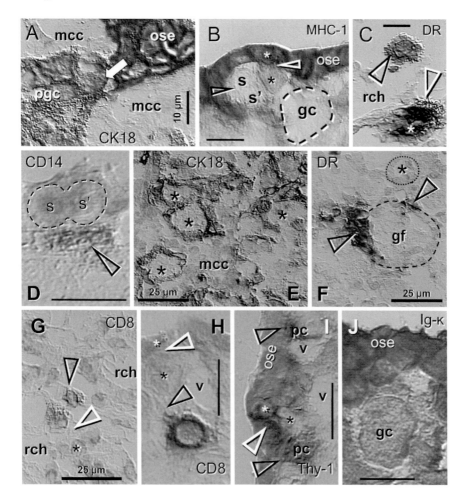

Fig. 16.2 Oocytes and granulosa cells originate from ovarian surface epithelium (OSE) stem cells (*ose*) in midpregnancy human fetuses. **A** Cytokeratin (CK) staining of a cluster of primitive granulosa cells (*pgc*) descending from the OSE (*arrow*) between mesenchymal cell cords (*mcc*). **B** Major histocompatibility complex class I (MHC-I) immunostaining, no hematoxylin counterstain. Asymmetric division (*white arrowhead*) gives rise to MHC-I-positive OSE (*white asterisk*) and MHC-I-negative germ (*black asterisk*) daughter cells. Symmetric division (*black arrowhead*) produces two MHC-negative (s and s') germ cells (crossing over). Larger (tadpole-like) germ cell (*dashed line*) enters the ovarian cortex. **C** HLA-DR+ (activated) monocyte-type cells (*black arrowhead*) migrate through the rete channels (*rch*) and interact (*white arrowhead*) with resident monocyte-derived cells (MDC; *asterisk*). **D** CD14 primitive MDCs associate (*arrowhead*) with symmetric division of germ cells (*dashed line*; hematoxylin counterstain). **E** Small (primordial) follicles (*asterisks*) with CK+ granulosa cells in the lower cortex. **F** HLA-DR cells intimately associate with growing primary follicles (*gf*) but not nongrowing ones (*asterisk*). **G** CD8 T cells (*black arrowhead*) in rete channels interacting with resident cells (*white arrowhead*) and beneath the OSE (**H**) associating (*arrowhead*) with emerging germ cell (*black asterisk*; *v*, vessel); (**I**) Thy-1 differentiation protein (*Thy-1*) is secreted (*arrowheads*) from vascular pericytes (*pc*) among OSE cells with emerging germ cells (*asterisks*). (**J**) Immunoglobulins (*Ig-k*) have a low affinity to germ-like cells (*gc*) but bind heavily to the OSE cells. Bars in **A–D** and **H–J** = 10 μm. (Adapted from **ref. 23** with permission of Humana Press Inc.)

Table 16.1 Working model of age-associated changes of ovary-committed MDCs (monocyte-derived cell) and T cells (OCMT) and hormonal signals (luteinizing hormone/human chorionic gonadotrophin [LH/hCG] and E2) required for the initiation and resumption of oogenesis in human ovaries

Period of life	OCMT[a]	LH/hCG[b]	E2[c]	Oogenesis
First trimester to midpregnancy	Yes	Yes	Yes	Yes[d]
Last trimester to newborn	Yes	No	Yes	No[d]
Postnatal to menarche	Yes	No	No	No[e]
Reproductive period[f]	Yes	Yes	Yes	Yes[d]
Premenopause[g]	No	Yes	Yes	No[e]
Postmenopause	No	Yes	No	No[d]

Source: Reprinted from ref. *23* with permission of Humana Press Inc.

[a] MDC and T cells with commitment for stimulation of OSE to germ cell transformation.

[b] Levels corresponding to the midcycle LH peak or more (hCG levels should be ten times more because it has a 10% affinity to the LH receptor compared to that of LH; *see* **ref. *52***).

[c] Levels corresponding to the preovulatory E_2 peak or more.

[d] Confirmed.

[e] Predicted.

[f] From menarche until 38 + 2 yr of age.

[g] From 38 + 2 yr until menopause.

the OSE appears not to be an accidental but OCMT-driven process. In other words, the number of OCMTs interacting with the OSE cells appears to determine the number of germ cells originating in the ovaries as well as the number of growing primary follicles.

16.1.1.1.2 Cessation of Oogenesis in Prenatal Human Ovaries

The origination of new human oocytes and primary follicles ceases after the second trimester of fetal intrauterine life, possibly because of the diminution of circulating human chorionic gonadotropin (hCG) in the fetal blood caused by the development of the placental "hCG barrier" *(23)*. Thereafter (perinatally), the layer of loose mesenchymal cells forming ovarian tunica albuginea develops, exhibiting some features of the OSE cells (cytokeratin expression) and hence possibly originating from epithelial–mesenchymal transition of the OSE cells *(12,24)*, as described in OSE cultures *(27)*. Under certain conditions *(23)*, these tunica albuginea mesenchymal cells could be transformed back into the OSE cells by mesenchymal–epithelial transition (i.e., into bipotent stem cells capable of differentiating into germ and granulosa cells in adult human ovaries). This, however, may not happen prior to puberty or around menarche *(23)*.

16.1.1.2 Oogenesis During the Prime Reproductive Period

During adulthood, the mesenchymal cells in the tunica albuginea are progenitors of bipotent OSE stem cells in human ovaries (*see* **Fig. 16.3A**). The advantage of mesenchymal cell progenitors is that they may be much more resistant to the epigenetic and environmental hazards compared to the highly sensitive oocytes, as evident from exponentially growing fetal abnormalities after termination of follicular renewal (*see* **Fig. 16.1**).

16.1.1.2.1 Follicular Renewal (see **Fig. 16.3**)

Follicular renewal in adult human ovaries from OSE stem cells is a two-step process, consisting of formation of epithelial cell nests (primitive granulosa cells) and origination of germ cells *(12)*. To form primitive granulosa cell nests, the segments of OSE cells directly associated with the ovarian cortex are overgrown by tunica albuginea and form solid epithelial cords, which fragment into small epithelial nests descending into the deep ovarian cortex, close to medulla. These nests associate with ovarian vessels to catch one of the circulating oocytes (*see* **Fig. 16.3B**). During follicle formation, extensions of granulosa cells penetrate the oocyte cytoplasm to form a single paranuclear Balbiani body (*see* **Fig. 16.3C**), which contains additional organelles the oocyte needs to develop later into the mature egg. Formation of primary follicles was also documented by double-color staining for cytokeratin of the nest cells and zona pellucida (ZP) glycoprotein staining of assembling oocytes (*see* **Fig. 16.3D**; **ref.** *12* provides a color view).

Germ cells originate by asymmetric division from OSE cells differentiating above the tunica albuginea (*see* **Fig. 16.3E**). The germ cells show meiotically expressed oocyte carbohydrate antigen (*see* **ref.** *12*) entering the tunica albuginea, where they undergo a symmetric division (*see* **Fig. 16.3F**) required for crossing over, and then entering the adjacent cortical vessels. During vascular transport, the germ cells increase to the small oocyte size, which is picked up by epithelial nests.

An alternative origin of germ cells in human ovaries is OSE crypts originating from OSE invaginations into the deep cortex. Such epithelial crypts (*see* **Fig. 16.3G**) show transformation of OSE cells into germ cells (inset), which are capable of saturating neighboring nests of primitive granulosa cells to form primary follicles. When the epithelial nests are not available, the tadpole-like germ cells, resembling thicker sperm (*see* **Fig. 16.3H**), attain cortical vessels and utilize vascular transport to reach distant targets (*see* **Fig. 16.3B**). Finally, the oocytes that were not utilized for the formation of new primary follicles degenerate in the medullary vessels (*see* **Fig. 16.3I**).

16.1.1.2.2 Association of Immune System-Related Cells with Follicular
            Renewal (see **Fig. 16.4**)

Like in fetal ovaries, the OCMTs show association with oogenesis in adult human ovaries. The primitive MDCs and CD8+ T cells accompany origination of germ cells from OSE stem cells during adulthood (*see* **Fig. 16.4A,B**). The MDCs are also associated with symmetric division of germ cells in the tunica albuginea and germ cells entering the upper cortex (*see* **Fig. 16.4C**). Activated MDCs are associated with migrating germ cells

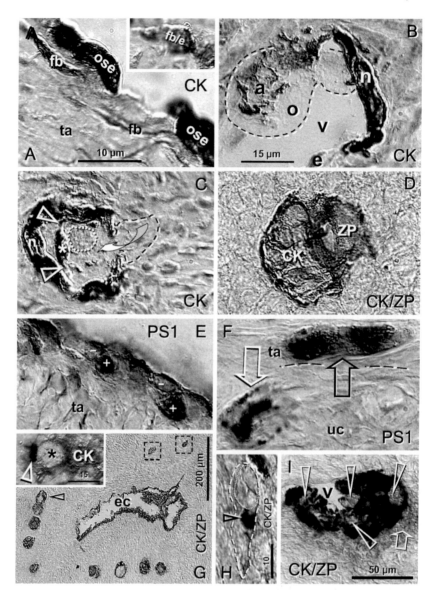

Fig. 16.3 Follicular renewal in adult human ovaries. **A** Cytokeratin positive cells of fibroblast type (*fb*) in tunica albuginea (*ta*) exhibit mesenchymal–epithelial transformation into ovarian surface epithelium (OSE) cells (*ose*). Inset shows a transitory stage (*fb/e*). **B** The CK+ (*dark color*) epithelial nest (*n*) inside of the deep cortical venule (*v*), which extends an arm (*a*) to catch the oocyte (*o, dashed line*) from the blood circulation. *e,* endothelial cell. **C** The nest body (*n*) and closing "gate." A portion of the oocyte (*dashed line*) still lies outside the complex and is expected to move inside (*arched arrow*). The oocyte contains intraooplasmic CK+ extensions from the nest wall (*arrowheads*), which contribute to the formation of CK+ paranuclear (Balbiani) body (*asterisk*). The oocyte nucleus is indicated by a *dotted line*. **D** Occupied "bird's nest" type indicates a halfway oocyte nest assembly. *CK* indicates cytokeratin staining of primitive granulosa cells, and *ZP* indicates zona pellucida expression in the assembling oocyte. **E** Segments of OSE show cytoplasmic PS1

(*see* **Fig. 16.4D**), and such cells associate with the cortical vasculature (*see* **Fig. 16.4E**) and utilize vascular transportation to reach distant destinations (*see* **Fig. 16.4F**).

The rete ovarii is absent in adult ovaries, and OCMTs during adulthood could originate from bone marrow and lymphoid tissues. It has been suggested that, during fetal immune adaptation, an "ovarian" memory is built within the developing immune system for support of follicular renewal by OCMT during adulthood *(24)*. With its utilization, such memory could be exhausted between 35 and 40 yr of age and follicular renewal terminated. It appears that termination of follicular renewal occurs at about 38 yr of age (*see* **Fig. 16.1**), resulting in a significant decline of oocyte numbers in human ovaries *(14)* and an abrupt change in the exponential rate of primary follicle loss *(28)*.

16.1.1.2.3   Summary on the Origin of New Primary Follicles in Adult Human
             Ovaries (*see* **Fig. 16.5**)

The origin of epithelial nests and germ cells and their assembly in adult human ovaries are schematically depicted in **Fig. 16.5**. Under the influence of local signaling (*see* **Fig. 16.5A**), and possibly immune system-related cells and neural signals and molecules *(11)*, such as tunica albuginea overgrows the upper ovarian cortex, and its mesenchymal cells attain cytokeratin expression and transform into OSE. In this way, the bilaminar OSE layer is formed, which descends into the cortex, and fragments into epithelial nests of primitive granulosa cells *(11,12)*. The epithelial nests move through stromal rearrangements into the lower cortex *(12)*, where they associate with cortical vessels (venules) to pick up circulating oocytes (*see* **Fig. 16.3B**).

Under the influence of hormonal (*see* **Fig. 16.5B**) and cellular signaling of OCMTs derived from immune system-related structures supposed to carry an "ovarian memory" (*see* **ref. 24**), the mesenchymal OSE precursors differentiate into the OSE cells covering the tunica albuginea layer and produce germ cells by asymmetric division (*see* **Figs. 16.3E, 4A,B**, and **ref. 12**), followed by a single symmetric division of germ cells (*see* **Figs. 16.3F and 4C**) required for crossing over. Subsequently, the germ cells enter the upper cortex (**Fig. 16.5B**) and vascular circulation (*see*

Fig 16.3 (continued) (meiotically expressed carbohydrate) expression. Asymmetric division of OSE cells gives rise to cells exhibiting nuclear PS1 (+ nuclei) and descending from the OSE into tunica albuginea (*ta*). **F** In tunica albuginea, the putative germ cells increase in size, show a symmetric division (*black arrow*), and exhibit development of cytoplasmic PS1 immunoexpression when entering (*white arrow*) the upper ovarian cortex (*uc*). **G** Association of primary follicles (*arrowhead*) with the cortical epithelial crypt (*ec*). Dashed boxes indicate unassembled epithelial nests. Inset shows origination of germ-like cells among CK+ cells (*CK*) in epithelial crypt. Note ZP+ segment (*white arrowhead*) associated with unstained round cell (*asterisk*). **H** Migrating germ cells with tadpole shape (*dashed line*) and nucleus (*dotted line*) and ZP staining of the intermediate segment (*arrowhead*). **I** Some medullary vessels (*v*) show accumulation of ZP+ (*dark color*) degenerating oocytes with unstained nuclei (*arrowheads*). Arrow indicates ZP release. (Adapted from **ref. 12**; see http://www.rbej.com/content/2/1/20 for color presentation and more details. Copyright Antonin Bukovsky.)

242

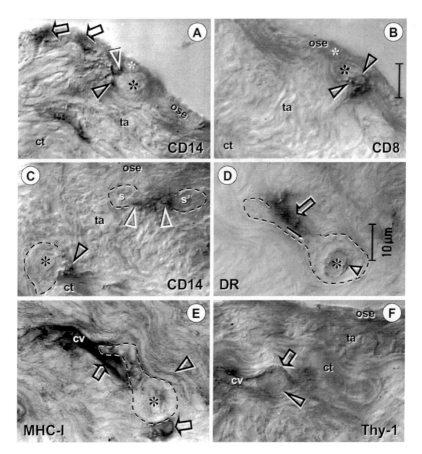

Fig. 16.4 Immune cells influence commitment of ovarian surface epithelium (OSE) stem cells. Staining of the adult human OSE (*ose*), tunica albuginea (*ta*), and adjacent cortex (*ct*) for CD14 of primitive monocyte-derived cell (MDC) and HLA-DR of activated MDC, CD8 of cytotoxic/suppressor T cells, major histocompatibility complex (MHC) class I heavy chain, and Thy-1 glycoprotein of pericytes, as indicated in panels. *Large asterisks* and *dashed lines* indicate putative germ cells. **A** Primitive MDC associate with OSE (*arrows*) and accompany (*arrowheads*) origination of germ cells by asymmetric division of OSE stem cells (*asterisks*). **B** Asymmetric division is also accompanied by extensions from T cells (*arrowheads*). **C** Primitive MDC accompany (*white arrowheads*) symmetric division (s–s') of germ cells in tunica albuginea and their migration into the adjacent cortex (*ct*). **D** Migrating tadpole-like germ cells are accompanied by activated MDC (*open arrow*), and HLA-DR material is apparent in the cytoplasm (*solid arrow*) and in the nuclear envelope (*arrowhead*). **E** The germ cells associate with cortical vasculature (*cv*) strongly expressing MHC-I (*arrows* vs *arrowhead*), enter and are transported (**F**) by the bloodstream. **F** (Adapted from **ref. *11*** with permission of Blackwell Publishing, Oxford, UK.)

Figs. 3F and **4C**), associate with blood vessels (*see* **Fig. 16.4E**), enter circulation (*see* **Fig. 16.4F**), and reach the small oocyte size before assembling with epithelial nests of primitive granulosa cells in the lower cortex (*see* **Fig. 16.3B**). The circulating oocytes may also contaminate other tissues, including bone marrow (*see* **Fig. 16.5**),

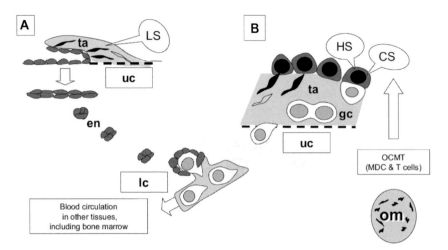

Fig. 16.5 Follicular renewal in adult ovaries is a two-step process based on mesenchymal–epithelial transition of tunica albuginea (*ta*) bipotent progenitor cells into ovarian surface epithelium (OSE) stem cells. **A** Epithelial nests: Segments of the OSE directly associated with the upper ovarian cortex (*uc*) are overgrown with tunica albuginea, which forms a solid epithelial cord that fragments into small epithelial nests (*en*) descending into the lower ovarian cortex (*lc*) and associating with the blood vasculature. Initiation of this process may require local signaling (*LS*), possibly both cellular and neural *(11)*. **B** Germ cells: Under the influence of cellular signaling (*CS*) of ovary-committed monocyte-derived cells (MDC) and T cells (OCMT) and hormonal signaling (*HS; see* **Table 16.1**), some OSE cells covering the tunica albuginea undergo asymmetric division and give rise to new germ cells. The germ cells subsequently divide symmetrically and enter adjacent cortical blood vessels. During vascular transport, they increase to oocyte size and are picked up by epithelial nests associated with the vessels. The OCMTs originate from bone marrow and lymphoid tissues carrying "ovarian" memory (*om*), which diminishes with utilization; when spent, the follicular renewal ceases in spite of persisting hormonal signaling

and eventually degenerate in ovarian vasculature (*see* **Fig. 16.3I**), to which they appear to have an exclusive affinity to home. For schematic description of an alternative origin of germ cells from epithelial crypts in adult human ovaries, *see* **Fig. 16.3G** and **ref. *12***.

16.1.1.2.4 Persisting Primary Follicles Lie in the Thy-1dp-Depleted Areas (*see* **Fig. 16.6**)

The role of Thy-1dp is associated with the stimulation of follicular growth and development *(11,29,30)*. Hence, the resting primary follicles should be devoid of the presence of Thy-1dp. Typically, the primary follicles occupy stromal areas low in Thy-1dp expression (*see* **Fig. 16.6A**), exhibiting an ovary within the ovary pattern *(12)*. It is apparent that growing primary follicles are associated with Thy-1dp secretion from accompanying vascular pericytes (*see* **Fig. 16.6B**), and resting follicles are not (see **Fig. 16.6C**).

16.1.2 Follicular Atresia and Renewal in Adult Rodents
(see Fig. 16.7)

Rat ovaries show some distinct general features compared to human ovaries. In contrast to humans, the adult rat ovaries are permanently covered by OSE cells. A distinct tunica albuginea layer of loose connective tissue and the deep cortex of compact stromal cells are not apparent. The ovaries are significantly smaller (5–7 vs 35–40 mm), and follicular turnover is more frequent—selection of approx. ten follicles in both ovaries and accompanying follicular atresia occurs during a short (4- to 5-d) cycle. Unfortunately, the granulosa cells in adult rat ovaries do not express cytokeratins (31).

Expression of ZP1,2,3 proteins in the ovary of a postpubertal rat is shown in **Fig. 16.7**. It is apparent that tadpole-like germ cells may also develop in adult rat ovaries. These cells may, but do not necessarily, originate from the OSE. An alternative germ cell origin site is the ovarian medulla or hilar region. In mice, the ovarian medulla was found to contain clusters of germ cells, which expressed some, but not all, markers of adult germ cells (32). Hence, the authors concluded that the ovarian medulla contains germ cells that are not oocytes. One may add that these germ cells are precursors of oocytes because that is a commitment of germ cells in the females. Indeed, we detected clusters of germ-like cells in adult rat ovaries expressing ZP4/ZPB, which was not detected in the follicular oocytes (13). It has been reported that

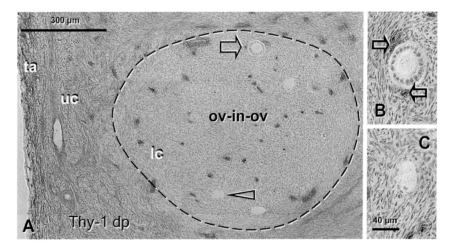

Fig. 16.6 Ovarian primary follicles and Thy-1 differentiation protein (*Thy-1 dp*). Thy-1dp was strongly expressed by tunica albuginea fibroblasts (*ta*) and moderately in the upper cortex (*uc*) and lower cortex (*lc*), excluding areas showing an "ovary-within-the-ovary" pattern (*ov-in-ov; dashed line*) with virtually no Thy-1dp immunoexpression, except for vascular pericytes (*solid arrow*). These areas characteristically contained primary follicles (*arrowhead* and **C**), some of which showed an increase in size accompanied by Thy-1dp+ pericytes (*open arrow* and **B**). (Adapted from ref. *12*, copyright Antonin Bukovsky.)

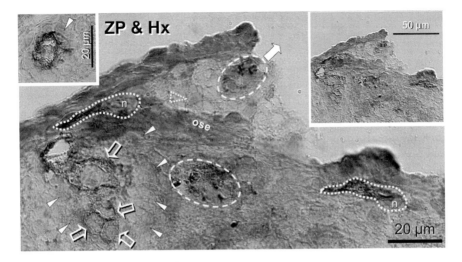

Fig. 16.7 The adult rat ovary stained for zona pellucida (*ZP*) (hematoxylin [*Hx*] counterstain of nuclei) shows two tadpole-like cells with ZP+ tails (*dotted lines*) and ZP unstained nuclei (*n*), two degenerating primary follicles (*dashed lines*) with possible expulsion (*solid arrow*), a cluster of ZP+ putative germ cells (*open arrows*), and adjacent granulosa-type cells (*arrowheads*), possibly in an epithelial cord descended from the surface epithelium (*ose*). The cells under the surface epithelium (SE) flap exhibit weaker staining of the nuclei (*dotted arrowhead*) compared to the ovarian surface epithelium (OSE) cells. The *right inset* shows the same area without symbols; the *left inset* shows ZP expression in the normal oocyte in the primary follicle. (Adapted from **ref. 23** with permission of Humana Press Inc.)

the rat ZP2 and ZP3 contain "minimum structures" to which rat sperm can bind, and rat ZP1 and ZP4/ZP4B are dispensable in the rats regarding their role in the oocyte–sperm interactions *(33)*. Hence, for instance, the ZP4/ZP4B glycoproteins could serve as molecules involved in early development of germ cells but are not functionally required for the developing and maturing oocytes.

In addition, in rat ovaries, the hilar region contains sex cords expressing an intrinsic bone morphogenetic protein (BMP) system replete with BMP ligands and receptors *(34)*. The tadpole putative germ cells expressing ZP1,2,3 appear to migrate under the ovarian surface, and they may settle and give rise to the clusters of dividing germ cells (*see* **Fig. 16.7**). The rat OSE cells may give rise to clusters of primitive granulosa cells available for the formation of primary follicles as in human fetuses.

The follicular renewal appears to be accompanied by atresia of primary follicles (*see* **Fig. 16.7**) degenerating either inside or leaving the ovary. It is possible that waves of follicular renewal induce atresia of resting follicles. Indeed, in premenopausal human ovaries lacking follicular renewal, the primary follicles appear to be depleted by their utilization only *(17)*; during the PRP with follicular renewal (*see* **Fig. 16.1**), the atresia affects 50–70% of primary follicles *(35)*.

16.1.3   Differentiation of Oocytes from Adult Stem Cells in Culture (see Fig. 16.8)

After studies of oogenesis in fetal and adult human ovaries, we wondered if a similar potential for OSE cells could be demonstrated in vitro. Our observations indicate that OSE cultures may differentiate into distinct cell types, including mesenchymal, epithelial, granulosa, and neural type cells, and oocytes (*see* **Fig. 16.8**). The presence of oocytes was not observed prior to d 4 of culture, indicating that they do not originate from follicular oocytes. Oocyte-type cells were found to develop in OSE cultures from premenopausal and postmenopausal ovaries and from ovaries of women exhibiting premature ovarian failure (POF) *(36,37)* because anovulatory ovaries are always covered with a continuous layer of the OSE *(22)* and often contain cortical epithelial crypts, an alternative source of germ cells *(12)*. On the other hand, a poor outcome was observed in cultures

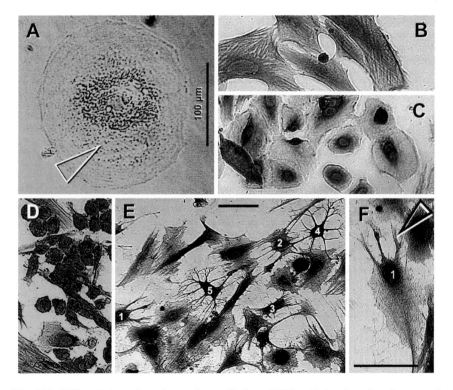

Fig. 16.8 Differentiation of ovarian surface epithelium (OSE) cells in vitro. Development of oocyte (**A**), fibroblasts (**B**), epithelial (**C**), granulosa (**D**), and neural-type cells (**E** and **F**) in primary OSE cultures. Arrowhead in **A** indicates initial breakdown of the germinal vesicle. Numbers in **E** show stages of neural cell differentiation. **F** A transition of an epithelial into a neural cell type (*arrowhead*) from **E** in detail. **A** A live culture in phase contrast; **B–F** Imunohistochemical staining for ZP1,2,3 glycoproteins. Bars = 100 μm. (Adapted from **ref. 36**, copyright Antonin Bukovsky)

derived from ovaries of ovulating women, for whom the occurrence of OSE is usually episodic during the cycle *(22)*.

16.1.4 Potential Treatment of Ovarian Infertility

It took 40 yr for the currently standard in vitro fertilization (IVF) technique to be utilized in clinical practice. Although the oocyte-type cells can be developed in OSE cultures for research purposes and confirmation that OSE cells are capable to be transformed into oocytes, it may take several years before utilization of such an approach for the treatment of ovarian infertility, which resists treatment by the current standard IVF techniques. Conditions for inclusion of patients with ovarian infertility into the first clinical trial were elaborated late in 2005 *(24)*, as discussed in the following subheadings.

16.1.4.1 Suitability of Patients for Clinical Trial

Patients with diagnosis of POF (aged less than or equal to 40–43 yr, depending on the institutional rules) could be included in the clinical trial. Optimally, these patients would have failed to conceive because of lack of functional oocytes during previous standard IVF therapy, or such therapy was impossible because of the lack of oocytes within ovaries, and they would be considering new options to have a genetically related child before undergoing conventional IVF with donated oocytes. Patients should provide a detailed medical history and available laboratory results for consideration for the trial. Ultrasound or magnetic resonance imaging (MRI) images of ovaries may be requested, and patients may be advised to utilize certain hormonal therapy, including replacement of the existing one, several weeks prior to admission.

Prospective patients and their partners should not carry any genetic alterations that can be transmitted to the child. Of particular importance is the exclusion of POF with fragile X permutation (>200 CGG trinucleotide repeats of FMR1 gene causing Fragile X-associated tremor/ataxia), because the birth of a child in such women may result in mental retardation of the progeny *(38)*. Genetic alterations are detected in a proportion of patients with POF, particularly those with primary amenorrhea *(39)*, and fragile X permutation has been detected in 4.8% of patients with POF *(40)*. Therefore, evidence of the lack of genetic abnormality should be provided, or the patients will be tested.

If needed, additional laboratory investigation of blood and urine as well as imaging procedures would be done after admission. All considered women should have a male partner with normal semen quality according to the World Health Organization (WHO) criteria. Women with infertile partners (i.e., with azoospermia) should be excluded.

Therapy of ovarian infertility with cultured ovarian stem cells should be explained to the patient by a specialist in gynecology and obstetrics who is familiar with the new technique. The medical documentation of each patient and her male partner should be evaluated by an interdisciplinary committee for IVF, which would decide inclusion into the trial.

16.1.4.2 Collection of Ovarian Stem Sells and Utilization of Oocytes

The OSE cells and a small ovarian biopsy are collected during laparoscopy. The OSE cells and cells collected by scraping of tissue biopsy are cultured for 5–10 d to determine whether they can produce new oocytes. If oocytes develop, they can be inseminated and fertilized by classical IVF or by intracytoplasmic sperm injection (ICSI) with the partner's semen. Embryos can be cultured to the blastocyst stage and, before transfer into the uterus, evaluated by preimplantation genetic diagnosis. All retrieved embryos should be cryopreserved because it is not possible to prepare endometrium and ovaries for a certain period of implantation. After appropriate hormonal preparation of the women, normal embryos could be thawed and at most one normal blastocyst transferred into the uterus. Supernumerary blastocysts are cryopreserved for the patient's potential later need. In case of a pregnancy, amniocentesis should be performed (genetic evaluation of embryo/fetus).

16.1.4.3 Potential Pitfalls

During the clinical trial, the following complications of cultured cells could occur: Oocytes will not develop, oocytes will not be appropriate for fertilization, oocytes will not be fertilized, fertilized oocytes will not develop into embryos, or embryos will not be transferred into the uterus because they will be genetically abnormal. If the oocyte culture is not successful, the infertility treatment could be continued by usual treatment with donated oocytes.

16.1.4.4 Initiation of the First Clinical Trial

Criteria for initiation of the clinical trial have been found appropriate and its initiation in the IVF laboratory, Department of Obstetrics and Gynecology, University Medical Center Ljubljana, approved by the Slovenian Committee for the Medical Ethic. Early in 2006, both authors involved (A.B. and I.V.-K.) met in the IVF laboratory, Ljubljana, Slovenia, to initiate the trial. The objective was to evaluate if there are OSE stem cells in infertile women with POF, and if these cells can develop into oocytes capable of fertilization in vitro. After the informed consent process, three patients with POF, and no oocytes in ovaries, aged 30, 38, and 40 yr, and their normospermic partners were selected.

The OSE cells were collected during diagnostic laparoscopy by scratching the ovarian surface, and ovarian biopsies were collected from both ovaries. From half of each biopsy and collected OSE, the cell cultures were set up in Dulbecco's modified Eagle's medium (DMEM)/F12 medium with phenol red (weak estrogenic action), supplemented with antibiotics and 20% comprehensively heat-inactivated serum (59°C, 60 min) of the corresponding patient. The culture was monitored daily.

Ovarian cells attached to the bottom of a dish began to differentiate into epithelial and fibroblast cell types and some of them into oocytes. On d 3 of culture, the initial medium was replaced with in vitro maturation medium (MediCult IVM, Copenhagen, Denmark) supplemented with follicle-stimulating hormone (FSH; 75 mIU/mL), hCG (5 IU/mL), and 10% heat-inactivated patient's serum. Prepared male partner's sperm were added several hours later.

Embryo-like structures developed in the OSE cultures of two POF women on the next day. The structures detached spontaneously and were transferred into the wells with standard medium for IVF, where they developed progressively to the morula-, preblastocyst-, (*see* **Fig. 16.9**) and blastocyst-like structures. They were frozen to be later genetically analyzed and transferred into the uterus, if normal.

Remaining material from biopsies was investigated by immunohistochemistry for the presence of OSE and granulosa cells of primary follicles (cytokeratin expression). No primary or other follicle types were found, and development of oocyte-like cells and embryo-like structures after in vitro insemination of cultures correlated with the presence of the OSE in the biopsies. In one woman with no OSE in both biopsies, no oocyte-like cells developed, and embryo-like structures were absent after utilization of IVM and sperm.

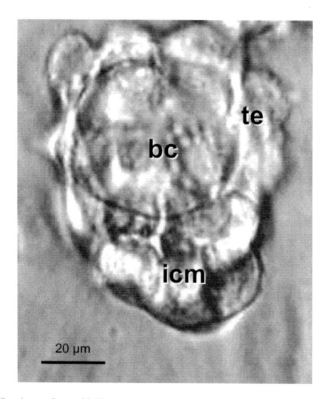

Fig. 16.9 Ovarian surface epithelium culture from premature ovarian failure (POF) patient after in vitro insemination. Optical section through preblastocyst-like structure with early formation of blastocoele (*bc*), trophectoderm (*te*), and inner cell mass (*icm*). (Adapted from **ref. *37*** with permission of Informa Healthcare, Informa UK Ltd.)

Results of this research confirmed the presence of OSE stem cells in some infertile women with POF, with these cells capable of developing into oocytes and able to be fertilized. These observations indicate that adult human ovaries are capable of producing new oocytes for follicular renewal, which, however, fails to occur in vivo in the POF patients. The OSE cultures offer new chances for infertile women with POF to have a genetically related offspring and are worthy of further investigation.

16.1.5 Conclusion

Altogether, the evolutionary point of view and available experimental and clinical data indicate that follicular renewal during PRP exists in human females and most probably without gaps throughout all animal kingdom species. We are convinced that it is just a matter of time before other investigators will realize this.

16.2 Materials

16.2.1 Human Ovaries in Vivo

16.2.1.1 Tissue Collection and Peroxidase Immunohistochemistry

1. Cryomold biopsy disposable vinyl specimens mold $10 \times 10 \times 5$ mm (Tissue-Tek, Miles Inc. Diagnostic Division, Elkhart, IN, cat. no. 4565).
2. Embedding medium for frozen tissue specimens to ensure optimal cutting temperature (OCT) compound (Tissue-Tek cat. no. 4583).
3. Fisherfinest* Premium Superfrost* Microscope Slides (Fisher Scientific International Inc., Pittsburgh, PA, cat. no. 12-544-7).
4. Fisherfinest® premium cover glasses (Fisher Scientific International Inc. cat. no. 12-548-5M).
5. Drierite anhydrous calcium sulfate (Hammond Drierite Co. Ltd., Xenia, OH, stock no. 23001).
6. Primary antibodies:

   a. Mouse-antihuman CK 18, clone CY-90 (Sigma-Aldrich Chemical Co., St. Louis, MO, cat. no. C8541) or CK 5, 6, 8, 17, clone MNF116 (Dako Corp., Carpinteria, CA, cat. no. M0821) ($5\,\mu$g/mL in phosphate-buffered saline [PBS]).
   b. Rabbit-anti-heat-solubilized porcine zona (HSPZ) *(41,42)*, kindly donated by Dr. Bonnie S. Dunbar, Department of Molecular and Cellular Biology, Baylor College of Medicine, Houston, Texas; diluted 1:20 and preabsorbed with rat kidney homogenate ($500\,$mg/mL; *see Note 1*) for 20 min at room temperature, spun ($14,000g$, 10 min, 4°C) and supernatant collected for immediate use or preserved frozen until required.

    c. Antibodies to HSPZ (*see* **Subheading** 16.*3.2.1.1*): R145, R146; diluted 1:80 and preabsorbed with rat kidney homogenate (500 mg/mL) for 20 min at room temperature, spun (14,000*g*, 10 min, 4°C), and supernatant collected for immediate use or preserved frozen until needed.

    d. Antibodies to human ZP4/ZPB (*see* **Subheading** *16.3.2.1.2*): R156, R 164; diluted 1:80 and preabsorbed with rat kidney homogenate (500 mg/mL; *see* **Note 1**) for 20 min at room temperature, spun (14,000*g*, 10 min, 4°C), and supernatant collected for immediate use or preserved frozen until needed.

    e. CD31 of endothelial cells, clone JC/70A, 5 µg/mL (Dako Corp. cat. no. M0823).

    f. HLA-DR of endothelial cells and activated tissue macrophages, clone MEM-12 *(43)*, 5 µg/mL (kindly donated by Drs. Ivan Hilgert and Vaclav Horejsi, Institute of Molecular Genetics, Academy of Sciences of the Czech Republic and Faculty of Sciences, Charles University, Prague).

    g. Thy-1dp of fibroblasts and pericytes, clone F15-42-01 *(44)* (kindly donated by Dr. Rosemarie Dalchau, Institute of Child Health, University of London).

7. Secondary antibodies:

    a. Peroxidase AffiniPure Goat Fab′ antirabbit immunoglobulin G (IgG) (H+L), Abs. human serum (GAR) (Protos Immunoresearch, Burlingame, CA, cat. no.761).

    b. Peroxidase conjugate swine antimouse IgG (SwAM/Px) (Sevapharma, Prague, Czech Republic; kindly provided by Dr. Jana Peknicova, Department of Biology and Biochemistry of Fertilization, Institute of Molecular Genetics, Academy of Sciences of the Czech Republic, Prague).

8. PBS: 1.06 g/L Na_2HPO_4, 0.35 g/L NaH_2PO_4.H_2O (Sigma cat. no. S9638), 8.5 g/L NaCl in distilled water. Adjust to pH 7.23. (10X stock solution is kept at −20°C: 53 g Na_2HPO_4, 17.5 g NaH_2PO_4.H_2O, 425 g NaCl in 5 L distilled water).

9. Tris-buffered saline Tween-20 (TBS-T): 9 g/L NaCl, 0.5 mL/L Tween-20® (Sigma cat. no. P7949), 10 mL/L 1*M* Tris-HCl (pH 7.4) (Fisher Scientific International Inc. cat. no. BP 152-1) in distilled water.

10. Acetone for fixation of tissue (Fisher Scientific International Inc. cat. no. HC 300-1 gal).

11. 100% reagent alcohol.

12. Methanol, Chromasolv®, for high-performance liquid chromatography, 99.9% (Sigma cat. no. 34860-2L-R) 42513.

13. Xylenes (Fisher Scientific International Inc. cat. no. HC 700-1 gal).

14. Sodium azide (Sigma cat. no. S2002).

15. Canada balsam, mounting medium for microscopy (Sigma cat. no. C1795).

16. Hematoxylin solution, Harris modified (Sigma cat. no. HHS16).

17. 3,3′-Diaminobenzidine tetrahydrochloride (DAB) (Sigma cat. no. D5637) 43018.

18. Hydrogen peroxide solution, 30% (w/w) in water (Sigma cat. no. H1009).

19. Vector SG substrate kit (Vector Laboratories cat. no. SK-4700).

20. Vector VIP substrate kit (Vector cat. no. SK-4600).

21. Cryostat microtome with specimen retraction during return travel (Carl Zeiss Microm HM 505 E; Microm Laborgeräte GmbH, Waldorf, Germany).
22. Leitz DM RB (Leica Inc., Wetzlar, Germany) microscope equipped with differential interference contrast and a DEI-470 CCD (charged coupled device) video camera system (Optronics Engineering, Goleta, CA) with detail enhancement. The video images were captured by CG-7 color frame grabber (Scion Corp., Frederick, MD) supported by Scion Image public software developed at the National Institutes of Health (Wayne Rasband, NIH, Bethesda, MD) and ported into Microsoft® Power-Point® 97 SR-2 (Microsoft Corp., Redmond, WA).

16.2.2   Rat Ovaries

16.2.2.1   Animals

1. Rats of the CII-ZV strain, born and maintained in the Laboratory of Biology of Reproduction, Zaragoza, UNAM, Mexico, were maintained on an light–dark cycle of 14 h light/10 h dark (lights on 0500 to 1900). The postnatal rats were weaned at 21 d of age, and they attained sexual maturity during the sixth week of life.
2. Female New Zealand white rabbits (R145, R146, R156, R164) were reared at the Small Animal Facility, National Institute of Immunology, New Delhi, India.

16.2.2.2   Antibodies to Zona Pellucida Glycoproteins

1. Complete Freund's adjuvant (CFA; Difco Laboratories, Detroit, MI, cat. no. 263810).
2. Incomplete Freund's adjuvant (IFA; Difco Laboratories cat. no. 263910).
3. Diphtheria toxoid (DT; Serum Institute, Pune, India, batch no. V6) 4554.
4. Glutaraldehyde (grade II, 25% aqueous solution, Sigma cat. no. A-8009).
5. Squalene (2,6,10,15,19,23-hexamethyl-2,6,10,14,18,22-tetracosahexaene; Sigma cat. no. S-3626).
6. Arlacel-A (mannide monooleate; Sigma cat. no. A-8009).
7. Antirabbit-IgG-HRPO conjugate (Pierce Biotechnology Inc., Rockford, IL, cat. no. 31460).

16.2.3   Tissue Culture for Research

16.2.3.1   Media and Supplements for Tissue Culture

1. DMEM nutrient mixture F12 Ham, with L-glutamine, 15 mM HEPES, without phenol red, without sodium bicarbonate (DMEM/F12, PhR−) (Sigma cat. no. D2902). Add sodium bicarbonate (3.7 g /L) (Sigma cat. no. S5761), adjust to pH 7.4, sterilize by filtration using a membrane with a porosity of 0.22 μ (Millipore

Corp., Billerica, MA, cat. no. SCGPS05RE). Supplement medium with antibiotics, 50 µg/mL gentamicin (Sigma cat. no. G1272), 100 U/mL penicillin, and 100 µg/mL streptomycin (Sigma cat. no. P0781), and eventually 20% fetal bovine serum (FBS; Gibco/BRL, Grand Island, NY, cat. no. 10099-141), as indicated in **Subheading 16.3**.

2. DMEM/F12 with L-glutamine, 15 mM HEPES, with phenol red, without sodium bicarbonate (DMEM/F12, PhR+) (Sigma cat. no. D-8900). Add sodium bicarbonate (3.7 g /L) (Sigma cat. no. S5761), adjust to pH 7.4, sterilize by filtration using a membrane with a porosity of 0.22 µ (Millipore Corp. cat. no. 475 SCGPS05RE). Supplement medium with antibiotics and FBS, as indicated above).

3. BD Falcon® 24-well tissue culture plates (Fisher Scientific International Inc. cat. no. 08-772-1; Falcon no. 353047).

4. BD Falcon 100 × 50 mm standard dishes (Petri dishes) (Fisher Scientific International Inc. cat. no. 08-757-100D; Falcon no. 351029).

5. Sterile stainless steel no. 22 surgical blades (Fisher Scientific International Inc. cat. no. 08-918-5C).

16.2.4 Tissue Culture for Clinical Trial

16.2.4.1 Media and Other Disposables for OSE Cultures

1. DMEM/F12, PhR+ (Sigma cat. no. D8900). Add sodium bicarbonate (3.7 g/L) (Sigma cat. no. S5761), adjust to pH 7.4, sterilize by filtration using a membrane with a porosity of 0.22 µ (Millipore Corp. cat. no. SCGPS05RE). Supplement medium with 20% comprehensively heat-inactivated human serum (HuS) of the corresponding patient and with the following antibiotics: 50 µg/mL gentamicin (Sigma cat. no. G1272), 100 U/mL penicillin, and 100 µg/mL streptomycin (Sigma cat. no. P0781).

2. In vitro maturation medium (MediCult IVM, Copenhagen, Denmark). Supplement medium with FSH, Gonal F, 75 NE/I.E. (Serono, Switzerland) and with hCG, Pregnyl®, 5000 I.E. (Organon, Oss, The Netherlands). First, add 1 ampule of Gonal F containing 75 IU (5.5 µg) FSH to 10 mL MediCult IVM medium (FSH solution). Second, add 1 ampule of Pregnyl containing 5000 IU hCG to another 10 mL of IVM medium (hCG solution). Finally, prepare IVM medium by mixing 9 mL IVM medium, 1 mL heat-inactivated patient's serum, 100 µL FSH solution, and 100 µL hCG solution. The final concentration of hormones is 75 mLU/mL FSH plus 5 IU/mL hCG.

3. Tissue culture plates; CE marked dishes for IVF, four-well dishes (Nalge Nunc International, New York, cat. no. 144444).

4. BD Falcon® 100 × 15 mm standard dishes (Petri dishes) (Fisher Scientific International Inc. cat. no. 08-757-100D; Falcon no. 351029).

5. Nylon cell strainer (70 µm; BD Falcon, Bedford, MA, cat. no. 08-771-2; Falcon no. 352350).

6. Glycerol (Sigma cat. no. G9012).

16.2.4.2 Sperm Preparation and Utilization

1. PureSperm (Nidacon, Goteborg, Sweden).
2. Sperm preparation medium (Medi-Cult, Jyllinge, Denmark).
3. Hypotaurine (Sigma cat. no. H1384).

16.2.4.3 Freezing and Thawing

1. Universal IVF medium (Medi-Cult).
2. First freezing solution: 5% glycerol in universal IVF medium.
3. Second freezing solution: 9% glycerol in universal IVF medium, $0.2M$ sucrose.
4. Plastic straw (Air Liquide, Bussy-Saint-George, France).
5. Freezing machine (Minicool 40, Air Liquide, France).
6. First thawing solutions free of cryoprotectant glycerol: $0.5M$ sucrose in universal IVF medium.
7. Second thawing solutions free of cryoprotectant glycerol: $0.2M$ sucrose in universal IVF medium.
8. Glass hood (K-System, Birkerød, Denmark).

16.2.4.4 Immunohistochemistry on Paraffin Sections

1. Primary antibody: Mouse-antihuman monoclonal antibody against high molecular weight cytokeratin, clone 34βE12 (Dako, Glostrup, Denmark, cat. no. M0630), 1:200 dilution.
2. Secondary antibody: Peroxidase-coupled rabbit antimouse immunoglobulins (Dako cat. no. 530P0161).
3. Formalin solution (10%; Fisher Scientific International Inc. cat. no. SF 98-4).
4. TissuePrep® embedding media (Fisher Scientific International Inc. cat. no. T565).
5. Citric acid for $0.01M$ citrate buffer, pH 6.0.
6. PBS: 1.06 g/L Na_2HPO_4 (Sigma cat. no. S9763), 0.35 g/L.
7. $NaH_2PO_4 \cdot H_2O$, 8.5 g/L NaCl in distilled water. Adjust to pH 7.23 (room temperature) prior to staining. (10X stock solutions are kept in −20°C: 53 g Na_2HPO_4, 17.5 g $NaH_2PO_4.H_2O$, 425 g NaCl in 5 L distilled water).
8. Diaminobenzidine liquid as recommended by the vendor (Dako K3465).
9. Canada balsam, mounting medium for microscopy.
10. Fisherfinest® Premium Superfrost® microscope slides.
11. Fisherfinest® premium cover glasses.
12. Xylenes.
13. 100% reagent alcohol.

16.3   Methods

16.3.1   Evaluation of the Ovaries

16.3.1.1   Tissue Processing for Peroxidase Immunohistochemistry: Cryostat Sections

1. On the bottom of marked $10 \times 10 \times 5$ mm Cryomold biopsy disposable vinyl specimen molds, drop 1 drop of OCT compound.
2. Make small pieces of tissue (ovaries, endometrium, and fallopian tube) and put it in the molds.
3. Pour OCT compound over the tissue so no spot is uncovered.
4. Freeze tissue samples in the molds by floating on liquid nitrogen (*see* **Note 2**) and store at −80°C until cutting.
5. Before cutting, clean the slides with acetone.
6. Cut the frozen samples. On every slide, place ten sections 7 μm thick each from distinct blocks (*see* **Note 3**).
7. Dry the slides at room temperature for 1 h and then fix in acetone at room temperature for 5 min. Dry in the hood at room temperature for 30 min. Keep the slides in the refrigerator in the box with Drierite anhydrous calcium sulfate (not longer than 1 wk) until the day of immunostaining.

16.3.1.2   Peroxidase Immunohistochemistry: Slides

1. Equilibrate the temperature of the slides by keeping them for 20 min in the closed box with Drierite anhydrous calcium sulfate in the hood at room temperature.
2. Incubate the slides with the specific primary antibodies (diluted to 5 μg/mL in PBS) in the wet chamber for 20 min at room temperature (*see* **Note 4**).
3. Rinse the slides three times (5 min each) with PBS.
4. Incubate the slides for 20 min at room temperature with the secondary antibody depending on the origin of the primary antibody (GAR, SwAM/Px) diluted 1:50. SwAM/Px has to be preabsorbed with rat kidney homogenate (500 mg/mL) for 20 min at room temperature (**5**), spun (14,000g, 10 min, 4°C) and supernatant collected.
5. Rinse the slides three times (5 min each) with PBS.
6. Dissolve 25 mg DAB in 1 mL PBS, filter the solution, and then dilute that in 50 mL PBS in the jar for the slides. Add 50 μL hydrogen peroxide solution.
7. Stain the slides in DAB for 5 min and then rinse in PBS.
8. Stain each slide with hematoxylin (diluted 1:20 in distilled water and filtered) for exactly 2 min. Subsequently, wash the slides as follows: distilled water, 1%

HCl, tap water, PBS (pH 8.0), and tap water. Wash the slides in the slowly running tap water for 3 min.

9. Dehydrate the sections in alcohol and xylenes as follows: 70% alcohol, 95% alcohol twice, 100% alcohol twice, xylenes three times.
10. Embed the specimens in Canada balsam diluted 1:1 in xylenes and cover with the cover glass.
11. Perform microscopic analysis using a microscope equipped with differential interference contrast.

16.3.1.2.1 Double-Color Immunohistochemistry

1. Finish **Subheading 16.3.1.2, step 7** (without dehydration and hematoxylin counterstain).
2. Rinse the slides in PBS.
3. Incubate the slides with the second specific primary antibody in the wet chamber for 20 min at room temperature (*see* **Note 5**).
4. Rinse the slides three times (5 min each) with PBS.
5. Incubate the slides for 20 min at room temperature with the secondary antibody depending on the origin of the primary antibody (GAR, SwAM/Px) diluted 1:50. SwAM/Px has to be preabsorbed with rat kidney homogenate (500 mg/mL) for 20 min at room temperature (*5*), spun (14,000g, 10 min, 4°C) and supernatant collected.
6. Rinse the slides three times (5 min each) with PBS.
7. Detect the antigen–antibody complex by the Vector SG detection kit according to the supplier's manual for a blue color.
8. Rinse the slides with PBS.
9. Wash the slides in tap water.
10. Dehydrate the slides in alcohol and xylenes as follows: 70% alcohol, 95% alcohol twice, 100% alcohol twice, xylenes three times.
11. Embed the specimens in Canada balsam diluted 1:1 in xylenes and cover the slides with the cover glass.
12. Perform microscopic analysis using a microscope equipped with differential interference contrast.

16.3.1.2.2 Triple-Color Immunohistochemistry

1. Finish **Subheading 16.3.1.2.1, step 8** (without dehydration and hematoxylin counterstain) for double-color immunohistochemistry.
2. Rinse the slides with PBS.
3. Incubate the slides with the third specific primary antibody in the wet chamber for 20 min at room temperature.
4. Rinse the slides three times (5 min each) with PBS.
5. Incubate the slides for 20 min at room temperature with the secondary antibody depending on the origin of the primary antibody (GAR, SwAM/Px) diluted

1:50. SwAM/Px has to be preabsorbed with rat kidney homogenate (500 mg/ mL) for 20 min at room temperature *(5)*, spun (14,000*g*, 10 min, 4°C) and supernatant collected.

6. Rinse the slides three times (5 min each) with PBS.
7. Detect the antigen–antibody complex using the Vector VIP detection kit according to the supplier's manual for a purple color.
8. Rinse the slides with PBS.
9. Wash the slides in tap water for 3 min.
10. Dehydrate the slides in alcohol and xylenes as follows: 70% alcohol, 95% alcohol twice, 100% alcohol twice, xylenes three times.
11. Embed the specimens in Canada balsam diluted 1:1 in xylenes and cover the slides with the cover glass.
12. Perform microscopic analysis using a microscope equipped with differential interference contrast.

16.3.2   Animals

After vaginal opening, the rats were monitored for cyclicity of ovarian function by evaluation of vaginal smears. Ovaries were collected from sexually mature rats, between d 45 and 60, and frozen for cryostat sections (*see* **Subheading 16.3.1.1**).

16.3.2.1   Generation of Antibodies to Zona Pellucida Glycoproteins

16.3.2.1.1   Generation of Antibodies to Porcine ZP1,2,3

The heat-solubilized isolated ZP (pSIZP) glycoproteins for antibodies to porcine ZP1,2,3 were prepared from pig ovaries collected from an abattoir as described previously *(45)*. Female New Zealand white rabbits (R145, R146) were used for the present study and were reared at the Small Animal Facility, National Institute of Immunology, New Delhi, India, per the guidelines and approval of the Institutional Animal Ethics Committee. Two rabbits were immunized intradermally with 100 µg pSIZP with complete CFA. Animals received two intramuscular boosters with same amount of pSIZP emulsified with IFA at monthly intervals. Animals were bled on the day of first injection and 15 d after the final booster.

16.3.2.1.2   Generation of Antibodies to Human Recombinant ZP4

In our experiment, purified recombinant hZP4/ZPB (10 mg) was conjugated to 5 mg DT using 0.1% glutaraldehyde as described previously *(46)*. Rabbits (n = 2; R156 and R 164) were immunized intradermally with the baculovirus-expressed recombinant hZP4 conjugated to DT. The animals received the conjugate equivalent to 250 µg of recombinant hZP4 emulsified using Squalene and Arlacel-A in a ratio of 4:1.

Two boosters were given intramuscularly at monthly intervals. Bleeds were collected at the time of the primary injection and 15 d after the second booster. The antibody titers were determined by direct-binding enzyme-linked immunosorbent assay (ELISA) as described previously except that instead of antimouse immunoglobulin-HRPO conjugate, antirabbit-IgG-HRPO conjugate at an optimized dilution of 1:2000 in PBS was employed *(47)*.

16.3.3   Tissue Culture for Research

1. Collect a 3 × 3 × 3 mm biopsy; mount it in OCT compound and freeze (*see* **Subheading 16.3.1.1**).
2. Scrape the surface epithelium of the ovaries gently with a sterile stainless steel surgical knife blade in the aseptic laminar flow hood into Petri dishes containing PhR– tissue culture medium without serum.
3. Cut the ovaries in half (one-half is kept by the Department of Pathology for the diagnostic evaluation). Scrape gently both the ovarian surface and the stroma of ovaries in the aseptic laminar flow hood into Petri dishes with PhR– tissue culture medium without serum.
4. Transfer the collected cells from the Petri dishes into 15-mL tubes and spin (1000*g*, 5 min at 25°C). Remove supernatant gently and resuspend the cells in culture medium (DMEM/F12, PhR– or DMEM/F12, PhR+) supplemented with inactivated 20% FBS.
5. Seed the cells into 24-well plates. The volume of medium is 350 µL per well, which is optimal for gas and nutrient exchange. Our experience led us to establish the following system (*see* **Fig. 16.10**): Rows A and B contain PhR– medium, and rows C and D contain PhR+ medium. The cells are initially seeded in columns 1–4, rows A and B, of the 24-well plate as indicated in **Fig. 16.10** and cultured in humidified atmosphere with 5% CO_2 at 37°C. The number of seeded cells is not possible to evaluate because of the strong contamination with erythrocytes, but varies between 100 and 1000 cells per well after medium replacement (*see also* **Note 6**).

   The medium with nonadherent cells and most erythrocytes is collected 48 h later and transferred into new wells (columns 1–4, rows C and D) as indicated in **Fig. 16.10**. Removal of erythrocytes is better if the plate is slightly shaken prior to collection of medium with floating cells. New medium is added to the rows A and B. Medium in rows C and D is also changed once 48 h later (*see* **Note 7**).

6. Incubate the cells in medium in a humidified atmosphere with 5% CO_2 at 37°C. Check medium content visually daily and add medium when necessary but do not replace medium completely because it may contain paracrine and autocrine substances and hormones secreted by the cultured cells. Finish cultures for evaluation between 5 and 10 d after seeding.

16.3.3.1 Peroxidase Immunohistochemistry: Tissue Culture

1. Aspirate medium from the wells and wash the wells with TBS-T for 10 min.
2. Fix the plates with cold methanol for 10 min.
3. Store wet in PBS with 0.01% sodium azide in the refrigerator until the immuno-histochemistry procedure. Prevent drying and wash and replace PBS when needed.
4. On the day of staining, remove the PBS from the wells and replace directly with primary antibody (300 µL/well). Depending on the antibody, the incubation time varies from 20 to 60 min at 37°C.
5. After incubation, wash the wells repeatedly three times with PBS at room temperature.
6. Apply the appropriate peroxidase-coupled secondary antibody (300 µL/well) and incubate for 20 min at room temperature.
7. Wash the wells extensively three times with PBS.

	1	2	3	4	5	6
A(-)	L1	R1	L2	R2		
B(+)	L1	R1	L2	R2		
C(-)	sL1	sR1	sL2	sR2		
D(+)	sL1	sR1	sL2	sR2		

Fig. 16.10 Seeding ovarian cultures in 24-well plates. The media are supplemented with 20% fetal bovine serum (FBS) and antibiotics, 350 µL per well; *rows A(−)* and *C(−)*:DMEM/F12, PhR−; *rows B(+)* and *D(+)*:DMEM/F12, PhR+. L, left ovary; R, right ovary. First-day seeding cells into rows A(−) and B(+): L1 and R1, scrapings from ovarian surfaces; L2 and R2, scrapings from ovarian surface plus cortex (mixed cultures). Third day 48 h after seeding: Supernatants (medium plus unattached ovarian cells and erythorcytes) from wells described above transfer one by one to rows C(−) and D(+). New media (PHR− or PhR+) put to the rows A(−) and B(+) immediately after each supernatant removal to prevent drying of adherent cells. Add 50 µL of corresponding medium to the supernatants in each well in rows C(−) and D(+) to adjust original volume. Remove supernatants with erythrocytes after 48 h and add fresh media. Adherent cells can eventually be washed with serum-free medium to remove erythrocytes. If the cells become overgrown, they can be passaged after trypsinization into remaining wells in columns 5 and 6 or another plate. Usually, confluent cells collected from one well are divided into five parts and seeded into one column (four new wells) and originating well

8. Detect the antigen–antibody complexes using either DAB, Vector VIP, or Vector SG.
9. Wash the wells in PBS and cover with PBS containing 0.01% sodium azide as a preservative.

16.3.4 Tissue Culture for Clinical Trial

16.3.4.1 Retrieval of Ovarian Stem Cells and Ovarian Biopsies

The women undergo diagnostic laparoscopy in general anesthesia and reverse Trendelenburg position. First, the surface of each ovary is gently scraped with laparoscopic scissors, and cells are collected by washing the scissor branches in tissue culture medium. Additional cells are similarly collected by laparoscopic brushes. In the next step, each ovary is washed with 10 mL saline solution, and the fluid is collected from the pouch of Douglas. Finally, a small biopsy (5 × 5 × 5 mm) is collected from each ovary, and the biopsy site is closed with a single suture.

16.3.4.2 Human Serum Heat Inactivation and OSE Cultures

1. Collect 30 mL venous blood into sterile tubes and let it coagulate at room temperature.
2. Spin the coagulated blood (3000g, 10 min, 25°C).
3. Transfer the HuS into new tubes and heat inactivate for 60 min in the 59°C water bath (*see* **Note 8**).
4. Spin the heat-inactivated serum (3000g, 10 min, 25°C).
5. Make aliquots of the supernatant and store them at −20°C until utilization.
6. Scrape the surface epithelium of half of each ovarian biopsy gently in the aseptic laminar flow hood with a sterile stainless steel surgical knife blade into Petri dishes containing tissue culture medium (DMEM/F12, PhR+).
7. Mince the tissue in medium with the blade into small pieces.
8. Pass the medium containing the cells and the remnants of the tissue through the 70-μm nylon cell strainer.
9. Spin the cell suspension (1000g, 10 min at 25°C), remove supernatant, and dilute the cells with DMEM/F12, PhR+ supplemented with 20% HuS.
10. Seed the cells into a four-well plate. Use 350 μL of the cell suspension per well. This is d 0.
11. Spin the cell suspension collected from the pouch of Douglas and seed the same way.
12. Incubate the cells in medium in a humidified atmosphere with 5% CO_2 at 37°C.
13. Collect the culture medium with unattached cells after 24 h (d 1) and transfer this medium into new wells.
14. Add new DMEM/F12, PhR+ medium supplemented with 20% HuS into the original wells to continue culture. Monitor daily under a converted microscope with a heated stage.

15. Replace the D-MEM/F12, PhR+ culture medium supplemented with 20% HuS with 350 µL in vitro maturation medium on d 3. Add sperm suspension (*see* **Subheading 16.3.4.3**) next or several days later.
16. Evaluate the ovarian cell cultures for the presence of embryo-, morula, and blastocyst-like structures.
17. Collect all these structures (embryo-, morula, and blastocyst-like) and freeze by a two-step method with glycerol as described in **Subheading 16.3.4.4**. Keep in liquid nitrogen at −196°C for later genetic analysis (*see* **Note 9**) and transfer to the uterus if normal.

16.3.4.3 Sperm Preparation and Utilization

1. Prepare the sperm using an 80%/40% gradient of PureSperm solution.
2. Spin the sperm at 600*g* for 30 min.
3. Wash the 80% fraction with sperm in 5 mL sperm preparation medium containing 50 m*M* antioxidant hypotaurine.
4. Spin the tube at 700*g* for 10 min.
5. Resuspend the pellet in 0.5 mL sperm preparation medium with hypotaurine and leave the sperm suspension in a CO_2 incubator at 37°C for 30 min to "swim-up."
6. For the fertilization, use the sperm from the top of the sperm suspension. Add 5 drops of sperm suspension into the wells with the ovarian cell culture. Leave one control well without sperm for comparison of the sperm effect.

16.3.4.4 Freezing of Embryos

1. Transfer embryos into the first freezing solution of the Blast Freeze kit for 10 min and incubate the embryos at 37°C in a CO_2 incubator.
2. Transfer embryos into the second freezing solution for 10 min and incubate at 37°C in a CO_2 incubator.
3. Place only one embryo into each plastic straw and start to cool in liquid nitrogen vapor using a freezing machine. Use a slow-cooling program at 35°C/min. At −6°C, manual seeding is performed to −40°C at 0.3°C/min, and from −40°C to −150°C at 35°C/min (*48*). Transfer the straws into liquid nitrogen (−196°C) and store until use.

16.3.4.5 Thawing of Embryos

1. Thaw the embryos on the day of the transfer.
2. Place the embryos into the first thawing solution of the Blast Thaw kit for 10 min at room temperature. Use a glass hood in which a 5% CO_2 stream can be regulated.

3. Place the embryos into the second thawing solution for 10 min at room temperature. Use a glass hood in which a 5% CO_2 stream can be regulated.
4. Wash the embryos in universal IVF medium and transfer them into fresh preincubated universal IVF medium.

16.3.4.6 Tissue Processing and Immunohistochemistry on Paraffin Slides

1. Fix the remaining half of each ovarian biopsy in formalin and embed in paraffin.
2. Cut the samples. Place 10-mm thick sections on every microscope slide.
3. Before staining, deparaffinize and rehydrate the sections by immersion into 0.01M citrate buffer, pH 6.0, at 98°C for 40 min. Cool the slides at room temperature.
4. Incubate the slides for 20 min with mouse monoclonal antibody against high molecular weight cytokeratin, clone 34βE12 (dilution 1:200).
5. Rinse the slides three times (5 min each) with PBS.
6. Incubate the slides with peroxidase-coupled rabbit antimouse immunoglobulins.
7. Rinse the slides three times (5 min each) with PBS.
8. Visualize the linked antibody by peroxidase diaminobenzidine solution as recommended by the vendor.
9. Dehydrate the slides in alcohol and xylenes as follows: 70% alcohol, 95% alcohol twice, 100% alcohol twice, xylenes three times.
10. Embed the specimens in Canada balsam diluted 1:1 in xylene and cover the slides with the cover glass.
11. The sections are analyzed under the light microscope for the presence of OSE cells and granulosa cells of primary follicles exhibiting brown staining for cytokeratin.

16.4   Notes

1. Preabsorption with rat kidneys efficiently removes nonspecific background. The kidneys are stored frozen, and an appropriate amount is minced several times with a stainless surgical knife just prior to antibody adsorption.
2. Remove molds from liquid nitrogen shortly before the freezing is completed and then place at −80°C. This will prevent cracking of samples from OCT overfreezing.
3. This will allow staining of several distinct samples at once. Make documentation of placement. After staining, you may draw a line net from the slide bottom with permanent color pens to easily distinguish the proper sample according to the documentation.
4. Most primary antibodies can be collected after staining, frozen, and reused several times. If sufficient concentration in some undiluted supernatant primary antibodies is not available, the efficacy can be increased by incubation overnight at 4°C, by incubation 1 h at 37°C, or a combination.
5. Because the diaminobenzidine (also SG or VIP) reaction product masks the antigen and catalytic sites of the first sequence of immunoreagents, preventing interaction with the reagents of the second sequence *(49)*, primary antibodies from the same species can be utilized for the double- and triple-color staining.

6. The number of wells seeded may be reduced to one-half to get a richer population of cells.
7. These so-called cell supernatants should first be spun down, liquid removed, and volume adjusted with new supplemented medium to 350 μL per well. They still may produce numerous clusters of proliferating OSE cells.
8. Such comprehensive heat inactivation of the patient's serum may be advantageous by eliminating not only complement but also other proteins with immunological activities that can be present particularly in POF patients because most women with POF have been found to have immune abnormalities *(39,50,51)*.
9. Preimplantation genetic analysis is essential to determine that the embryo is normal. If pregnancy develops, the amniocentesis should be performed for genetic evaluation of embryo/fetus.

Acknowledgments We thank Drs. Bonnie S. Dunbar and Sarvamangala V. Prasad of the Department of Molecular and Cellular Biology, Baylor College of Medicine, Houston, Texas, for kindly supplying additional ZP1,2,3 antibodies and Dr. A. Neil Barclay of the Sir William Dunn School of Pathology, University of Oxford, Oxford, United Kingdom, who kindly provided mouse antirat Ia and LCA monoclonal antibodies. Peroxidase conjugate of swine antimouse IgG was kindly provided by Dr. Jana Peknicova of the Department of Biology and Biochemistry of Fertilization, Institute of Molecular Genetics, Academy of Sciences of the Czech Republic, Prague, and HLA-DR was kindly donated by Drs. Ivan Hilgert and Vaclav Horejsi of the Institute of Molecular Genetics, Academy of Sciences and Faculty of Sciences, Charles University, Prague, Czech Republic. Clone F15-42-01 to human Thy-1dp was kindly donated by Dr. Rosemarie Dalchau, Institute of Child Health, University of London. We also thank Drs. Helena Meden-Vrtovec, Tomaž Tomaževič, Jasna Šinkovec, and Andrej Vogler of the Department of Obstetrics and Gynecology, University of Ljubljana Medical Center, for stimulating suggestions and support, as well as excellent laparoscopic collection of OSE cells and ovarian biopsies by Dr. Andrej Vogler. This work was supported by the Physicians' Medical Education and Research Foundation award to A.B.

References

1. Allen, E. (1923) Ovogenesis during sexual maturity. *Am. J. Anat.* **31,** 439–481.
2. Pearl, R., and Schoppe, W. F. (1921) Studies on the physiology of reproduction in the domestic fowl. XVIII. Further observations on the anatomical basis of fecundity. *J. Exp. Zool.* **34,** 101–1189.
3. Zuckerman, S. (1951) The number of oocytes in the mature ovary. *Recent Prog. Horm. Res.* **6,** 63–109.
4. Franchi, L. L., Mandl, A. M., and Zuckerman, S. (1962) The development of the ovary and the process of oogenesis, in *The ovary* (Zuckerman, S., ed.)–, Academic Press, London, pp. 1–88.
5. Zuckerman, S., and Baker, T. G. (1977) The development of the ovary and the process of oogenesis, in *The ovary*, Vol. 1 (Zuckerman, S., and Weir, B. J., eds.)–, Academic Press, New York, pp. 41–67.
6. Kingery, H. M. (1917) Oogenesis in the white mouse. *J. Morphol.* **30,** 261–315.
7. Evans, H. M., and Swezy, O. (1931) Ovogenesis and the normal follicular cycle in adult mammalia. *Mem. Univ. Calif.* **9,** 119–224.
8. Gerard, P. (1920) Contribution a l'etude de l'ovarie des mammiferes. L'ovaire de Galago mossambicus(Young). *Arch. Biol.* **43,** 357–391.
9. Rao, C. R. N. (1928) On the structure of the ovary and the ovarian ovum of Loris lydekkerianus Cabr. *Q. J. Micr. Sci.* **71,** 57–73.
10. Zuckerman, S., and Weir, B. J. (1977) *The ovary*, 2nd ed., Vol. I, –Academic Press, New York.
11. Bukovsky, A., Keenan, J. A., Caudle, M. R., Wimalasena, J., Upadhyaya, N. B., and Van Meter, S. E. (1995) Immunohistochemical studies of the adult human ovary: possible contribution of immune and epithelial factors to folliculogenesis. *Am. J. Reprod. Immunol.* **33,** 323–340.

12. Bukovsky, A., Caudle, M. R., Svetlikova, M., and Upadhyaya, N. B. (2004) Origin of germ cells and formation of new primary follicles in adult human ovaries. *Reprod. Biol. Endocrinol.* **2**, 20. Available at: http://www.rbej.com/content/2/1/20.

13. Bukovsky, A., Ayala, M. E., Dominguez, R., Svetlikova, M., and Selleck-White, R. (2007). Bone marrow derived cells and alternative pathways of oogenesis in adult rodents. *Cell Cycle* 6(18), 2306–2309.

14. Block, E. (1952). Quantitative morphological investigations of the follicular system in women. Variations at different ages. Acta Anat. (Basel). 14, 108–123.

15. Kerr, J. B., Duckett, R., Myers, M., Britt, K. L., Mladenovska, T., and Findlay, J. K. (2006) Quantification of healthy follicles in the neonatal and adult mouse ovary: evidence for maintenance of primordial follicle supply. Reproduction. 132, 95–109.

16. Johnson, J., Canning, J., Kaneko, T., Pru, J. K., and Tilly, J. L. (2004) Germline stem cells and follicular renewal in the postnatal mammalian ovary. Nature. 428, 145–150.

17. Gougeon, A., Echochard, R., and Thalabard, J. C. (1994) Age-related changes of the population of human ovarian follicles: increase in the disappearance rate of non-growing and early-growing follicles in aging women. Biol. Reprod. 50, 653–663.

18. Peters, H., and McNatty, K. P. (1980) The ovary. A correlation of structure and function in mammals, University of California Press, Berkeley, CA.

19. Waldeyer, W. (1870) Eierstock und Ei, Engelmann, Leipzig.

20. Simkins, C. S. (1928) Origin of the sex cells in man. Am. J. Anat. 41, 249–253.

21. Brambell, F. W. R. (1927) The development and morphology of the gonads of the mouse. Part 1. The morphogenesis of the indifferent gonad and of the ovary. Proc. R. Soc. 101, 391–409.

22. Motta, P. M., Van Blerkom, J., and Makabe, S. (1980) Changes in the surface morphology of ovarian "germinal" epithelium during the reproductive cycle and in some pathological conditions. J. Submicrosc. Cytol. 12, 407–425.

23. Bukovsky, A., Caudle, M. R., Svetlikova, M., Wimalasena, J., Ayala, M. E., and Dominguez, R. (2005) Oogenesis in adult mammals, including humans: a review. Endocrine. 26, 301–316.

24. Bukovsky, A., Copas, P., and Virant-Klun, I. (2006) Potential new strategies for the treatment of ovarian infertility and degenerative diseases with autologous ovarian stem cells. Expert Opin. Biol. Ther. 6, 341–365.

25. Williams, A. F., and Barclay, A. N. (1988) The immunoglobulin superfamily—domains for cell surface recognition. Annu. Rev. Immunol. 6, 381–405.

26. Bukovsky, A., Caudle, M. R., Keenan, J. A., et al. (2001) Association of mesenchymal cells and immunoglobulins with differentiating epithelial cells. BMC Dev. Biol. 1, 11.

27. Auersperg, N., Wong, A. S., Choi, K. C., Kang, S. K., and Leung, P. C. (2001) Ovarian surface epithelium: biology, endocrinology, and pathology. Endocr. Rev. 22, 255–288.

28. Faddy, M. J. (2000) Follicle dynamics during ovarian ageing. Mol. Cell Endocrinol. 163, 43.

29. Bukovsky, A., Caudle, M. R., Keenan, J. A., Wimalasena, J., Foster, J. S., and Van Meter, S. E. (1995) Quantitative evaluation of the cell cycle-related retinoblastoma protein and localization of Thy-1 differentiation protein and macrophages during follicular development and atresia, and in human corpora lutea. Biol. Reprod. 52, 776–792.

30. Bukovsky, A. (2006) Immune system involvement in the regulation of ovarian function and augmentation of cancer. Microsc. Res. Tech. 69, 482–500.

31. Czernobilsky, B., Moll, R., Levy, R., and Franke, W. W. (1985) Co-expression of cytokeratin and vimentin filaments in mesothelial, granulosa and rete ovarii cells of the human ovary. Eur. J. Cell Biol. 37, 175–190.

32. Johnson, J., Bagley, J., Skaznik-Wikiel, M., et al. (2005) Oocyte generation in adult mammalian ovaries by putative germ cells in bone marrow and peripheral blood. Cell. 122, 303–315.

33. Boja, E. S., Hoodbhoy, T., Garfield, M., and Fales, H. M. (2005) Structural conservation of mouse and rat zona pellucida glycoproteins. Probing the native rat zona pellucida proteome by mass spectrometry. Biochemistry. **44**, 16445–16460.

34. Erickson, G. F., and Shimasaki, S. (2003) The spatiotemporal expression pattern of the bone morphogenetic protein family in rat ovary cell types during the estrous cycle. *Reprod. Biol. Endocrinol.* **1**, 9.

35. Ingram, D. L. (1962) Atresia, in *The ovary* (Zuckerman, S., ed.), Academic Press, London, pp. 247–273.
36. Bukovsky, A., Svetlikova, M., and Caudle, M. R. (2005) Oogenesis in cultures derived from adult human ovaries. *Reprod. Biol. Endocrinol.* **3,** 17.
37. Bukovsky, A., and Virant-Klun, I. (2006) Adult stem cells in the human ovary, in *Stem cells in reproductive medicine: basic science and therapeutic potential* (Simon, C., and Pellicer, A., eds,), Informa Healthcare, London, pp. 53–70.
38. Corrigan, E. C., Raygada, M. J., Vanderhoof, V. H., and Nelson, L. M. (2005) A woman with spontaneous premature ovarian failure gives birth to a child with fragile X syndrome. *Fertil. Steril.* **84,** 1508.
39. Rebar, R. W. (2000) Premature ovarian failure, in *Menopause biology and pathobiology* (Lobo, R. A., Kesley, J., and Marcus, R., eds,), Academic Press, San Diego, CA, pp. 135–146.
40. Gersak, K., Meden-Vrtovec, H., and Peterlin, B. (2003) Fragile X premutation in women with sporadic premature ovarian failure in Slovenia. *Hum. Reprod.* **18,** 1637–1640.
41. Sehgal, S., Gupta, S. K., and Bhatnagar, P. (1989) Long-term effects of immunization with porcine zona pellucida on rabbit ovaries. *Pathology.* **21,** 105–110.
42. Miller, C. C., Fayrer-Hosken, R. A., Timmons, T. M., Lee, V. H., Caudle, A. B., and Dunbar, B. S. (1992) Characterization of equine zona pellucida glycoproteins by polyacrylamide gel electrophoresis and immunological techniques. *J. Reprod. Fertil.* **96,** 815–825.
43. Horejsi, V., Hilgert, I., Kristofova, H., Bazil, V., Bukovsky, A., and Kulhankova, J. (1986) Monoclonal antibodies against human leucocyte antigens. I. Antibodies against beta-2-microglobulin, immunoglobulin kappa light chains, HLA-DR-like antigens, T8 antigen, T1 antigen, a monocyte antigen, and a pan-leucocyte antigen. *Folia Biol. (Praha).* **32,** 12–25.
44. McKenzie, J. L., and Fabre, J. W. (1981) Human Thy-1: unusual localization and possible functional significance in lymphoid tissues. *J. Immunol.* **126,** 843–850.
45. Yurewicz, E. C., Sacco, A. G., and Subramanian, M. G. (1987) Structural characterization of the Mr = 55,000 antigen (ZP3) of porcine oocyte zona pellucida. Purification and characterization of alpha- and beta-glycoproteins following digestion of lactosaminoglycan with endo-beta-galactosidase. *J. Biol Chem.* **262,** 564–571.
46. Govind, C. K., Srivastava, N., and Gupta, S. K. (2002) Evaluation of the immunocontraceptive potential of *Escherichia coli* expressed recombinant non-human primate zona pellucida glycoproteins in homologous animal model. *Vaccine.* **21,** 78–88.
47. Govind, C. K., Hasegawa, A., Koyama, K., and Gupta, S. K. (2000) Delineation of a conserved B cell epitope on bonnet monkey (*Macaca radiata*) and human zona pellucida glycoprotein-B by monoclonal antibodies demonstrating inhibition of sperm-egg binding. *Biol. Reprod.* **62,** 67–75.
48. Virant-Klun, I., Tomazevic, T., Bacer-Kermavner, L., Mivsek, J., Valentincic-Gruden, B., and Meden-Vrtovec, H. (2003) Successful freezing and thawing of blastocysts cultured in sequential media using a modified method. *Fertil. Steril.* **79,** 1428–1433.
49. Sternberger, L. A., and Joseph, S. A. (1979) The unlabelled antibody method. Contrasting color staining of paired hormones without antibody removal. *J. Histochem. Cytochem.* **27,** 1424–1429.
50. Hoek, A., van Kasteren, Y., de Haan-Meulman, M., Schoemaker, J., and Drexhage, H. A. (1993) Dysfunction of monocytes and dendritic cells in patients with premature ovarian failure. *Am. J. Reprod. Immunol.* **30,** 207–217.
51. Hoek, A., van Kasteren, Y., de Haan-Meulman, M., Hooijkaas, H., Schoemaker, J., and Drexhage, H. A. (1995) Analysis of peripheral blood lymphocyte subsets, NK cells, and delayed type hypersensitivity skin test in patients with premature ovarian failure. *Am. J. Reprod. Immunol.* **33,** 495–502.
52. Bousfield, G. R., Butnev, V. Y., Gotschall, R. R., Baker, V. L., and Moore, W. T. (1996) Structural features of mammalian gonadotropins. *Mol. Cell Endocrinol.* **125,** 3–19.

Index

Printed in the United States of America